Advanced Electric Power System 2023

Advanced Electric Power System 2023

Editor

Ying-Yi Hong

MDPI

Basel • Beijing • Wuhan • Barcelona • Belgrade • Novi Sad • Cluj • Manchester

Editor
Ying-Yi Hong
Chung Yuan Christian University
Taoyuan
Taiwan

Editorial Office
MDPI AG
Grosspeteranlage 5
4052 Basel, Switzerland

This is a reprint of articles from the Special Issue published online in the open access journal *Energies* (ISSN 1996-1073) (available at: https://www.mdpi.com/journal/energies/special_issues/TVD92H72EC).

For citation purposes, cite each article independently as indicated on the article page online and as indicated below:

Lastname, A.A.; Lastname, B.B. Article Title. *Journal Name* **Year**, *Volume Number*, Page Range.

ISBN 978-3-7258-1615-6 (Hbk)
ISBN 978-3-7258-1616-3 (PDF)
doi.org/10.3390/books978-3-7258-1616-3

Contents

About the Editor . . . . **vii**

Yuan-Kang Wu, Deng-Yue Gau and Trinh-Duc Tung
Overview of Various Voltage Control Technologies for Wind Turbines and AC/DC Connection Systems
Reprinted from: *Energies* **2023**, *16*, 4128, doi:10.3390/en16104128 . . . . **1**

Hakim Bennani, Ahmed Chebak and Abderrazak El Ouafi
Unique Symbolic Factorization for Fast Contingency Analysis Using Full Newton–Raphson Method
Reprinted from: *Energies* **2023**, *16*, 4279, doi:10.3390/en16114279 . . . . **18**

Joaquín Pedra, Luis Sainz and Lluís Monjo
DC Admittance Model of VSCs for Stability Studies in VSC-HVDC Systems
Reprinted from: *Energies* **2023**, *16*, 5457, doi:10.3390/en16145457 . . . . **35**

Stefano Quaia, Alessandro Mauri, Alessio Marchesin and Paolo Manià
Actions to Handle Increasing Distributed Generation and Rising Voltage during Low-Demand Periods: Tap Staggering Practices and Field Tests in the Italian Transmission Network
Reprinted from: *Energies* **2023**, *16*, 6258, doi:10.3390/en16176258 . . . . **52**

Oluwafemi Emmanuel Oni and Omowunmi Mary Longe
Analysis of Secondary Controller on MTDC Link with Solar PV Integration for Inter-Area Power Oscillation Damping
Reprinted from: *Energies* **2023**, *16*, 6295, doi:10.3390/en16176295 . . . . **65**

Sheng Li, Can Zhang and Jili Zuo
Long-Term Voltage Stability Bifurcation Analysis and Control Considering OLTC Adjustment and Photovoltaic Power Station
Reprinted from: *Energies* **2023**, *16*, 6383, doi:10.3390/en16176383 . . . . **83**

Chandu Valuva, Subramani Chinnamuthu, Tahir Khurshaid and Ki-Chai Kim
A Comprehensive Review on the Modelling and Significance of Stability Indices in Power System Instability Problems
Reprinted from: *Energies* **2023**, *16*, 6718, doi:10.3390/en16186718 . . . . **100**

Prottay M. Adhikari, Luigi Vanfretti, Hao Chang and Koushik Kar
Real-Time Control of a Battery Energy Storage System Using a Reconfigurable Synchrophasor-Based Control System
Reprinted from: *Energies* **2023**, *16*, 6909, doi:10.3390/en16196909 . . . . **145**

Guoqing Zhang, Wengen Gao, Yunfei Li, Xinxin Guo, Pengfei Hu and Jiaming Zhu
Detection of False Data Injection Attacks in a Smart Grid Based on WLS and an Adaptive Interpolation Extended Kalman Filter
Reprinted from: *Energies* **2023**, *16*, 7203, doi:10.3390/en16207203 . . . . **165**

Eduardo Tavares Silvério and Jose Rubens Macedo Junior
Measuring and Modeling the Skin Effect for Harmonic Power Flow Studies
Reprinted from: *Energies* **2023**, *16*, 7913, doi:10.3390/en16237913 . . . . **185**

Tianmeng Yang, Suhua Lou, Meng Zhang, Yanchun Li, Wei Feng and Jicheng Liu
Research on a Two-Layer Optimal Dispatching Method Considering the Mutual Aid of Peak Regulating Resources among Regional Power Grids
Reprinted from: *Energies* **2024**, *17*, 667, doi:10.3390/en17030667 . . . . **204**

About the Editor

Ying-Yi Hong

Ying-Yi Hong (Senior Member, IEEE) received a B.S.E.E. degree from Chung Yuan Christian University (CYCU), Taiwan, in 1984, an M.S.E.E. degree from National Cheng Kung University (NCKU), Taiwan, in 1986, and a Ph.D. degree from the Department of Electrical Engineering, National Tsing Hua University (NTHU), Taiwan, in December 1990. Sponsored by the Ministry of Education, Taiwan, he conducted research in the Department of Electrical Engineering, University of Washington, Seattle, from August 1989 to August 1990. He has been with CYCU since 1991, where he was the Dean of the College of Electrical Engineering and Computer Science from 2006 to 2012. He was promoted to Chair Professor in 2024 due to his exceptional research, leadership, teamwork, and international collaboration. From 2012 to 2018, he was the General Secretary of CYCU, where he is currently the Vice President. He received the Outstanding Professor of Electrical Engineering Award from the Chinese Institute of Electrical Engineering (CIEE), Taiwan, in 2006 and the Outstanding Professor of Engineering Award from the Chinese Institute of Engineers (CIE), Taiwan, in 2024. He was the Chair of the IEEE PES Taipei Chapter in 2001. His research interests include power system analysis and artificial intelligence applications.

energies

MDPI

Article

Overview of Various Voltage Control Technologies for Wind Turbines and AC/DC Connection Systems

Yuan-Kang Wu *, Deng-Yue Gau and Trinh-Duc Tung

Department Electrical Engineering, National Chung-Cheng University, No. 168, University Rd., Chiayi 62102, Taiwan; willie890486@yahoo.com.tw (D.-Y.G.); tungtd@alum.ccu.edu.tw (T.-D.T.)
* Correspondence: allenwu@ccu.edu.tw

Abstract: Wind power generation is one of the mainstream renewable energy resources. Voltage stability is as important as the frequency stability of a power system with a high penetration of wind power generation. The advantages of high-voltage direct current (HVDC) transmission systems become more significant with the increase of both installed capacity and transmission distance in offshore wind farms. Therefore, this study discusses various voltage control methods for wind turbines and HVDC transmission systems. First, various voltage control methods of a wind farm were introduced, and they include QV control and voltage droop control. The reactive power of a wind turbine varies with active power, while the active power from each wind turbine may be different owing to wake effects. Thus, QV and voltage droop control with varying gain values are also discussed in this paper. Next, the voltage control methods for an HVDC transmission system, such as power factor control, voltage control, and V_{ac}-Q control, are also summarized and tested in this study. When a three-phase short circuit fault occurs or a sudden reactive power load increases, the system voltage would drop immediately. Thus, various voltage control methods for wind turbines or HVDC can make the system's transient response more stable. Therefore, this study implemented the simulation scenarios, including a three-phase short circuit fault at the point of common coupling (PCC) or a sudden increase of reactive power load, and adopted various voltage control methods, which aim to verify whether additional voltage control methods are effective to improve the performance of transient voltage. The voltage control method has been implemented in PSCAD/EMTDC, and the simulation results show that the QV control performs better than the droop control. In addition, when applying the voltage control technique during a three-phase fault, transient voltage nadir can be improved through either an HVDC transmission system or an AC transmission system.

Keywords: renewable energy; voltage stability; high-voltage direct current (HVDC); voltage control; voltage droop control

Citation: Wu, Y.-K.; Gau, D.-Y.; Tung, T.-D. Overview of Various Voltage Control Technologies for Wind Turbines and AC/DC Connection Systems. *Energies* **2023**, *16*, 4128. https://doi.org/10.3390/en16104128

Academic Editor: Juan-José González de la Rosa

Received: 15 April 2023
Revised: 13 May 2023
Accepted: 15 May 2023
Published: 16 May 2023

1. Introduction

With global warming and sustainable development, renewable energy has developed rapidly during the last few decades. Wind energy is one of the primary types of renewable energy, and the installation is more concentrated than solar energy. The high penetration of offshore wind farms causes concern about frequency and voltage stability. Thus, many countries have requested wind farms to provide voltage support at the point of connection in their grid codes [1,2]. Using the power electronic converter, the doubly-fed induction generator (DFIG) and full converter wind turbine (FCWT) can provide reactive power support to the system's voltage [3]. To connect an offshore wind farm, the HVDC transmission system is much more suitable than the HVAC transmission system, with a transmission distance of over 100 km and a capacity larger than 100 MVA [4]. Besides, voltage source converter-based high voltage direct current (VSC-HVDC) can control active and reactive power independently [5], which makes it control the voltage easily. Without the consideration of reactive power compensations like Static Synchronous Compensators (STATCOM) or Static Var Compensators (SVC), this study first tested various control methods for wind turbines and VSC-HVDC systems.

With a high penetration of wind power generation, the voltage issues become important. The Power-Voltage analysis [6] about static voltage stability indicated that the voltage would collapse without any contingency if a transfer of wind power generation increases. Meanwhile, it only raises a small transfer limit by installing more reactive compensation resources. In the aspect of dynamic voltage stability, compared to a stiff system, a weak grid is generally concerned about overvoltage after a grid fault owing to a higher sensitivity of dV/dQ [7]. The voltage variation to reactive power injection is defined as the grid strength, presented by the Short Circuit Ratio (SCR). The voltage at a strong grid with a high SCR fluctuates less than at a weak grid with a small SCR [8]. Ref. [9] provides a study about the influence of SCR on the voltage control of wind power plants. It shows that it is easier to impact the voltage at the point of common coupling (PCC) with a lower SCR value.

In Ref. [10], it divided the voltage control method of a large-scale wind farm into three categories: decentralized, centralized and hierarchical controls. The reactive power capability of different wind turbines varies according to wake effects; thus, Ref. [11] proposed an adaptive Q-V method that allows wind turbines with more reactive power capability to provide more reactive power. Then, the adaptive and fixed Q-V schemes were simulated by considering different grid stiffness and disturbance types. Ref. [12] presented both variable voltage droop control and constant droop control to reduce the voltage fluctuation caused by varying loads at PCC. A reactive power coordination control strategy was proposed in [13] to optimize voltage quality and minimize power loss using a genetic algorithm; moreover, the control method was confirmed better than the unit power factor by investigating the maximum delta voltage and the voltage characteristic coefficient in a test system integrated with three wind farms. To sum up, two factors that influence the voltage regulation in an offshore wind farm include the capability of reactive power supported by wind turbines and the grid strength at PCC.

Ref. [14] proposed a coordinated control scheme of active and reactive power oscillation damping controllers for a utility-scale wind power plant to support voltage stability and provide fast voltage recovery after grid faults. In [15], the work presented a communication-free coordinated fault ride-through (FRT) control between the wind farm-side converters and the wind turbine inverters based on the injected sequence harmonics; the proposed method enables the DC voltage to be recovered quickly within an accepted range, which helps improve the decline of AC voltage. In [16], a reactive voltage control technique for PMSG-based wind farms was suggested. It considers sufficient reactive power and balanced terminal voltage. Thus, the method can regulate the PCC voltage of wind farms while still maintaining the balance of the terminal voltage of a wind farm. In [17], a multi-objective Q-V coordinated control technique for wind farms was suggested, which considers voltage variation, converter junction temperature, and power loss. In [18], a coordinated voltage regulation technique based on model predictive control was suggested for enabling wind farms to support black start. Additionally, the coordinated voltage regulation can precisely consider the effect of active power on voltage variation, and the reactive and active power from wind turbines and energy storage can be coordinated to increase the capability of wind farms. To increase the voltage stability of power systems at the PCC, Ref. [19] suggested a synchrophasor data-based Q-V droop (SQVD) control technique for wind farms. The suggested method enhances the Q-V droop performance of wind farms by coupling the corresponding wide-area signals with the local signals at each wind farm. A distributed reactive power control (DVC) scheme based on the consensus alternating direction method of multipliers (ADMM) was introduced in [20]. This scheme was applied to wind farms to optimize reactive power utilization. To reduce voltage deviation and active power losses, a decentralized voltage control based on the standard ADMM was also created in [20] for wind farm operation.

Besides wind turbines, reactive power compensation devices also play a vital role in voltage regulation. Ref. [21] utilized STATCOM with CMC-based topology, and the effect was examined to resolve voltage fluctuation. Ref. [22] suggested potential methods that can improve the voltage stability of wind farms: one is to install a static var compensator (SVC) to provide dynamic reactive power support, and the other is to select a doubly-fed induction generator (DFIG) that can control reactive power flexibly without installing reactive power compensation devices. After a three-phase short circuit fault, the problem of overvoltage could cause wind turbines to trip off. Thus, a coordinated control between SVC and DFIG can decrease the magnitude of overvoltage compared to the use of SVC only [23].

Compared to traditional AC transmission, VSC-HVDC can enhance voltage stability by providing extra reactive power [24]. In [25], the stability analysis was carried out for a 100 MW solar plant through a connection of HVDC. Ref. [26] implemented a system simulation and showed the voltage at PCC during a three-phase short circuit fault. The simulation considered different levels of wind power penetration with a VSC-HVDC connection. The results showed that the voltage at the PCC is proportional to the wind power integration. In Ref. [27], the control strategy that considers wind farm, STATCOM and HVDC was tested by load switching and three-phase short circuit fault, where the offshore wind farm is connected by the line commutated converter high-voltage direct current (LCC-HVDC) transmission system. Ref. [28] showed that the control scheme of VSC-HVDC can influence the maximum transfer of active power. In Refs. [29–31], some voltage control methods for a VSC-HVDC-connected weak grid were proposed. Two control modes were proposed in [32,33] by coordinating wind generators with VSC-HVDC based on model predictive control. The normal operation mode can maintain a stable voltage and reduce power loss, while the corrective mode can help the voltage reach 1.0 p.u. rapidly when wind turbines were reconnected to the grid after a storm. The control method presented in [34] can minimize power loss of offshore wind farms, increase the amount of active power transfer, and improve voltage stability during system transients. From the above literature reviews, the HVDC-connected system can provide better performance than the AC-connected system.

In addition to reviewing various up-to-date voltage control methods, this paper also compared the performances of transient voltage responses based on different voltage control techniques. To demonstrate the superior performance of the QV control, this paper also carried out the control methods for the voltage droop control and the QV control, and it compared all of these control methods with AC or DC transmission systems. The simulation results reveal that the voltage nadir can be improved by the above control methods. Additionally, the performance of voltage responses with HVDC transmission systems is better than that with AC transmission systems.

The paper is organized as follows: Section 2 introduces various voltage control methods utilized for wind turbines and VSC-HVDC. Section 3 presents the used AC and HVDC test systems and demonstrates the simulation results under various operating scenarios. Finally, the conclusions are drawn in Section 4.

2. Various Voltage Control Methods

In this study, the control methods are separated into two parts. The first part is about the control methods for wind turbines, which is introduced in Section 2.1, while the second part is about the control methods for VSC-HVDC, which is presented in Section 2.2.

2.1. Wind Farm Control

In this study, the DFIG-based wind turbine was used based on the generic model, and its block diagram is listed in Figure 1. There are four main blocks inside the generic model: the converter control model, the generator/converter model, the pitch control model, and the drive train model. The reactive power reference (Q_{ref}) generated by QV, voltage droop or other controls is sent to the converter control model to obtain the required voltage. A detailed description of each model can be found in [35,36].

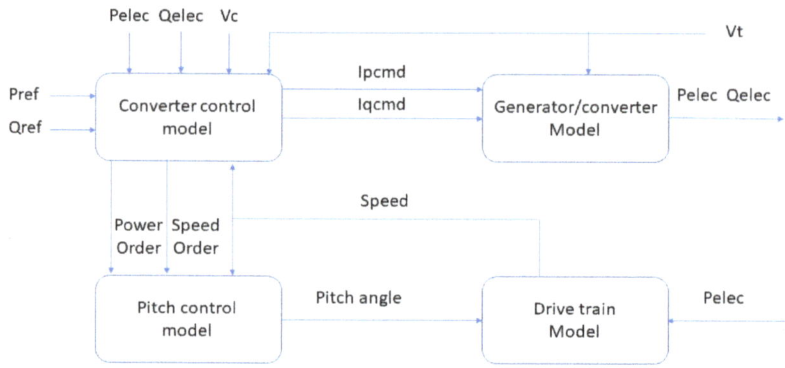

Figure 1. The block diagram of the generic model of a wind turbine.

2.1.1. PQ Diagram

The main restriction of the reactive power supported by a wind turbine is its stator current and rotor current. Figure 2 shows a PQ diagram from the stator current of a DFIG [9], which indicates the maximum reactive power absorbed or provided by a DFIG is based on its active power. For instance, Table 1 shows the relationship between active and reactive power for a DFIG-based wind turbine. As the active power is maximal, the reactive power can be supported or absorbed by 20% and 30% of capacity, respectively. However, if active power is below the maximum, the supported or absorbed reactive power can be increased. That is, the controller of a wind turbine can control its reactive power according to its PQ diagram.

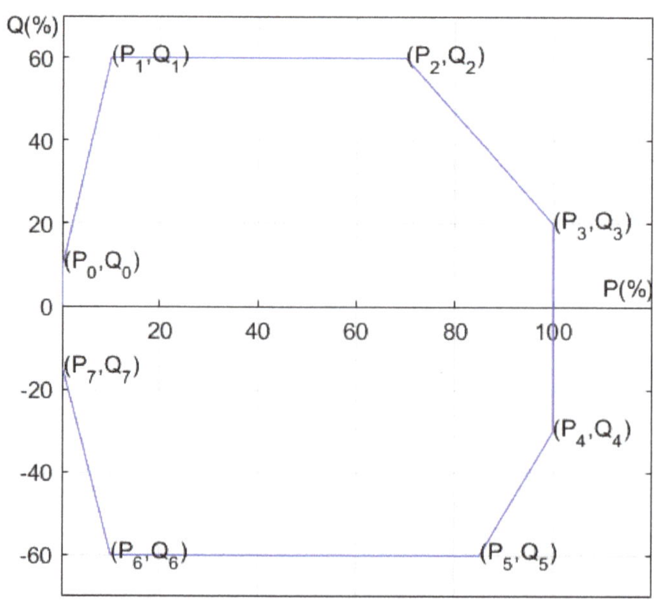

Figure 2. The PQ diagram of a DFIG-based wind turbine.

Table 1. The PQ chart of a DFIG.

Point	P (%)	Q (%)
0	0	10
1	10	60
2	70	60
3	100	20
4	100	−30
5	85	−60
6	10	−60
7	0	−15

2.1.2. QV Control

The adaptive QV control enables a wind turbine to provide more reactive power. Figure 3 shows the block diagram to obtain the gain k_{q_i} used in Figure 4. First, the upper Q_{max_i} and lower limit Q_{min_i} of a wind turbine are determined in the adaptive QV controller based on the desired PQ diagram. Next, the gain k_{q_i} at each wind turbine is obtained by considering its reactive power limit. This control method can control the voltage at the PCC and the terminal of each wind turbine. For example, in Figure 4, the limit of ±0.1 (i.e., the block of "limit" in Figure 4) restricts the difference between the reference voltage (V_{ref_pcc}) and the measured voltage (V_{meas_pcc}) at the PCC [11]. Next, the difference between the reference voltage and the measured voltage at the wind turbine is added to the output signal from the limit to obtain the error signals. Finally, the required Q_{ref} during the fault can be computed. In contrast, the QV gain of the fixed QV control is set to 5 in this study.

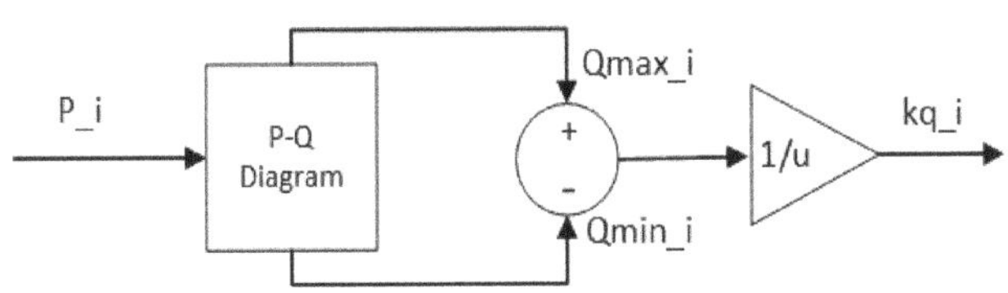

Figure 3. The block diagram to obtain the kq_i.

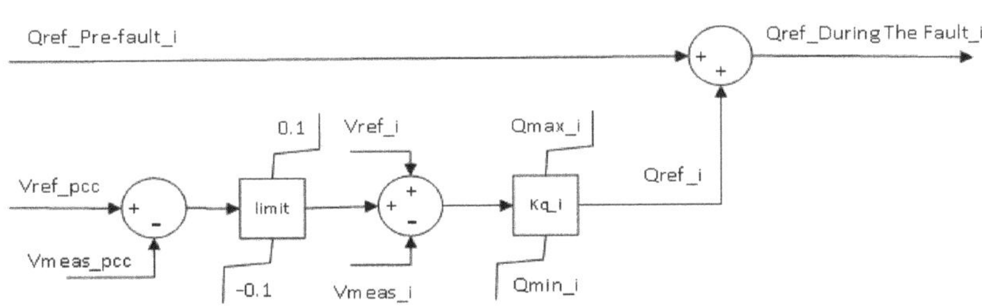

Figure 4. The control block of the adaptive QV control.

2.1.3. Voltage Droop Control

Voltage droop control is a kind of reactive control method. The functional theory of voltage droop control is similar to the frequency control of a synchronous generator. It can stabilize the voltage by providing or absorbing reactive power from wind turbines. The conventional voltage droop is expressed as

$$Q_{ref-Duringthefault} = Q_{refPre-fault} - \frac{1}{R_i}\left(V_{measpcc} - V_{refpcc}\right) \quad (1)$$

where $V_{measpcc}$ is the voltage that measures at the PCC; V_{refpcc} is the voltage reference at the PCC; $Q_{refPre-fault}$ is the original reference value of reactive power from a wind turbine; $Q_{ref-Duringthefault}$ is the new reference value of reactive power from a wind turbine. The coefficient $1/R_i$ is the reciprocal of k_{q_i}. The structure of voltage droop control is shown in Figure 5.

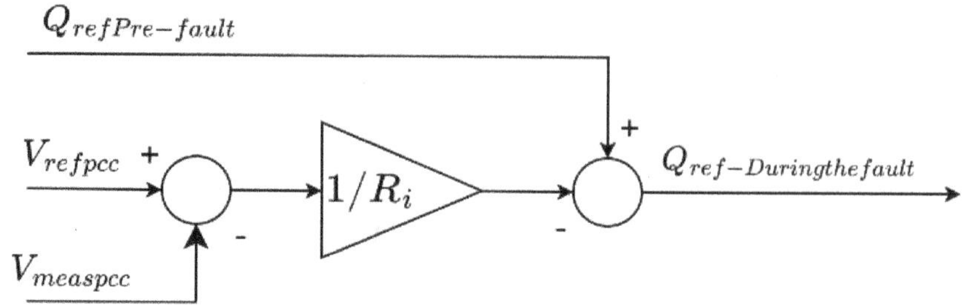

Figure 5. The control block of the voltage droop control.

2.2. VSC-HVDC Control

Figure 6 shows the control strategy of a VSC-HVDC-connected offshore wind farm (OWF). The I_{dref} of inner control is derived from either V_{DC} control or P control at the GSVSC or WFVSC. The I_{qref} of inner control is derived from the proposed control method to control the voltage at PCC. A detailed description of the inner control loop, V_{DC} control and P control can be obtained from [37,38].

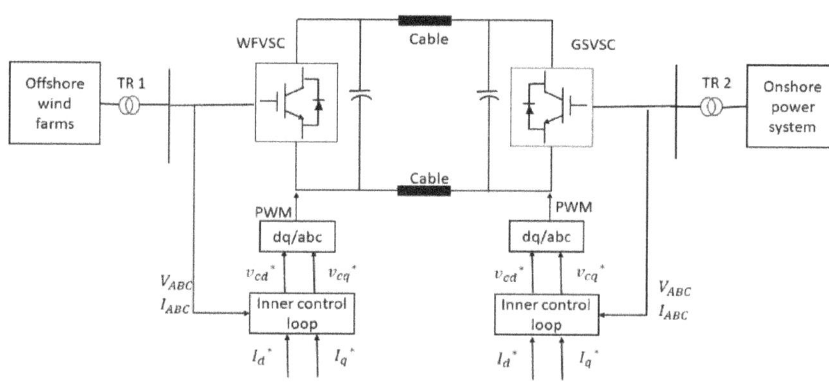

Figure 6. The control strategy of VSC-HVDC connected OWF.

2.2.1. Power Factor Control

Figure 7 shows the block diagram of power factor control, where both power factor and active power are used to obtain the required reactive power. The sign of the power factor can identify whether the reactive power is provided or absorbed. Finally, Q_{ref} is sent to the Q controller to compare with the measured reactive power Q_{meas} at the PCC, and the error between them is sent to the PI controller to obtain I_{qref}. Figure 8 shows the relationship between reactive power and power factor. To identify the leading or lagging power factor, Figure 8 uses different signs to present them. For instance, a negative sign indicates a leading power factor that a wind turbine absorbs reactive power; the sign is positive for a lagging power factor.

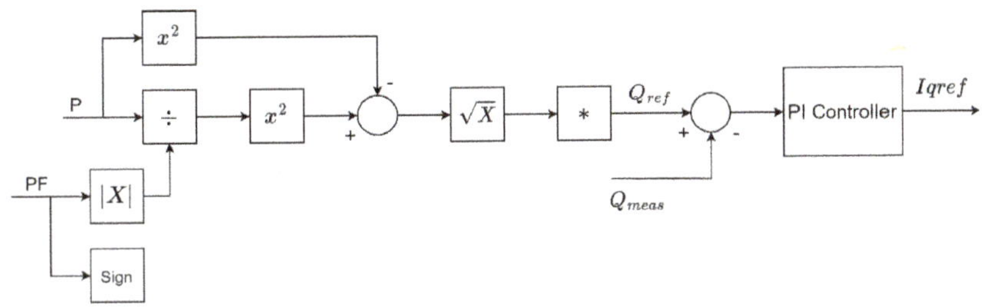

Figure 7. Power factor control.

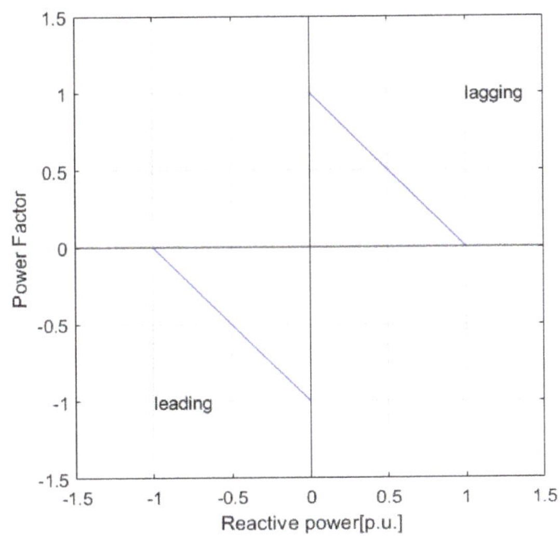

Figure 8. Presentation of power factor.

2.2.2. V_{ac}-Q Droop Control

Figure 9 shows the block diagram for the V_{ac}-Q droop control. In this figure, the deadband can set an acceptable voltage margin that the controller does not trigger, which only provides the desired reactive power. While the voltage exceeds the range, the controller can provide or absorb reactive power for voltage support. The $Q_{ref-Duringthefault}$ is sent to Q_{ref} and Q_{meas} to get I_{qref}.

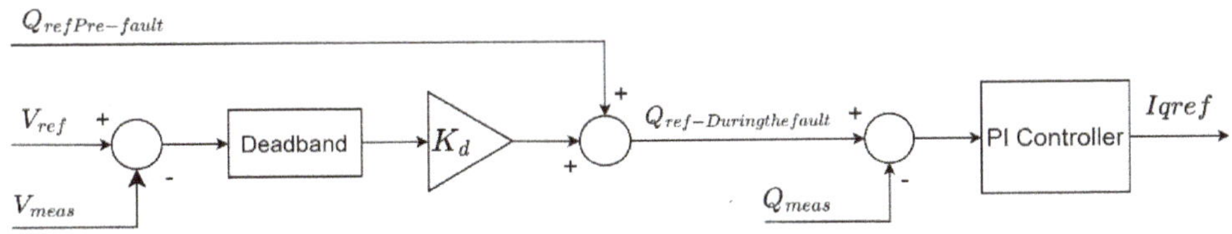

Figure 9. V_{ac} -Q droop control.

2.2.3. Voltage Control

Voltage control uses the error between the reference voltage V_{ref} and the measured voltage V_{meas}; then the error between them is sent to the PI controller. The output of the PI controller is I_{qref} that can adjust reactive power to maintain the voltage at the reference value. The topology of voltage control is shown in Figure 10.

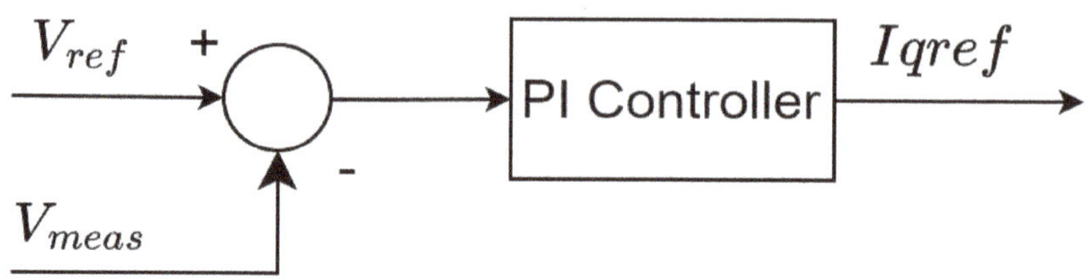

Figure 10. Voltage control.

3. Results and Discussions

3.1. Simulation Topology of the Used AC System and HVDC System

To investigate the effect of the proposed methods on the system voltage, the simulations have been done in an AC test system, which consisted of equivalent wind farms and an onshore power system, as shown in Figure 11; in addition, Figure 12 shows the HVDC test system. Each equivalent wind farm includes four wind turbines with a rated power of 8 MW each. The total rated power of the offshore wind farm is 160 MW. On the onshore grid, there are six synchronous generators with a rated power of 125 MW each. Table 2 shows the assumed wind speeds for WTs. It is assumed to give a steady-state (constant) wind-speed scenario for each wind farm, and different wind farms would have different wind speeds because of wake loss or other reasons. For example, the wind speed to the first and fifth wind farms are assumed as 11.6 m/s and 9 m/s, respectively. Power Generation from wind turbines will be affected by different wind speeds.

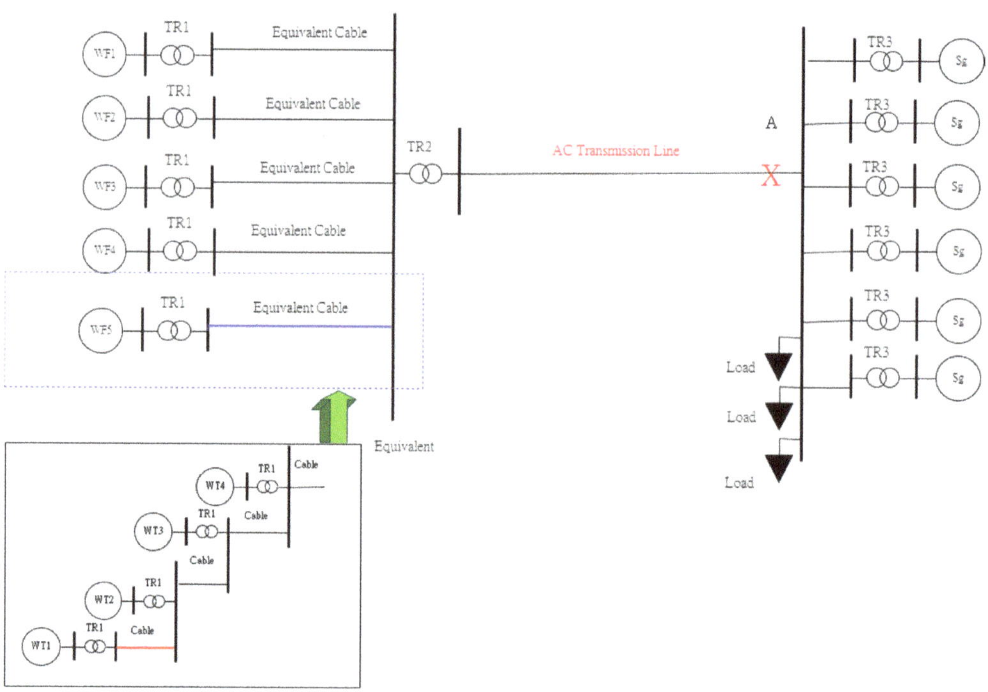

Figure 11. AC Test system.

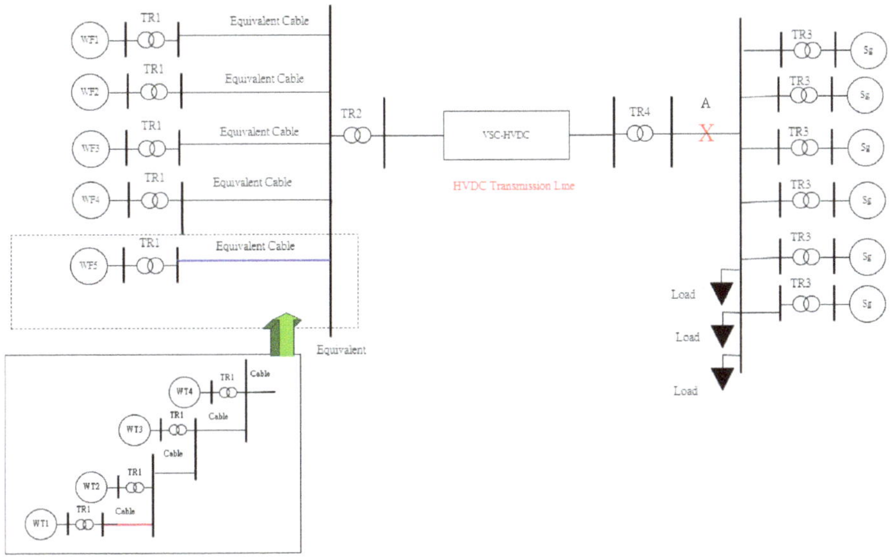

Figure 12. HVDC Test system.

Table 2. Wind speed of equivalent WFs.

No. of Wind Farm	Wind Speed (m/s)
1	11.6
2	11.4
3	10.6
4	10
5	9

Table 3 indicates the parameter of the cable. The turns ratio of the transformer in TR1 and TR2 is 0.69 KV/66 KV and 66 KV/161 KV, respectively. In a large-scale wind farm with multiple wind turbines, the equivalent wind-farm modeling technique from multiple wind turbines with many connection lines to an equivalent wind turbine with an equivalent connection line is significant, which greatly reduces simulation time. Thus, this work established an equivalent wind farm modeling based on [39,40]. In Table 3, the cable parameters before equivalence indicate the parameters of an original cable from a wind turbine to the connection point, as shown in the red line of Figure 11 or Figure 12. However, the parameters after equivalence mean the equivalent impedance of the cable from an equivalent wind turbine to the connection point, as shown in the blue line of Figure 11 or Figure 12.

Table 3. Parameters of cable.

Parameters	Before Equivalent	After Equivalent
R (Ohms)	0.045	0.084375
L (mH)	0.371362	0.696303
C (uF)	0.24	0.96
Rate voltage (kv)	66	66

The duration of the experiment is 12 s. However, most of the figures (Figures 13–24) only show the runtime within 10 s because the time length is enough. In Sections 3.2 and 3.3, an additional reactive power load of 180 MVar is assumed to be increased into the onshore grid at 8 s. In Sections 3.4 and 3.5, a three-phase short-circuit fault occurs at 8 s at point A of Figures 11 and 12, and the fault lasts for 100 ms.

3.2. A Reactive Power Load Is Added to the System with an AC Transmission System

To investigate whether the proposed methods can work adequately in the AC system or not, an increase of reactive load (180 MVar) is added into the system in 8 s.

Figure 13 shows the transient voltage curve at the PCC using the Adaptive QV, Fixed QV, Variable droop and Constant droop controls. Figure 14 shows the transient reactive power that inserts into the PCC with the corresponding control methods. The voltage nadir using the fixed QV is higher than that using the variable droop control, although the peak reactive power using the fixed QV is lower than that using the variable droop control because the fixed QV method has the characteristics of rapid response. That is, the response time of reactive power is quick.

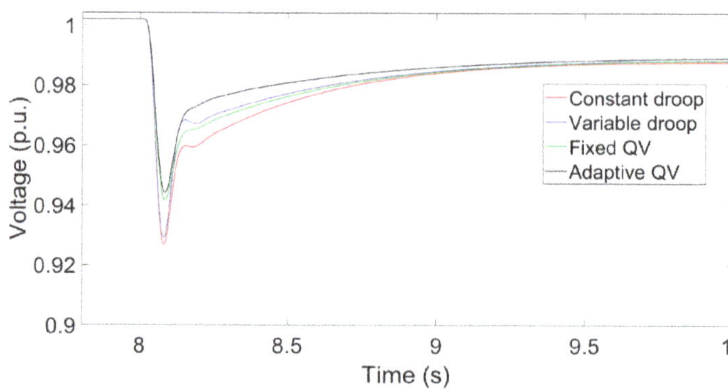

Figure 13. Transient voltage curve with an AC transmission system.

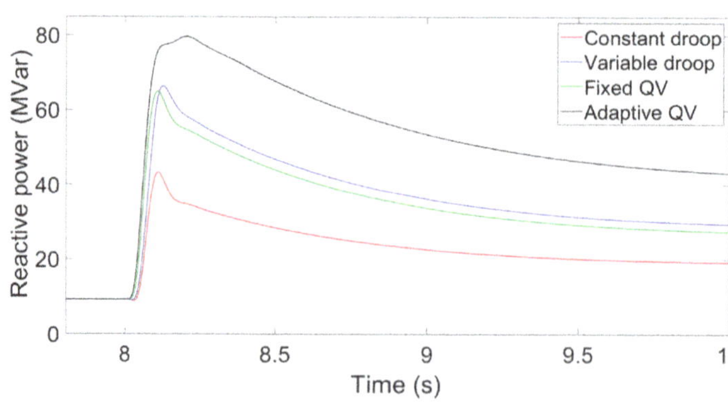

Figure 14. Transient Reactive power with an AC transmission system.

3.3. A Reactive Power Load Is Added to the System with a VSC-HVDC Transmission System

To investigate whether the proposed methods can work appropriately in the HVDC-connected wind power system, a reactive load (180 MVar) is assumed to be added into the system in 8 s. The power factor of power factor control and V_{ac}-Q droop control is set to 0.97 leading. Figure 15 shows the transient voltage curve at the PCC using the Voltage control, V_{ac}-Q Droop and Power factor controls. The voltage by using the voltage control recovers quickly because it provides a rapid reactive power response after the reactive power load is added. The transient voltage using the power factor control drops the most because it maintains a constant power factor without providing extra reactive power.

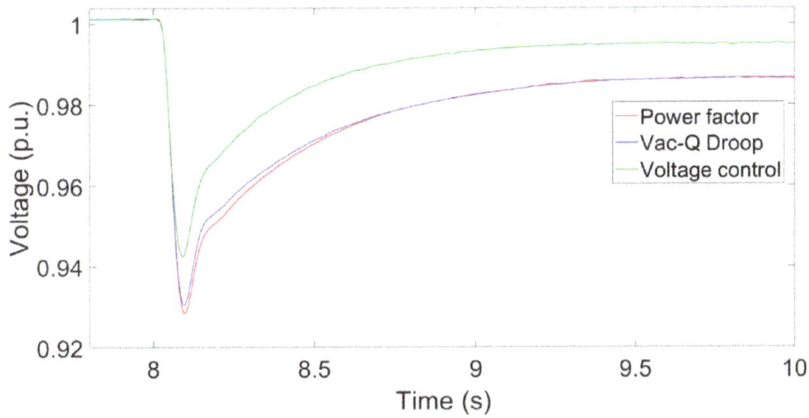

Figure 15. Transient voltage curve with a VSC-HVDC transmission system.

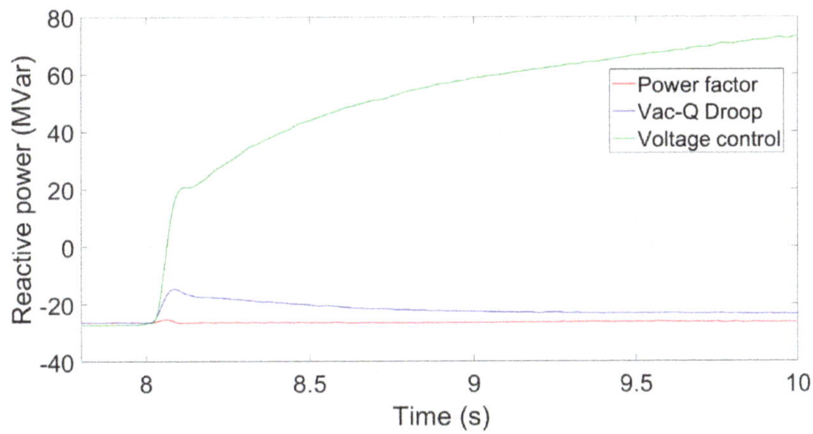

Figure 16. Transient reactive power with a VSC-HVDC transmission system.

3.4. Three-Phase Short Circuit Fault at the System with an AC Transmission System

In this study, a three-phase short circuit fault for 100 ms occurs at 8 s in point A (Figures 11 and 12). Figure 17 shows the transient voltage curve at the PCC using Adaptive QV, Fixed QV, Variable droop and Constant droop controls. Figure 18 shows the maximum reactive power that inserts into the PCC using Variable droop control, Adaptive QV, Fixed QV, and Constant droop control. Like the case in Section 3.2, the characteristics of the QV method with a rapid response can cause a higher voltage nadir than the droop control. The transient voltage using adaptive QV or fixed QV is similar, while the voltage using the variable droop or the constant droop control is similar. The voltage nadir at the PCC can be improved slightly using adaptive QV or fixed QV, which is like the simulation results in [11].

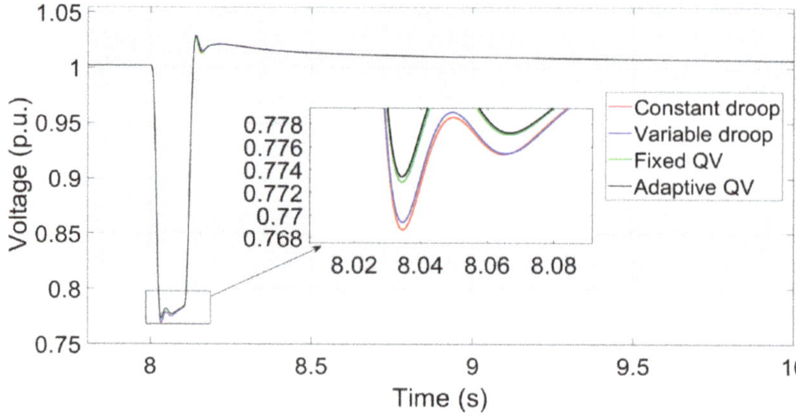

Figure 17. The transient voltage at the PCC using AC transmission.

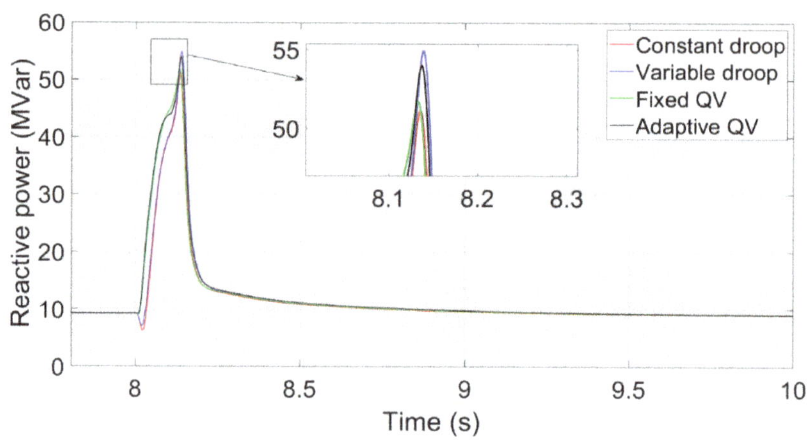

Figure 18. The transient reactive power from the wind farm using AC transmission.

3.5. Three-Phase Short Circuit Fault at the System with a VSC-HVDC Transmission System

A three-phase short circuit fault for 100 ms occurs at 8 s in point A. Figure 19 shows the transient voltage curves at the PCC using Voltage control, V_{ac}-Q Droop and Power factor control. The voltage at the PCC using V_{ac}-Q droop control recovers quicker than that using the power factor control because it provides extra reactive power after the fault. A large kd value in the V_{ac}-Q droop control can provide more reactive power. The voltage at the PCC using the voltage control becomes slightly higher after the fault because this control method can provide much reactive power.

3.6. Comparison of an Increase of Reactive Power Load at Both Systems with Different Transmission Types

Figure 21 summarizes the transient voltage curves at PCC using Adaptive QV, Voltage control, Power factor control, and no control. The red line and black line represent the transient voltage based on the AC and HVDC transmission systems, respectively. Notably, there is no control for the power factor-based method in the HVDC transmission system. The voltage drop is less if any control method is implemented at both AC and HVDC transmission systems because they can provide extra reactive power after the fault. Without a doubt, the voltage nadir with controls is higher than those without any control. The voltage nadir using the adaptive QV is higher than that using the voltage control because of the the parameters of PI controller. The PI parameter can be adjusted to avoid providing much reactive power during a three-phase short circuit fault, which causes overvoltage after a fault.

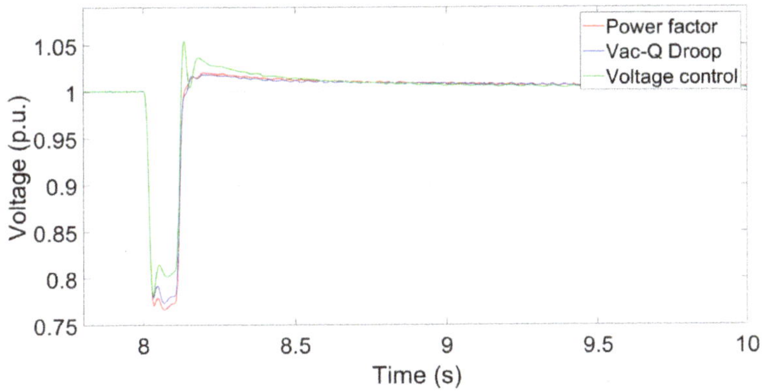

Figure 19. The transient voltage at the PCC using DC transmission.

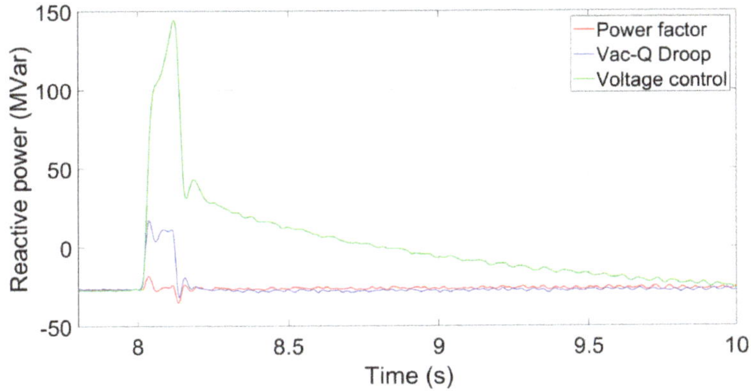

Figure 20. The transient reactive power from the wind farm using DC transmission.

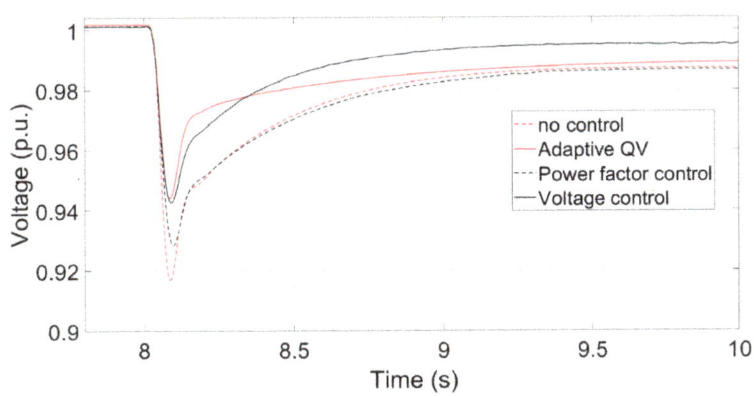

Figure 21. Comparison of transient voltage curves.

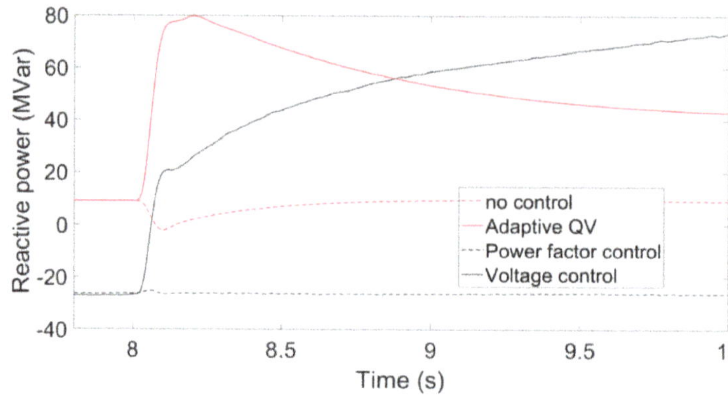

Figure 22. Comparison of reactive power curves.

3.7. Comparison of Three-Phase Short Circuit Fault at Both Transmission Systems

Figure 23 shows the transient voltage curves at the PCC using the Voltage control, Adaptive QV, Power factor control, and no control. The voltage nadir using the voltage control is higher than that using the adaptive QV control because the reactive power using the Voltage control can be increased quickly.

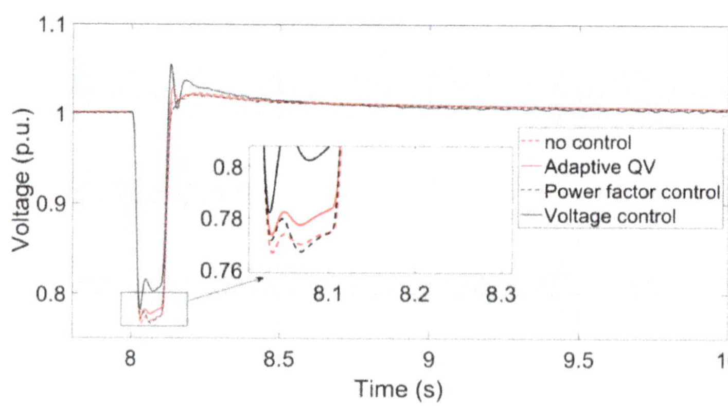

Figure 23. Comparison of transient voltage curves.

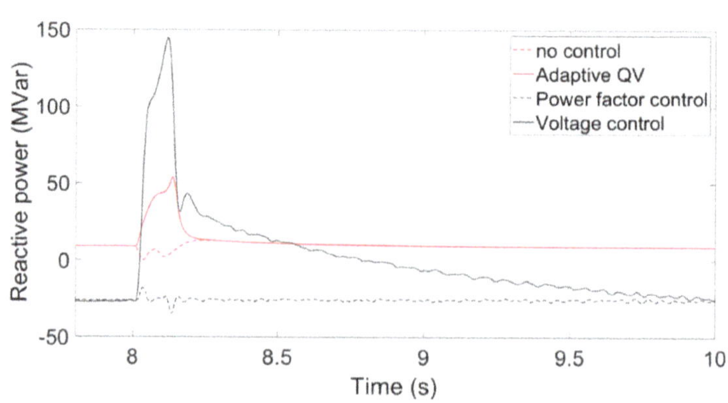

Figure 24. Comparison of reactive power curves.

3.8. Comparison of Tripping of Synchronous Generators at Both Transmission Systems

It is assumed that a synchronous generator is tripped at 8 s, and Figure 25 shows the simulation results, which reveals that the influence on voltage is less.

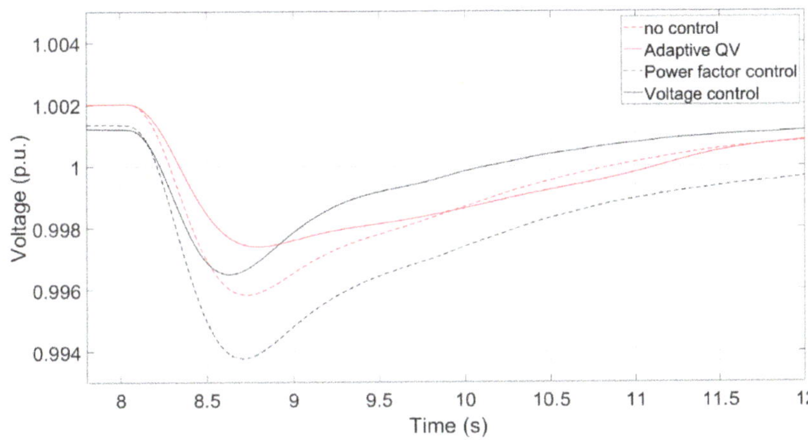

Figure 25. Transient voltage in Section 3.8.

4. Conclusions

With a high penetration of wind power generation in a power system, wind turbines should provide more ancillary services like traditional synchronous generators. Thus, some voltage control methods, such as voltage droop control and QV control, have been proposed recently. Besides, many offshore wind farms will be installed in the world, and high voltage direct current-based transmission systems should be utilized to connect wind farms and the main power grid, especially when the installed capacity or the transmission distance of offshore wind farms is increased. Therefore, this study compared both voltage droop control and QV control for wind turbines, and both voltage control and V_{ac}-Q control for HVDC transmission systems. In addition, equivalent wind farms are also established.

From the simulation results based on the test system, the transient voltage nadir using any control is higher than that without any control, and the QV control can achieve a better result compared to the droop control in both scenarios. Additionally, the voltage nadir using the voltage control with an HVDC transmission system is higher than that with an AC transmission system during a three-phase fault.

The fluctuation of voltage at the connected point varies based on the grid strength. That is, the voltage fluctuation is high if the wind farm is connected to a weak grid, and vice versa. As a result, applying an appropriate voltage control method should consider the grid strength. For instance, if the grid strength at the PCC is weak, additional reactive power compensation devices could be needed to maintain the voltage stability. Additionally, coordinated control schemes combining wind turbines with other devices should be considered in the future.

Author Contributions: Conceptualization, Y.-K.W. and D.-Y.G.; methodology, Y.-K.W. and D.-Y.G.; validation, Y.-K.W.; formal analysis, D.-Y.G. and T.-D.T.; investigation, Y.-K.W.; resources, Y.-K.W.; data curation, Y.-K.W.; writing—original draft preparation, D.-Y.G. and T.-D.T.; writing—review and editing, Y.-K.W.; visualization, Y.-K.W.; supervision, Y.-K.W.; project administration, Y.-K.W.; funding acquisition, Y.-K.W. All authors have read and agreed to the published version of the manuscript.

Funding: This research was funded by the Ministry of Science and Technology (MOST) of Taiwan, grant number MOST 110-2221-E-194-029-MY2.

Conflicts of Interest: The authors declare no conflict of interest.

References

1. Wu, Y.-K.; Chang, S.-M.; Mandal, P. Grid-connected wind power plants: A survey on the integration requirements in modern grid codes. *IEEE Trans. Ind. Appl.* **2019**, *55*, 5584–5593. [CrossRef]
2. Ellis, A.; Nelson, R.; Von Engeln, E.; MacDowell, J.; Casey, L.; Seymour, E.; Peter, W.; Williams, J.R.; Walling, R.; Barker, C.; et al. Review of existing reactive power requirements for variable generation. In Proceedings of the 2012 IEEE Power and Energy Society General Meeting, San Diego, CA, USA, 22–26 July 2012; pp. 1–7. [CrossRef]
3. Martin, J.A.; Hiskens, I.A. Reactive power limitation due to wind-farm collector networks. In Proceedings of the 2015 IEEE Eindhoven PowerTech 2015, Eindhoven, The Netherlands, 29 June–2 July 2015; pp. 1–6. [CrossRef]
4. Sharma, R. *Electrical Structure of Future Off-Shore Wind Power Plant with a High Voltage Direct Current Power Transmission*; Technical University of Denmark: Kongens Lyngby, Denmark, 2012.
5. Silva, B.; Moreira, C.L.; Seca, L.; Phulpin, Y.; Lopes, J.A.P. Provision of Inertial and Primary Frequency Control Services Using Offshore Multiterminal HVDC Networks. *IEEE Trans. Sustain. Energy* **2012**, *3*, 800–808. [CrossRef]
6. Nawir, M.; Adeuyi, O.; Wu, G.; Liang, J. Voltage stability analysis and control of wind farms connected to weak grids. In Proceedings of the 13th IET International Conference on AC and DC Power Transmission (ACDC 2017), Manchester, UK, 14–16 February 2017. [CrossRef]
7. Schmall, J.; Huang, S.-H.; Li, Y.; Billo, J.; Conto, J.; Zhang, Y. Voltage stability of large-scale wind plants integrated in weak networks: An ERCOT case study. In Proceedings of the 2015 IEEE Power & Energy Society General Meeting, Denver, CO, USA, 26–30 July 2015; pp. 1–5.
8. Huang, S.-H.; Schmall, J.; Conto, J.; Adams, J.; Zhang, Y.; Carter, C. Voltage control challenges on weak grids with high penetration of wind generation: ERCOT experience. In Proceedings of the 2012 IEEE Power and Energy Society General Meeting, San Diego, CA, USA, 22–26 July 2012; pp. 1–7. [CrossRef]
9. Martínez, J.; Kjær, P.C.; Rodriguez, P.; Teodorescu, R. Design and Analysis of a Slope Voltage Control for a DFIG Wind Power Plant. *IEEE Trans. Energy Convers.* **2011**, *27*, 11–20. [CrossRef]
10. Li, Q.; Zhang, Y.; Ji, T.; Lin, X.; Cai, Z. Volt/Var Control for Power Grids with Connections of Large-Scale Wind Farms: A Review. *IEEE Access* **2018**, *6*, 26675–26692. [CrossRef]
11. Kim, J.; Seok, J.-K.; Muljadi, E.; Kang, Y.C. Adaptive *Q–V* Scheme for the Voltage Control of a DFIG-Based Wind Power Plant. *IEEE Trans. Power Electron.* **2015**, *31*, 3586–3599. [CrossRef]
12. Li, Y.; Xu, Z.; Zhang, J.; Meng, K. Variable Droop Voltage Control for Wind Farm. *IEEE Trans. Sustain. Energy* **2017**, *9*, 491–493. [CrossRef]
13. Liu, B.; Zhang, X.; Kong, D.; Liu, X.; Chang, X. Research on Reactive Power and Voltage Coordination Control Strategy for Multi-wind Farm Connecting to Power Grid. In Proceedings of the 2019 IEEE Innovative Smart Grid Technologies-Asia (ISGT Asia), Chengdu, China, 21–24 May 2019; pp. 1633–1637. [CrossRef]
14. Basu, M.; Kim, J.; Nelms, R.M.; Muljadi, E. Coordination of Utility-scale PV Plant and Wind Power Plant in In-ter-area-Oscillation Damping. In Proceedings of the IEEE Transactions on Industry Applications 2023, Nashville, TN, USA, 29 October–2 November 2023; pp. 1–9.
15. Jia, K.; Dong, X.; Wen, Z.; Wu, W.; Bi, T. Harmonic Injection Based Fault Ride-through Control of MMC-HVDC Connected Offshore Wind Farms. *IEEE Trans. Sustain. Energy* **2023**, 1–11. [CrossRef]
16. Dai, J.; Wan, L.; Chang, P.; Liu, L.; Zhou, X. Reactive Voltage Control Strategy for PMSG-Based Wind Farm Considering Reactive Power Adequacy and Terminal Voltage Balance. *Electronics* **2022**, *11*, 1766. [CrossRef]
17. Tan, H.; Li, H.; Yao, R.; Zhou, Z.; Liu, R.; Wang, X.; Zheng, J. Reactive-voltage coordinated control of offshore wind farm considering multiple optimization objectives. *Int. J. Electr. Power Energy Syst.* **2021**, *136*, 107602. [CrossRef]
18. Liu, W.; Liu, Y.; Wu, L. Model Predictive Control Based Voltage Regulation Strategy Using Wind Farm as Black-Start Source. *IEEE Trans. Sustain. Energy* **2023**, *14*, 1122–1134. [CrossRef]
19. Mahish, P.; Mishra, S. Synchrophasor Data Based Q-V Droop Control of Wind Farm Integrated Power Systems. *IEEE Trans. Power Syst.* **2022**, *38*, 358–370. [CrossRef]
20. Huang, S.; Wu, Q.; Guo, Y.; Chen, X.; Zhou, B.; Li, C. Distributed Voltage Control Based on ADMM for Large-Scale Wind Farm Cluster Connected to VSC-HVDC. *IEEE Trans. Sustain. Energy* **2019**, *11*, 584–594. [CrossRef]
21. Han, C.; Huang, A.Q.; Baran, M.E.; Bhattacharya, S.; Litzenberger, W.; Anderson, L.; Johnson, A.L.; Edris, A.-A. STATCOM impact study on the integration of a large wind farm into a weak loop power system. *IEEE Trans. Energy Convers.* **2008**, *23*, 226–233.
22. Mi, Z.; Tian, H.; Yu, Y.; Su, X.; Fan, X.; Feng, J. Study on voltage stability of power grid with large scale wind farm inter-connected. In Proceedings of the 2009 International Conference on Sustainable Power Generation and Supply, Nanjing, China, 6–7 April 2009; pp. 1–6.
23. Ding, R.; Meng, C.; Qiao, Y. The coordinating control of voltage and reactive power between SVC and DFIG after LVRT. In Proceedings of the 2015 IEEE Eindhoven PowerTech 2015, Eindhoven, The Netherlands, 29 June–2 July 2015; pp. 1–5. [CrossRef]
24. Latorre, H.F.; Ghandhari, M. Improvement of voltage stability by using VSC-HVdc. In Proceedings of the 2009 Transmission & Distribution Conference & Exposition: Asia and Pacific, Seoul, Republic of Korea, 26–30 October 2009; pp. 1–4.
25. Hussein, I.I.; Essallah, S.; Khedher, A. Improvement of the Iraqi Super Grid Performance Using HVDC/HVAC Links by the Integration of Large-Scale Renewable Energy Sources. *Energies* **2022**, *15*, 1142. [CrossRef]

26. Liu, H.; Chen, Z. Impacts of large-scale offshore wind farm integration on power systems through VSC-HVDC. In Proceedings of the 2013 IEEE Grenoble Conference, Grenoble, France, 16–20 June 2013; pp. 1–5. [CrossRef]
27. Xu, J.; Liu, B.; Torres-Olguin, R.E.; Undeland, T. Grid integration of large offshore wind energy and oil & gas installations using LCC HVDC transmission system. In Proceedings of the SPEEDAM 2010, Pisa, Italy, 14–16 June 2010; pp. 784–791. [CrossRef]
28. Bernat, J.O.; Preece, R. Impact of VSC-HVDC Reactive Power Control Schemes on Voltage Stability. In Proceedings of the 2019 IEEE PowerTech, Milan, Italy, 23–27 June 2019; pp. 1–6. [CrossRef]
29. Arani, M.F.M.; Mohamed, Y.A.-R.I. Analysis and Performance Enhancement of Vector-Controlled VSC in HVDC Links Connected to Very Weak Grids. *IEEE Trans. Power Syst.* **2016**, *32*, 684–693. [CrossRef]
30. Egea-Alvarez, A.; Fekriasl, S.; Hassan, F.; Gomis-Bellmunt, O. Advanced vector control for voltage source converters connected to weak grids. *IEEE Trans. Power Syst.* **2015**, *30*, 3072–3081. [CrossRef]
31. Zhao, M.; Yuan, X.; Hu, J.; Yan, Y. Voltage dynamics of current control time-scale in a VSC-connected weak grid. *IEEE Trans. Power Syst.* **2015**, *31*, 2925–2937. [CrossRef]
32. Guo, Y.; Gao, H.; Wu, Q.; Zhao, H.; Østergaard, J. Coordinated voltage control scheme for VSC-HVDC connected wind power plants. *IET Renew. Power Gener.* **2018**, *12*, 198–206. [CrossRef]
33. Guo, Y.; Gao, H.; Wu, Q.; Zhao, H.; Ostergaard, J.; Shahidehpour, M. Enhanced Voltage Control of VSC-HVDC-Connected Offshore Wind Farms Based on Model Predictive Control. *IEEE Trans. Sustain. Energy* **2017**, *9*, 474–487. [CrossRef]
34. Sakamuri, J.N.; Rather, Z.H.; Rimez, J.; Altin, M.; Göksu, Ö.; Cutululis, N.A. Coordinated voltage control in offshore HVDC connected cluster of wind power plants. *IEEE Trans. Sustain. Energy* **2016**, *7*, 1592–1601. [CrossRef]
35. Pourbeik, P. *Proposed Changes to the WECC WT3 Generic Model for Type 3 Wind Turbine Generators*; Electric Power Research Institute, Inc.: Palo Alto, CA, USA, 2013; Volume 12, p. 11.
36. Ellis, A.; Kazachkov, Y.; Muljadi, E.; Pourbeik, P.; Sanchez-Gasca, J.J. Description and technical specifications for generic WTG models—A status report. In Proceedings of the 2011 IEEE/PES Power Systems Conference and Exposition, Phoenix, AZ, USA, 20–23 March 2011; pp. 1–8. [CrossRef]
37. Lu, G.-L.; Lin, C.-H.; Wu, Y.-K. Comparison of Communication-Based and Coordination-Based Frequency Control Schemes for HVdc-Connected Offshore Wind Farms. *IEEE Trans. Ind. Appl.* **2021**, *57*, 3352–3365. [CrossRef]
38. Lin, C.-H.; Wu, Y.-K. Overview of Frequency-Control Technologies for a VSC-HVDC-Integrated Wind Farm. *IEEE Access* **2021**, *9*, 112893–112921. [CrossRef]
39. Muljadi, E.; Butterfield, C.; Ellis, A.; Mechenbier, J.; Hochheimer, J.; Young, R.; Miller, N.; Delmerico, R.; Zavadil, R.; Smith, J. Equivalencing the collector system of a large wind power plant. In Proceedings of the IEEE Power Engineering Society General Meeting 2006, Montreal, QC, Canada, 18–22 June 2006. [CrossRef]
40. Wu, Y.-K.; Zeng, J.-J.; Lu, G.-L.; Chau, S.-W.; Chiang, Y.-C. Development of an equivalent wind farm model for frequency regulation. *IEEE Trans. Ind. Appl.* **2020**, *56*, 2360–2374. [CrossRef]

energies

MDPI

Article

Unique Symbolic Factorization for Fast Contingency Analysis Using Full Newton–Raphson Method

Hakim Bennani [1,*], Ahmed Chebak [2] and Abderrazak El Ouafi [1]

[1] Mathematic Computer and Engineering Department, University of Quebec at Rimouski, Rimouski, QC G5L 3A1, Canada; abderrazak_elouafi@uqar.ca
[2] Green Teck Institute, Mohammed VI Polytechnic University, Benguerir 43150, Morocco; ahmed.chebak@um6p.ma
* Correspondence: hakim1201@hotmail.com

Abstract: Contingency analysis plays an important role in assessing the static security of a network. Its purpose is to check whether a system can operate safely when some elements are out of service. In a real-time application, the computational time required to perform the calculation is paramount for operators to take immediate actions to prevent cascading outages. Therefore, the numerical performance of the contingency analysis is the main focus of this current research. In power flow calculation, when solving the network equations with a sparse matrix solver, most of the time is spent factorizing the Jacobian matrix. In terms of computation time, the symbolic factorization is the costliest operation in the LU (Lower-upper) factorization process. This paper proposes a novel method to perform the calculation with only one symbolic factorization using a full Newton–Raphson-based generic formulation and modular approach (GFMA). The symbolic factorization retained can be used during the iterations of any power flow contingency scenario. A computer study demonstrates that reusing the same symbolic factorization greatly reduces computation time and improves numerical performance. Power system security assessment under N-1 and N-2 contingency conditions is performed for the IEEE standard 54-bus and 108-bus to evaluate the numerical performance of the proposed method. A comparison with the conventional power flow method shows that the time required for the analysis is shortened considerably, with a minimum gain of 228%. The comparative analysis demonstrates that the proposed solution has better numerical performance for large-scale networks.

Keywords: contingency analysis; symbolic factorization; generic formulation; KLU solver

Citation: Bennani, H.; Chebak, A.; El Ouafi, A. Unique Symbolic Factorization for Fast Contingency Analysis Using Full Newton–Raphson Method. *Energies* **2023**, *16*, 4279. https://doi.org/10.3390/en16114279

Academic Editors: Ying-Yi Hong and Sérgio Cruz

Received: 3 April 2023
Revised: 3 May 2023
Accepted: 17 May 2023
Published: 23 May 2023

1. Introduction

Contingency analysis is regarded as an integral part of power system security analysis as it determines the integrity of the power system and verifies the post-contingency equilibrium state in terms of overloads and voltage deviations [1,2]. Its main purpose is to check whether the system can operate safely without any limit violation (i.e., equipment overloading, under- or over-voltage) after the occurrence of a contingency [3].

In N-1 contingency analysis, the main focus is to evaluate the effects of a single equipment outage on power system operating conditions once the single component (such as the loss of a generator, line, and transformer) has been retrieved from the network. Traditional N-1 contingency analysis performs numerous power-flow runs with a total number of N scenarios. After solving the power flow problem for each contingency scenario, the limits are verified for branches and nodes. The system is N-1 secure if, for all scenarios, the following constraints are fulfilled:

$$I \leq I_{max}$$
$$U_{\min} \leq U \leq U_{\max}$$

The extension of N-1 contingency analysis considers simultaneous component outages (N-k contingencies). N-k security refers to the ability of the system to maintain secure operation after k components have failed in a system with N components [4,5]. For a large-scale network, strict N-k checking requires a large amount of power flow calculation. However, in real-time security assessment, the time window for system operators to analyze the problem and take corrective actions is quite limited. Therefore, the main challenge is to find an efficient method to significantly reduce the resolution time without compromising solution accuracy [6].

In contingency analysis, the power flow method used to perform the contingency analysis has a direct impact on the solution time. The Newton–Raphson power flow-based approach is generally preferred for its superior convergence characteristics and accuracy. The most computationally demanding part of this method is solving the linear equations with LU (Lower-upper) factorization at each iteration. In terms of computation time, the symbolic factorization is the costliest operation in the LU factorization process. This paper focuses on using the Newton–Raphson method to solve contingency analysis. The central question is whether it is possible to avoid the repetitive symbolic factorizations to effectively shorten the analysis time.

We will now highlight the strengths and weaknesses of the most important methods used to reduce the computational time of contingency analysis. This critical literature review will make evident the effectiveness of the proposed solution in simulating the contingency analysis of large-scale networks (Transmission and distribution systems).

The issue of improving computational speed in contingency analysis has been discussed in many works. In [7,8], the compensation theorem is applied to simulate the changes in the passive elements of the network without changing the Jacobian matrix structure. However, this method is well suited to applications involving linear network equations, which is not the case for analysis based fully on Newton's method.

The main alternative to the compensation method for network matrix modification is to perform partial refactorization [9]. This method consists of updating the LDU (Lower-diagonal-upper) factors to reflect changes in some elements of the Jacobian matrix. The weakness of this method is that the refactorization effort is affected by the position of the modified matrix elements; thus, little or no savings can be obtained if one of the modified elements is near the top of the matrix.

The DC power flow method has been widely used to accelerate the computation of contingency analysis and improve its numerical performance [10]. The DC power flow simplifies the AC power flow to a linear circuit problem. This method is convergent and non-iterative but less accurate than AC power flow solutions.

A reduction in computational time can be achieved by reducing the number of contingency cases to be tried [11,12]. This approach involves ranking contingencies in descending order of severity index. The Contingencies can then be simulated, starting with the most severe and continuing until there is no overloading caused by the outage of branches. The main shortcoming of this method is that the contingency ranking by the index may introduce some errors due to the screening effect.

In [13], the graph theory was used to demonstrate that the base case symbolic factorization can be reused to solve contingency cases. The key assumption of this method is that there is no branch to be put in service during a contingency. This limitation prevents practical use in distribution systems where the introduction of new lines to reroute flows occurs in the most interesting contingencies.

Recent research in the area includes optimization-based methods [14–20]. These methods do not enumerate every contingency but attempt to find the most severe contingency to reduce the computational burden.

Table 1 demonstrates a summary of the reviewed classical methods for managing contingency analysis, with the major strengths and weaknesses of each one.

Table 1. Summary of the classical methods of contingency analysis.

Method	**Strengths**	**Weaknesses**	**Reference**
Compensation method	• Fast linearized power flow outage analysis. • Requires less computer storage. • Include the assessment of simultaneous outages of k-branches (N-k contingencies).	• Nonlinear networks are not supported. • Simulate only the outage of passive elements.	[7,8]
Partial refactorization	• Factorization is performed only on the affected part of the matrix.	• Numerical performance depends on the position of the modified elements. • Numerical performance decreases if the number of changes in the matrix is small.	[9]
DC power flow	• Very fast. • Non-iterative approach.	• Risk of divergence. • Less accurate.	[10]
Severity index	• Number of scenarios to be evaluated is greatly reduced.	• Errors due to the screening effect. • Do not enumerate every contingency.	[11,12]
Graph theory	• Possibility of re-using the base case symbolic factorization. • Fast.	• Unsuited to simulate remedial actions. • Efficient for contingencies involving only a few branch elements.	[13]
Optimization methods	• Reduction in CPU time. • Remedial actions can be included in the optimization function. • Include the assessment of simultaneous outages of k-branches.	• Do not enumerate every contingency. • High Complexity.	[14–20]

Additionally, the aforementioned methods have only addressed cases where the change in the sparsity pattern of the Jacobian matrix is caused by the change in network topology. However, in an automatic adjusting solution [21,22] a new symbolic factorization is not only required for a change involving network topology but also during Newton's iterative process (due to control adjustment); hence, the symbolic factorization of the base case Jacobian matrix cannot be systematically reused in contingency cases.

In this paper, an original modeling technique has been proposed to avoid the occurrence of change in the sparsity pattern within the same power flow run (due to control adjustments) and after evaluating a new contingency scenario (due to a change in network topology). The proposed solution will overcome the limitations identified in classical methods (Table 1) and allow for the systematic and fast solving of contingency analysis. More specifically, we propose to perform the contingency analysis using only one symbolic factorization. The proposed approach is completely generic and can easily accommodate the simultaneous outages of k-branches in the network (i.e., N-k contingency analysis) with the possibility to include any remedial actions that consist of putting in new service elements.

This paper uses the power flow solver based on GFMA formulation to solve the power flow problem [23]. The algorithm of GFMA is based on the rigorous full Newton–Raphson approach and relies on the automatic adjustment technique, which provides an accurate and fast power flow solution. The high flexibility of this formulation offers the possibility of re-using the base case symbolic factorization of the Jacobian matrix to recompute the power flow of any other scenario. The basic idea behind the proposed approach is to imbue the GFMA power-flow algorithm with the concept of the dynamic parameters to preserve the same sparsity pattern of the Jacobian matrix. The dynamic parameters concept is the main contribution of this research paper.

This paper starts with a brief presentation of the GFMA method for solving power flow problems. The second part presents the modeling concept of dynamic parameters. In the third part, the dynamic parameters are introduced into the component models to perform a unique symbolic factorization in the contingency analysis. The last part presents illustrative simulation examples to evaluate saved time for relatively large transmission networks.

2. GFMA Formulation

The formulation of GFMA is recalled here to establish the basis for the following algorithms. GFMA provides a generic power flow formulation capable of handling complex component models and arbitrary network topology. In this method, each component of a system is modeled autonomously as a subsystem using a non-causal mathematical representation. The implicit representation of the system and the modular approach avoid many theoretical modeling complications, as the variables and load flow constraints can be arbitrarily defined. The GFMA formulation is generic, extensible, and can handle arbitrary component models without any known limitation.

Each component is described using the model representation in the form of Equation (1). For the convenience of the reader, matrices and vectors are indicated in bold type characters.

$$\mathbf{g_k}(\mathbf{u_k}, \mathbf{y_k}, \mathbf{x_k}, \boldsymbol{\lambda}_\mathbf{k}) = 0 \tag{1}$$

where:

- $\mathbf{u_k}$ are the model input variables.
- $\mathbf{y_k}$ are the model output variables.
- $\mathbf{x_k}$ are the model internal variables.
- $\boldsymbol{\lambda}_\mathbf{k}$ are the model variables.

Each model introduces m inputs, n outputs, and h internal state variables. These variables are expressed in real units and can be arbitrarily defined. The voltages are defined as inputs to establish the electrical connection with the connected bus and the currents as outputs to formulate Kirchhoff's Current Law. The inputs and outputs represent the interface between the model and the system. The readers can refer to [23] for more details.

The component models are aligned along the diagonal to form the equipment Jacobian matrix. This block diagonal matrix is augmented with the linking equations to establish the connection between components. These extra-equations are external to the components model and have a linear form as follows:

$$\sum_{i,j} (c_{i,k} \cdot y_{i,k} + d_{j,l} \cdot u_{j,l}) = 0 \tag{2}$$

where $y_{i,k}$ is the ith output variable of the kth model; $u_{j,l}$ is the jth input variable of the lth model; $c_{i,k}$ and $d_{j,n}$ are the coefficients associated with $y_{i,k}$ and $u_{j,l}$, respectively.

Herein, the system is initialized using the single-iteration FP solution [24]. The estimation of the initial guess of the state vector is obtained by converting (1) to the linearized version and solving the resulting linear system of equations in the form of

$$\mathbf{A}.\mathbf{x} = \mathbf{b} \tag{3}$$

A comprehensive example is shown in Figure 1 to explore the structure of the Jacobian matrix. As can be seen from Figure 2, the upper and lower block of the Jacobian matrix represent the network equipment and linking equations, respectively. The obtained Jacobian matrix is sparse and unsymmetrical. It is important to notice that the system buses do not contribute to the Jacobian matrix of the system with additional nonzero elements (Figure 2).

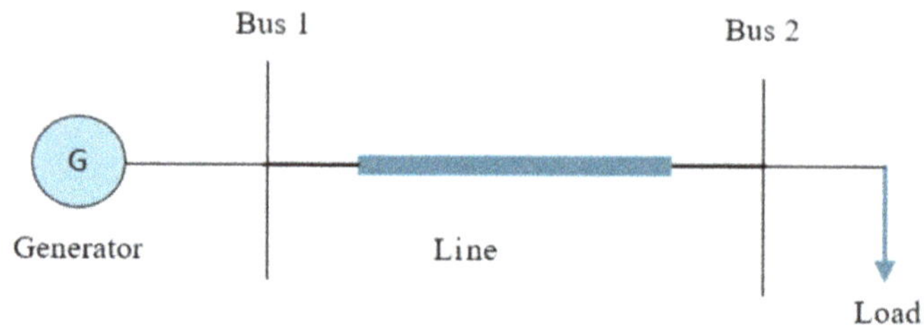

Figure 1. Test system example.

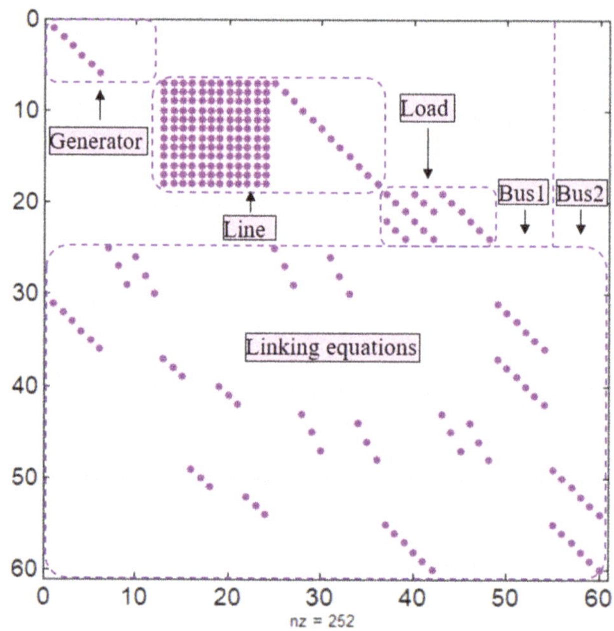

Figure 2. Matrix structure.

3. Dynamic Parameters Approach

A sparse matrix solver provides efficient numerical solutions to solve linear equations. The KLU solver is suitable for solving unsymmetrical and sparse linear systems [25] and is considered one of the most efficient packages designed for electrical circuits. Roughly speaking, the KLU solver performs two steps in sequence:

1. Symbolic factorization: performs optimal permutations and pre-ordering to reduce the fill-ins (number of non-zero elements).
2. Numerical factorization: Computes the factorization sub-matrices L and U.

The change in the Jacobian matrix structure is due to the presence of conditional statements ("if-else") in the code corresponding to different operating conditions. These conditions are evaluated, and the appropriate constraint equation is executed (see Equation (4)).

$$\begin{array}{ccc} g = f_1 & if & Condition(1) \\ \vdots & \vdots & \vdots \\ g = f_i & else\,if & Condition(i) \\ \vdots & \vdots & \vdots \\ g = f_n & else & Condition(n) \end{array} \quad (4)$$

The equation can be formulated as one compact mismatch equation.

$$g = \lambda_0 f_0 + \sum_{i=1}^{i=n} \lambda_i f_i \quad (5)$$

where:

- f_i and λ_i are the non-linear constraint equation and the dynamic parameter for the *ith* condition, respectively.
- f_0 and λ_0 are the linear constraint equation and the dynamic parameter used for the ini-tialization. It is worth noting that the term $\lambda_0 f_0$ can be dropped from Equation (5) if the model is linear.

The dynamic parameters approach consists of updating the parameters λ_i from iteration to iteration to enforce the appropriate mismatch equation. Specifically, when a new condition is flagged, the parameter associated with the enforced constraint equation is set to one, while the other parameters are all set to zero. This modeling technique prevents a change in the sparsity pattern of the Jacobian matrix since the same mathematical expression holds for all operating conditions.

Equations (4) and (5) are mathematically equivalent but numerically different. The main advantage of formulation (5) resides in its high computational performance since the repetitive symbolic factorizations are systematically avoided.

The dynamic parameters approach has been successfully applied to the power flow problem and will be extended here for contingency analysis. It will be demonstrated in the next section that the use of this modeling technique finds useful applications in network problems involving a change in network topology.

4. Outage Modeling

In this section, the dynamic parameters approach presented in [23] is extended to avoid the symbolic factorizations related to the change in network topology. Herein, the constraint equation enforcing the outage of the equipment (f_c) and its associated parameter (λ_c) are incorporated in the original formulation as follows:

$$g = \lambda_c f_c + (1 - \lambda_c).(\lambda_0 f_0 + \sum_{i=1}^{i=n} \lambda_i f_i) \quad (6)$$

The first term of Equation (6) prevents change in the nonzero pattern of the Jacobian matrix caused by the change in the network topology. To simulate a new contingency scenario, the parameter λ_c associated with the localized equipment outage must be set to one. For the base case scenario, the same parameter is set to zero for all elements in the network.

A flowchart depicting the updates in the dynamic parameters in the contingency algorithm is shown in Figure 3.

From a computational point of view, the constraint equation f_c is considered a judicious choice if the following conditions are fulfilled:

- Condition 1: Ideally, the constraint equation f_c should not introduce additional non-zero elements in the Jacobian matrix other than those initially generated from f_i.
- Condition 2: A linear constraint equation is preferred, as the Newton–Raphson converges more robustly due to its more linear formulation.

In the following sections, the equations for each model are developed using a compact mismatch equation according to the dynamic parameter approach. For the sake of clarity, the model equations are expressed in complex form, and the partial derivatives are not detailed here, as they can be obtained using the symbolical computation [26].

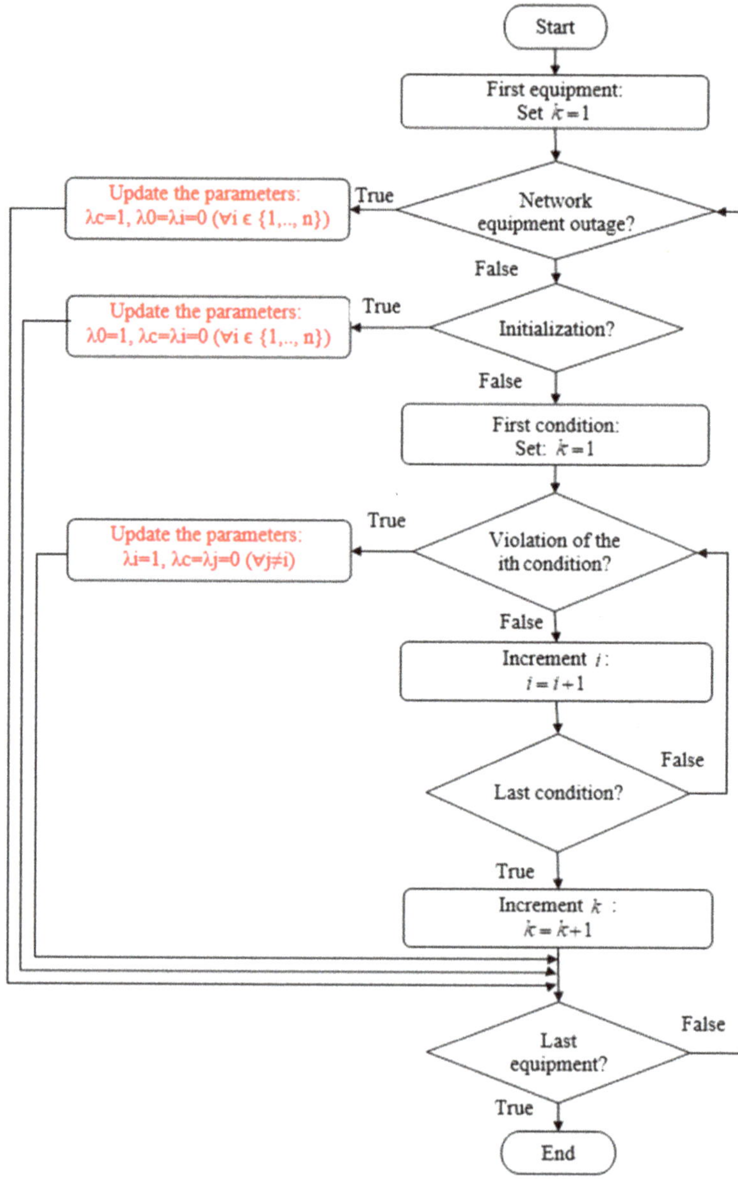

Figure 3. Updates in the dynamic parameters in the contingency case power flow solution.

4.1. Line Outage

The outage of the line is represented by a linear constraint equation forcing the entering currents to zero. For the sake of clarity, the line model of Figure 4 is presented as an example.

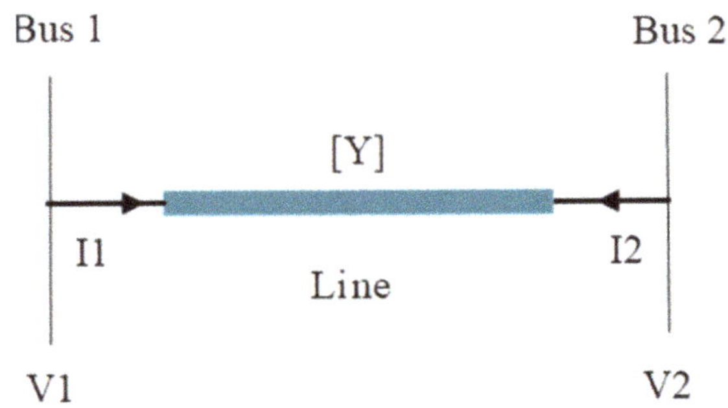

Figure 4. Line model.

The vector mismatch equation of the line formulated with the dynamic parameters is given by

$$g = \lambda_c.\underbrace{\begin{bmatrix} I_1 \\ I_2 \end{bmatrix}}_{'f_c} + (1-\lambda_c).(\underbrace{\begin{bmatrix} I_1 \\ I_2 \end{bmatrix} - Y.\begin{bmatrix} V_1 \\ V_2 \end{bmatrix}}_{'f_1}) \tag{7}$$

It should be noted that the linear constraint equation f_0 used for the initialization is not required here since the line model is inherently linear.

4.2. Generator Outage

The synchronous generator (SG) is modeled as variable susceptance and conductance (G and B) [23]. The latter two (G and B) are defined in the model as internal state variables which are solved to satisfy the power flow constraints of the machine. This model is numerically more efficient and converges faster than the conventionnel internal voltage behind reactance [27,28].

The SG state variables added to the system's variables are as follows:

$$X_{SG} = [\underbrace{\begin{matrix} V_k^{R,a}, V_k^{R,b}, V_k^{R.c} \\ V_k^{I,a}, V_k^{I.b}, V_k^{I,c} \end{matrix}}_{inputs}, \underbrace{G, B}_{Internal}, \underbrace{\begin{matrix} I_k^{R,a}, I_k^{R,b}, I_k^{R.c} \\ I_k^{I,a}, I_k^{I.b}, I_k^{I,c} \end{matrix}}_{outputs}] \tag{8}$$

This model introduces eight unknown variables (six for output currents and two for internal variables) required for the solvability of the same number of equations.

For a machine with PV constraint, the mismatch equation written with dynamic parameters is given by

$$g = \lambda_c.\underbrace{\begin{bmatrix} G \\ B \end{bmatrix}}_{f_c} + (1-\lambda_c).(\lambda_0.\underbrace{\begin{bmatrix} G - G_0 \\ B - B_0 \end{bmatrix}}_{f_0} + \lambda_1.\underbrace{\begin{bmatrix} P_G - P_{des} \\ V_G - V_{des} \end{bmatrix}}_{'f_1} + \lambda_2.\underbrace{\begin{bmatrix} 0 \\ Q_G - Q_{\min} \end{bmatrix}}_{'f_2} + \lambda_3.\underbrace{\begin{bmatrix} 0 \\ Q_G - Q_{\max} \end{bmatrix}}_{'f_3}) \tag{9}$$

where the active and reactive power is expressed as a function of the state variables

$$P_G + jQ_G = \mathbf{I}^*_{\mathbf{abc}}\mathbf{V}_{\mathbf{abc}} \tag{10}$$

The sign * indicates the complex-conjugate operator.

This mismatch equation can capture all possible operating conditions of the machine, including the initialization and equipment outage.

In the initialization stage, the parameter λ_0 is set to one while all other parameters are set to zero, and the non-linear constraint equations of the generator are reverted to the linearized version. The linear constraint equation f_0 is specifically introduced in the mismatch equation to estimate the initial value of the susceptance and conductance. For a machine with PV constraint, G_0 and B_0 are estimated, assuming a nominal power factor of the machine and the desired active power.

$$G_0 + jB_0 = \frac{conj(P_{des} + jQ_{des})}{|V_{LLnom}|^2} \tag{11}$$

The parameters λ_1,λ_2,λ_3 are dynamically updated during the iterative power flow process to flag all operating conditions of the generator. This allows the machine to switch from PV to PQ when its reactive power limit is violated.

The outage of the generator is represented by an equation forcing the susceptance and conductance to zero. This constraint equation (f_c) is linear and does not introduce any additional nonzero elements in the mismatch equation other than those generated from f_0,f_1,f_2, and f_3. The generator is considered effectively out of service if the parameter is set to one.

Six equations are also introduced from the equation relating the terminal voltages and currents.

$$\mathbf{I_{abc}} = (\mathbf{A}\mathbf{Y_{012}}\mathbf{A}^{-1}).\mathbf{V_{abc}} \tag{12}$$

where $\mathbf{A}$ is the Fortescue's transformation matrix; $\mathbf{V_{abc}}$ and $\mathbf{I_{abc}}$ are the vector of generator voltages and current in phase domain, respectively.

$$\mathbf{Y_{012}} = \begin{bmatrix} Y_0 & 0 & 0 \\ 0 & G + jB & 0 \\ 0 & 0 & Y_2 \end{bmatrix} \tag{13}$$

With the introduction of the dynamic parameters, the constraint Equation (13) becomes

$$\mathbf{I_{abc}} = \mathbf{A}\begin{bmatrix} Y_0 & 0 & 0 \\ 0 & \lambda_c.(G_0 + jB_0) + (1-\lambda_c).(\lambda_0.(G_0 + jB_0) + (\lambda_1 + \lambda_2 + \lambda_3).(G + jB)) & 0 \\ 0 & 0 & Y_2 \end{bmatrix} \mathbf{A}^{-1}\mathbf{V_{abc}} \tag{14}$$

4.3. Transformer Outage

In this section, the transformer outage model is implemented in a three-phase power-flow solution method based on the GFMA formulation. Therein, an ideal transformer connected between nodes k and n is modeled by describing the voltage and current equations relating the primary and secondary windings. The node m is inserted to include the three-phase impedance transformer. This solution is generic and can be used to handle arbitrary transformer connections. The transformer model depicted in Figure 5 is used to derive the constraint equations in the phase domain for an arbitrary three-phase transformer connection.

The constraint equations for the transformer are given by

$$g = \begin{cases} \underbrace{\lambda_c \cdot \begin{bmatrix} -\mathbf{I}_{\mathbf{m}_{abc}} \\ \mathbf{I}_{\mathbf{n}_{abc}} \end{bmatrix}}_{'f_c} + (1-\lambda_c) \cdot \Bigg(\underbrace{\begin{bmatrix} -\mathbf{I}_{\mathbf{m}_{abc}} \\ \mathbf{I}_{\mathbf{n}_{abc}} \end{bmatrix} - \begin{bmatrix} \mathbf{Y_{tr}} & -\mathbf{Y_{tr}} \\ -\mathbf{Y_{tr}} & \mathbf{Y_{tr}} \end{bmatrix} \begin{bmatrix} \mathbf{V}_{\mathbf{m}_{abc}} \\ \mathbf{V}_{\mathbf{n}_{abc}} \end{bmatrix}}_{f_1} \Bigg) \\ \mathbf{V}_{\mathbf{m}_{abc}} - \mathbf{D}(\ ratio\ ).\mathbf{V}_{\mathbf{k}_{abc}} \\ \mathbf{I}_{\mathbf{k}_{abc}} + \mathbf{D}^{\mathrm{T}}(\ ratio\ ) \cdot \mathbf{I}_{\mathbf{m}_{abc}} \end{cases} \tag{15}$$

where $\mathbf{Y_{tr}}$ is the three-phase transformer admittance matrix; *ratio* is the turn ratio of the master transformer; D is the dependency that is expressed as a function of turn ratio; and the sign ^{T} indicates the transpose matrix operator.

The transformer outage is simulated by forcing the output currents at both sides of the transformer to zero. This is achieved by setting the parameter λ_c to one.

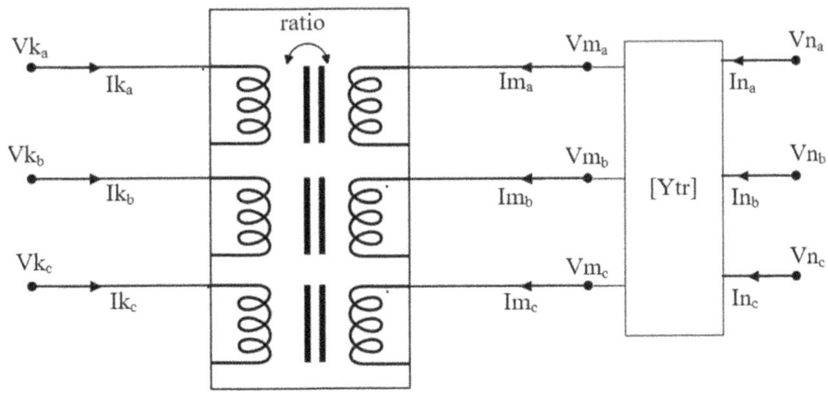

Figure 5. Three-phase transformer.

4.4. Switching Device Outage

For an ideal switch connected between nodes k and m, the model equation of the switch device depicted in Figure 6 is given by

$$g = \lambda_c.\underbrace{I_{km}}_{'f_c} + (1-\lambda_c).\underbrace{(V_k - V_m)}_{'f_1} \quad (16)$$

Figure 6. Switch device.

In this equation, the switch is considered in-service if λ_c is set to zero (i.e., closed position). The same equation can be used to simulate the outage of the ideal switch by setting λ_c to one (i.e., opened position).

4.5. Load Shedding

An optimal load-shedding scheme is necessary under contingency conditions to prevent cascade outages and complete black-out [29]. The load is formulated using the current-mismatch equations [30,31]. The load mismatch equation is given by

$$g = \lambda_c.\underbrace{I_{km}}_{'f_c} + (1-\lambda_c).(\lambda_0.\underbrace{(I_{km} - Y_L(V_k - V_m))}_{f_0} + \lambda_1.\underbrace{(I_{km} - \frac{P_{des} - jQ_{des}}{V_k^* - V_m^*})}_{'f_1}) \quad (17)$$

where V_k and V_m are the voltages at nodes k and m, respectively; I_{km} is the injected load current; and Y_L is the estimated load admittance used for the initialization.

5. General Aspects of the Algorithm

With the introduction of the dynamic parameters in the formulation of mismatch equations, the sparsity pattern of the Jacobian matrix corresponding to the base case scenario becomes identical to that of any contingency case. The unique symbolic factorization retained in all scenarios of contingency analysis corresponds to that performed for the system initialization of the base case scenario.

In the base case scenario, all the network equipment is initially in service. However, any equipment that was initially out of service but expected to be connected as part of remedial action must be included in the base case scenario and modeled as an out-of-service element. From a modeling point of view, adding and removing an element from the system are treated similarly.

Simultaneous outages (N-2) are treated as single outages (N-1). Both can reuse the symbolic factorization of the base case scenario.

The main drawback of the proposed approach is that the nonzero elements and the size of the Jacobian matrix in the contingency cases are slightly overestimated. To illustrate this point, let us consider an outage of a transmission line that is modeled using dynamic parameters. The Jacobian elements are obtained by replacing the dynamic parameter λ_c in Equation (7) with one and then deriving the partial derivatives with respect to the state variables. Expressed in real form, the Jacobian elements of the positive sequence model of the line are given by

$$\begin{array}{ll} \frac{\partial g}{\partial I1_{R,I}} = 1 & \frac{\partial g}{\partial I2_{R,I}} = 1 \\ \frac{\partial g}{\partial V1_{R,I}} = 0 & \frac{\partial g}{\partial V2_{R,I}} = 0 \end{array} \tag{18}$$

As can be seen from Equation (18), eight nonzero elements contribute to removing a branch element from the system. These elements will exist only if the symbolic factorization of the base case scenario is reused for the contingency scenario. This is evident from the fact that all elements in the base case are initially considered in service. It is worth noting that the KLU solver does not drop numerically zero entries from its sparse matrix; therefore, all the entries that are present in the data structure of the Jacobian matrix are not dropped, even though they are numerically zero.

The contingency analysis is performed with the algorithm depicted in Figure 7.

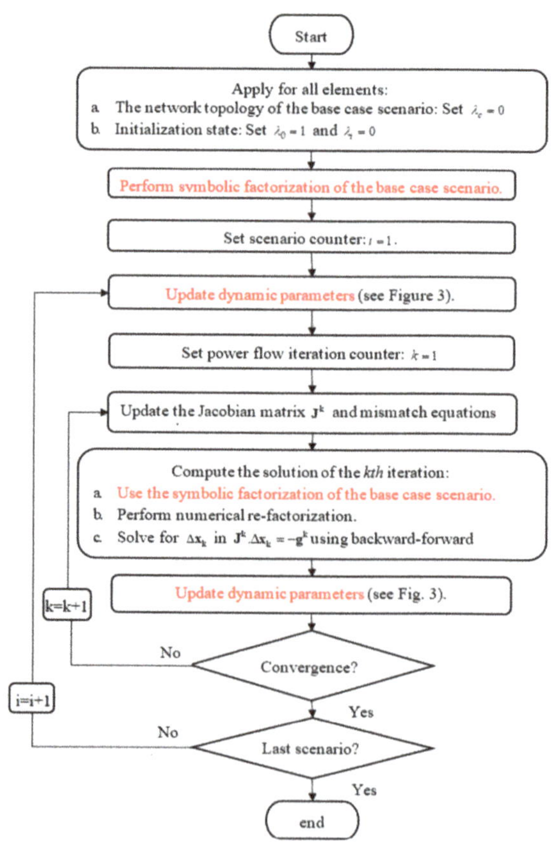

Figure 7. Contingency algorithm.

6. Validation

The proposed solution has been tested for the IEEE 57-bus and the IEEE 118-bus test systems. These networks are used to evaluate the numerical performance of the proposed solution. The objective is to compare the computational cost of the proposed solution (GFMA) with the contingency analysis method implemented in CYME power engineering software. This conventional method consists of performing extensive simulations and the strict checking of all possible contingencies using a MANA (Modified-augmented-nodal-analysis)-based power flow solution [32,33]. MANA and GFMA are two different formulations, but both are implemented in a-b-c reference frame and solved using the full-Newton algorithm and KLU solver.

Herein, the simulations are performed for the N-1 and N-2 security criteria. The N-1 criterion considers a unique contingency that is cleared by the primary protection, in this case, the faulted element is disconnected from the system without any further impact on the system. However, the N-2 criterion considers that the first outage in the system results in a second component failure. It is worth noting that the N-k security criterion leads to C_k^N possible contingencies ($C_k^N = \frac{N!}{k!(N-k)!}$).

To evaluate the resolution time of the contingency analysis, timers were inserted into the code sections, where the symbolic and numerical factorizations are performed. The normalized CPU time with respect to the proposed solution is presented at the end of this section, and a comparison between the numerical efficiency of these two methods serves as a closing statement.

All algorithms have been programmed and executed using the MATLAB Platform, which runs on a dedicated machine (3.4 GHz i7-2600 CPU with 16 GB of RAM). A tolerance of 0.01% on the voltage mismatch is used as the stopping criterion for each power flow run.

6.1. Case-1: IEEE 57-Bus Test System

The IEEE 57-bus system shown in Figure 8 is used to test the numerical performance of the proposed method and to assess the gain in computation time. The network contains 57 buses, 63 lines, 7 generators, 15 two-winding transformers, and 42 loads. The total number of components that can potentially fail in the system is 85 ($N = 63 + 7 + 15$). Contingency is carried out using the GFMA formulation proposed in this paper and MANA formulation implemented in CYME software.

The strict N-l of the IEEE 57-bus system needs 85 (C_1^{85}) outage analyses. Table 2 shows that both methods require the same number of numerical factorizations; however, the number of symbolic factorizations required in the MANA is 190, while the proposed solution performs the same calculation with only one symbolic factorization. Moreover, when the nonzero pattern is already known, the depth-first search used in Gilbert/Peierls method can be skipped. This means that numerical factorization becomes computationally less expensive when the sparsity pattern is not changed. Therefore, the cost of the numerical factorization is smaller for GFMA than MANA. In this case, the GFMA is about 2.30 faster than MANA (Table 2).

For N-2 contingency analysis, the IEEE 57-bus system needs 3570 scenarios (C_2^{85}). As expected, the CPU time is significantly reduced since one symbolic factorization is performed to evaluate this large number of outage scenarios. In this case, the GFMA is about 2.28 faster than MANA (Table 3). Although the performance ratio depends on the system size and the number of controller devices, this case study clearly demonstrates that the time required for the analysis is shortened considerably with the use of the proposed method.

Table 2. CPU Timings for the N-1 contingency analysis using MANA and GFMA.

	IEEE 57-Bus System		IEEE 108-Bus System	
	Proposed Solution (GFMA)	**CYME (MANA)**	**Proposed Solution (GFMA)**	**CYME (MANA)**
Number of Scenarios	85		240	
Reading and Treatment of network file	Not included		Not included	
Build and Update Jacobian Matrix	1.84 s	1.76 s	5.57 s	5.13 s
	Symbolic Factorizations		Symbolic Factorizations	
Number of Symbolic Factorizations	1	190	1	531
Timing of Symbolic Factorizations	0.010 s	2.09 s	0.013 s	6.90 s
	Numerical Factorizations		Numerical Factorizations	
Number of Numerical Factorizations	425 *	425	1177 *	1177
Timing of Numerical Factorizations	1.06 * s	2.97 s	3.01 s *	8.43 s
Solve	0.084 s	0.082 s	0.12 s	0.10 s
Total CPU Time	2.99 s (100%)	6.91 s (230%)	4.24 s (100%)	20.56 s (484%)

* Numerical Refactorization.

Table 3. CPU Timings for the N-2 contingency analysis using the MANA and GFMA.

	IEEE 57-Bus System		IEEE 108-Bus System	
	Proposed Solution (GFMA)	**CYME (MANA)**	**Proposed Solution (GFMA)**	**CYME (MANA)**
Number of Scenarios	3570		28,680	
Reading and Treatment of network file	Not included		Not included	
Build and Update Jacobian Matrix	76.28 s	72.46 s	687.40 s	651.43 s
	Symbolic Factorizations		Symbolic Factorizations	
Number of Symbolic Factorizations	1	8925	1	71,690
Timing of Symbolic Factorizations	0.011 s	93.71 s	0.013 s	931.97 s
	Numerical Factorizations		Numerical Factorizations	
Number of Numerical Factorizations	16,950 *	16,950	136,170 *	136,170
Timing of Numerical Factorizations	42.37 s *	110.17 s *	267.00 s *	695.00 s
Solve	3.65 s	3.54 s	31.86 s	29.45 s
Total CPU Time	122.31 s (100%)	279.89 s (228%)	986.27 s (100%)	2308.00 s (234%)

* Numerical Refactorization.

The sparsity pattern of the Jacobian matrix of the base case scenario has a dimension of $\mathbb{R}^{3078X3078}$ with 16,242 nonzero elements (see Figure 9). The matrix structure of the base case scenario is identical to any contingency case.

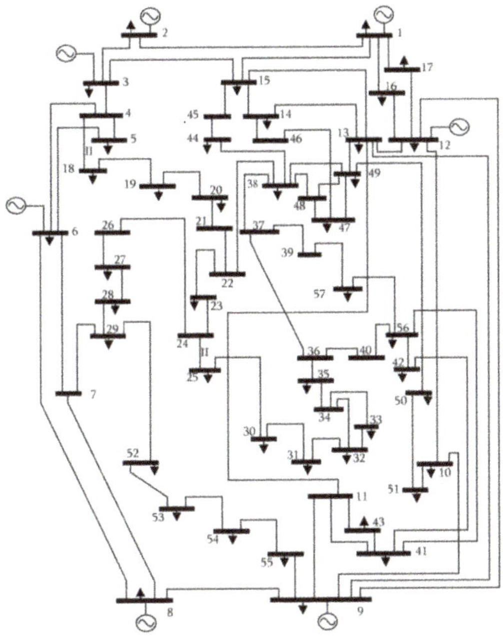

Figure 8. IEEE 57-bus System.

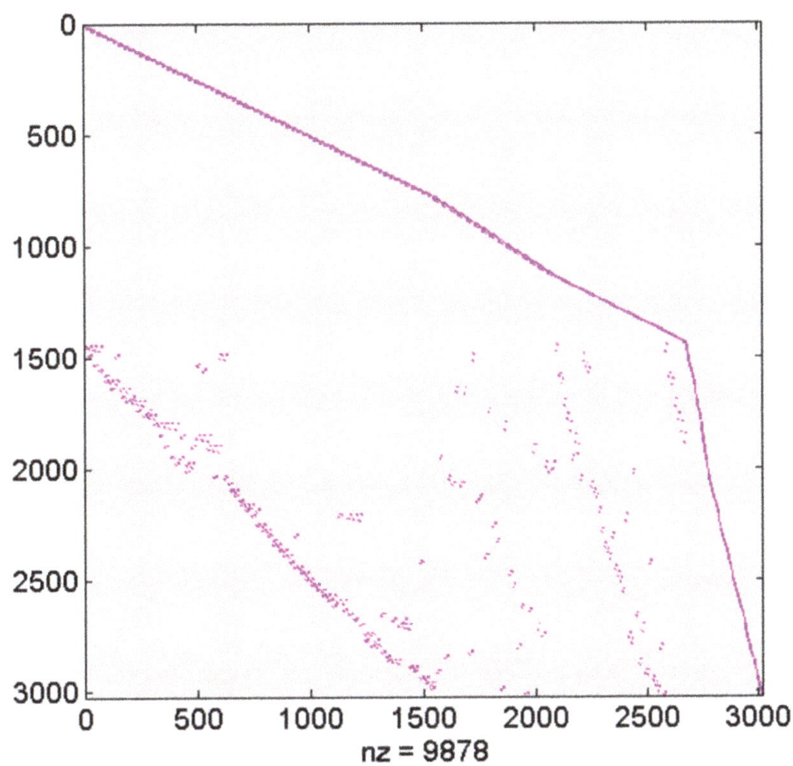

Figure 9. Sparsity pattern of the Jacobian matrix of the base case scenario (IEEE 57-bus System).

6.2. Case-2: IEEE 108-Bus Test System

This IEEE 108-bus test system was used in various studies for contingency evaluation and voltage security assessment. This system contains 19 generators, 35 synchronous condensers, 177 lines, 9 transformers, and 91 loads (Figure 10). The total number of components is 240 ($N = 19 + 35 + 177 + 9$). The strict N-l and N-2 checking of the IEEE 108-bus system needs 240 and 3570 outage analyses, respectively.

The comparative analysis presented in Tables 2 and 3 leads to the same findings as those observed in the IEEE 54-bus system. Once more, the proposed solution proves to be faster than MANA, with a significant reduction in resolution time.

In addition, the proposed method performs better for larger systems and its efficiency increases as the system increases in size. This can be observed from the results of Tables 2 and 3, where the detailed CPU time is provided for both networks (IEEE-54 and IEEE-108 systems). This is explained by the fact that the number of symbolic factorizations required to cover all contingencies increases drastically with the size of the network.

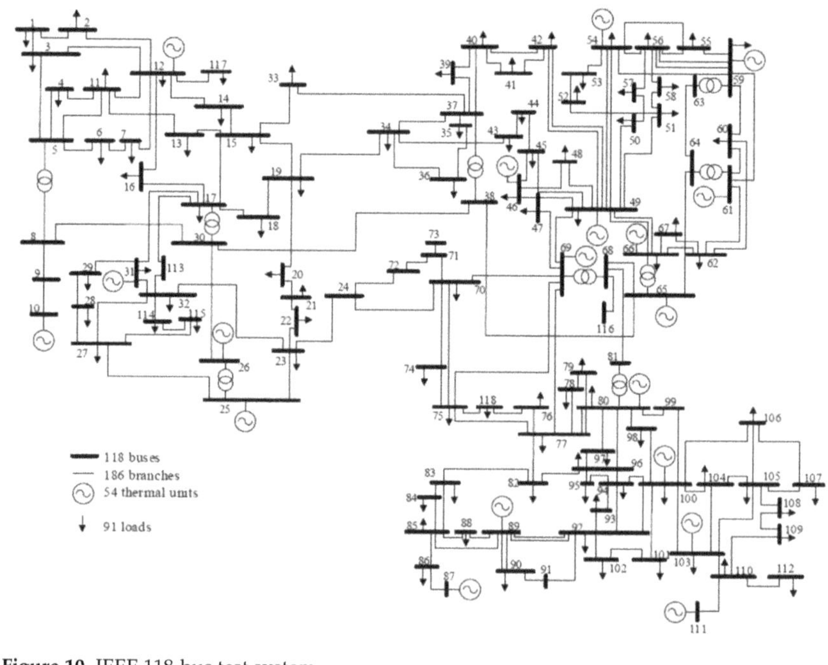

Figure 10. IEEE 118-bus test system.

7. Conclusions

The proposed modeling technique presented in this paper exploits the flexibility of the GFMA formulation to avoid the repetitive symbolic factorizations required during the iterative power flow process and after the modification of network topology. Herein, the contingency analysis is performed with only one symbolic factorization using a Newton-like method and KLU sparse matrix solver.

The proposed approach is completely generic and can be applied to simulate simultaneous outages of k-branches elements in the network and any remedial actions that consist of putting in service new elements. The proposed solution is generic and can be applied to the strict N-k checking without any known limitation.

The new method has been tested on IEEE 57-bus and IEEE 108-bus test systems, with all the details represented and simulated. The power system security assessment under N-1 and N-2 contingency conditions demonstrates that a significant computational gain has been achieved by applying the concept of dynamic parameters. The results also showed that the method performs better for larger systems and that its efficiency increases as the system increases in size.

Author Contributions: Conceptualization, H.B.; methodology, H.B.; software, H.B.; validation, H.B.; formal analysis, H.B.; investigation, H.B.; resources, H.B.; data curation, H.B.; writing—original draft preparation, H.B.; writing—review and editing, H.B. and A.C.; visualization, H.B.; supervision, A.C. and A.E.O.; project administration, A.C. and A.E.O.; funding acquisition, A.E.O. All authors have read and agreed to the published version of the manuscript.

Funding: Financial support from the University of Quebec at Rimouski under grant AEO-760700.

Data Availability Statement: Not applicable.

Acknowledgments: The financial support from the University of Quebec at Rimouski under grant AEO-760700 is gratefully acknowledged.

Conflicts of Interest: The authors declare no conflict of interest.

References

1. Kip, M.; Lei, W.; Prabha, K. Power system security assessment. *IEEE Power Energy Mag.* **2004**, *2*, 30–39.
2. Mohammed, S.; William, T.; Yong, F. Impact of Security on Power Systems Operation. *Proc. IEEE* **2005**, *93*, 2013–2025.
3. Bulat, H.; Franković, D.; Vlahinić, S. Enhanced Contingency Analysis—A Power System Operator Tool. *Energies* **2021**, *14*, 923. [CrossRef]
4. Liana, C.; Xuan, L.; Zuyi, L. Preventive Mitigation Strategy for the Hidden N-k Line Contingencies in Power Systems. *IEEE Trans. Reliab.* **2018**, *67*, 1060–1070.
5. Babalola, A.; Belkacemi, R.; Zarrabian, S. Real-time cascading failures prevention for multiple contingencies in smart grids through a multi-agent system. *IEEE Trans. Smart Grid* **2018**, *9*, 373–385. [CrossRef]
6. Hassan, H.; Mehrdad, T.; Pedram, S. Accurate, simultaneous and Real-Time screening of N-1, N-k, and N-1-1 contingencies. *Int. J. Electr. Power Energy Syst.* **2022**, *136*, 107592. [CrossRef]
7. William, F.T. Compensation Methods for Network Solutions by Optimally Ordered Triangular Factorization. *IEEE Trans. Power Appar. Syst.* **1972**, *PAS-91*, 123–127.
8. Ongun, A.; Brian, S.; William, F.T. Sparsity-Oriented Compensation Methods for Modified Network Solutions. *IEEE Trans. Power Appar. Syst.* **1983**, *PAS-102*, 1050–1060.
9. Sherman, M.C.; Vladimir, B. Partial Matrix Refactorization. *IEEE Trans. Power Syst.* **1986**, *1*, 193–199.
10. John, K.; Felix, W. Analysis of linearized decoupled power flow approximations for steady-state security assessment. *IEEE Trans. Circuits Syst.* **1984**, *31*, 623–636.
11. Mikolinnas, T.A.; Bruce, W. An Advanced Contingency Selection Algorithm. *IEEE Power Eng. Rev.* **1981**, *2*, 25–26. [CrossRef]
12. Halpin, T.F.; Robert, F.; Fink, R. Analysis of automatic contingency selection algorithms. *IEEE Power Eng. Rev.* **1984**, *5*, 29–30.
13. Elliott, M.C.; Jingjin, W.; Yiting, Z.; Renchang, D.; Guangyi, L. Symbolic Factorization Re-utilization for Contingency Analysis. In Proceedings of the IEEE Power & Energy Society General Meeting (PESGM), Portland, OR, USA, 5–10 August 2018; pp. 1–5.
14. Pudi, S.; Sanjeeb, M. Power system contingency ranking using Newton Raphson load flow method. In Proceedings of the Annual IEEE India Conference (INDICON), Mumbai, India, 13–15 December 2013; pp. 1–4.
15. Rocco, C.M.; Ramirez-Marquez, J.E.; Salazar, D.E.; Yajure, C. Assessing the Vulnerability of a Power System Through a Multiple Objective Contingency Screening Approach. *IEEE Trans. Reliab.* **2011**, *60*, 394–403. [CrossRef]
16. Ding, T.; Li, C.; Yan, C.; Li, F.; Bie, Z. Bilevel Optimization Model for Risk Assessment and Contingency Ranking in Transmission System Reliability Evaluation. *IEEE Trans. Power Syst.* **2017**, *32*, 3803–3813. [CrossRef]
17. Street, A.; Oliveira, F.; Arroyo, J.M. Contingency-Constrained Unit Commitment with n–K Security Criterion: A Robust Optimization Approach. *IEEE Trans. Power Syst.* **2011**, *26*, 1581–1590. [CrossRef]
18. Wang, Q.; Watson, J.; Guan, Y. Two-stage robust optimization for N-k contingency-constrained unit commitment. *IEEE Trans. Power Syst.* **2013**, *28*, 2366–2375. [CrossRef]
19. Levitin, G. Optimal Defense Strategy Against Intentional Attacks. *IEEE Trans. Reliab.* **2007**, *56*, 148–157. [CrossRef]
20. Levitin, G.; Hausken, K. Redundancy vs. Protection vs. False Targets for Systems Under Attack. *IEEE Trans. Reliab.* **2009**, *58*, 58–68. [CrossRef]
21. Stott, B. Review of load-flow calculation methods. *Proc. IEEE* **1974**, *62*, 916–929. [CrossRef]
22. Norris, M.P.; Scott, M. Automatic Adjustment of Transformer and Phase-Shifter Taps in the Newton Power Flow. *IEEE Trans. Power Appar. Syst.* **1971**, *PAS-90*, 103–108.
23. Hakim, B.; Ahmed, C.; Abderrazak, E. A generic three-phase power flow formulation for flexible modeling and fast solving for large-scale unbalanced networks. *Int. J. Electr. Power Energy Syst.* **2023**, 148.
24. Wilsun, X.; Hermann, W.D.; José, R.M. Generalized three-phase power flow method for the initialization of EMTP simulations. In Proceedings of the POWERCON '98. 1998 International Conference on Power System Technology. Proceedings (Cat. No.98EX151), Beijing, China, 18–21 August 1998; Volume 2, pp. 875–879.

25. Timothy, A.D.; Eknathan, P.N. Algorithm 907: KLU, a direct sparse solver for circuit simulation problems. *ACM Trans. Math. Softw.* **2010**, *37*, 3.
26. Canizares, C.A. Applications of symbolic computation to power system analysis and teaching. In Proceedings of the 2006 PES Power Systems Conference and Exposition, Atlanta, GA, USA, 29 October–1 November 2006; p. 361.
27. Chen, T.H.; Chen, M.S.; Inoue, T.; Kotas, P.; Chebli, E.A. Three-phase cogenerator and transformer models for distribution system analysis. *IEEE Trans. Power Deliv.* **1991**, *6*, 1671–1681. [CrossRef]
28. Tatianna, B.; Penido, D.R.R.; Araujo, L.R. A new synchronous DG model for unbalanced multiphase power flow studies. *IEEE Trans. Power Syst.* **2020**, *35*, 803–813.
29. Bhuvanagiri, R.; Mohan, K.; Surya, V. Priority Based Optimal Load Shedding in a Power System Network under Contingency Conditions. In Proceedings of the International Conference for Advancement in Technology (ICONAT), Goa, India, 21–22 January 2022; pp. 1–5.
30. Sereeter, B.; Vuik, K.; Witteveen, C. Newton Power Flow Methods for Unbalanced Three-Phase Distribution Networks. *Energies* **2017**, *10*, 1658. [CrossRef]
31. Garcia, P.; Pereira, J.; Carneiro, S.; Da Costa, V.; Martins, N. Three-phase power flow calculations using the current injection method. *IEEE Trans. Power Syst.* **2000**, *15*, 508–514. [CrossRef]
32. Ilhan, K.; Jean, M.; Ulas, K.; Gurkan, S.; Omar, S. Multiphase Load-Flow Solution for Large-Scale Distribution Systems Using MANA. *IEEE Trans. Power Deliv.* **2014**, *29*, 908–915.
33. Jean, M. Régimes transitoires électromagnétiques: Simulation. In *Réseaux Électriques et Applications;* TECHNIQUES DE L'INGENIEUR: Saint-Denis, France, 2015.

energies

Article

DC Admittance Model of VSCs for Stability Studies in VSC-HVDC Systems

Joaquín Pedra [1,*], Luis Sainz [1] and Lluís Monjo [2]

[1] Department of Electrical Engineering (ETSEIB—UPC), Universitat Politècnica de Catalunya, Av. Diagonal 647, 08028 Barcelona, Spain; luis.sainz@upc.edu
[2] Department of Electrical Engineering (EPSEVG—UPC), Universitat Politècnica de Catalunya, Av. Victor Balaguer s/n, 08800 Vilanova i La Geltrú, Spain; lluis.monjo@upc.edu
* Correspondence: joaquin.pedra@upc.edu

Abstract: High-voltage direct current (HVDC) systems linked to AC grids with converters are promising energy transmission systems. These systems present complex AC- and DC-side dynamic interactions. Impedance-based stability studies have recently been proposed to assess DC-side dynamics from DC-side characterization of voltage source converters (VSCs) considering AC-side dynamics. However, the existing approaches used for stability studies in VSC-HVDC systems do not completely model VSCs because they do not consider together the VSC delay, the grid voltage feedforward filter, and all the *d*- and *q*-reference current controls. Moreover, these approaches are analytically characterized from *dq*-real space vectors (less related to circuit theory than *dq*-complex space vectors), and some work with simple AC grids. The main contribution of this paper is a detailed and complete DC admittance model of VSCs from *dq*-complex space vectors, which considers the VSC delay, feedforward filter, and *d*- and *q*-reference current controls, and also a general AC grid. The proposed model can be used for DC-side stability studies in VSC-HVDC systems considering AC grid dynamics. The capabilities and drawbacks of impedance-based stability methods for DC-side stability assessment were analyzed, and the positive-net-damping criterion was validated as a robust approach. The model was validated by PSCAD/EMTDC simulations and applied to a stability study in a VSC-HVDC system.

Keywords: impedance modeling; voltage source converters; HVDC transmission

Citation: Pedra, J.; Sainz, L.; Monjo, L. DC Admittance Model of VSCs for Stability Studies in VSC-HVDC Systems. *Energies* **2023**, *16*, 5457. https://doi.org/10.3390/en16145457

Academic Editor: Ying-Yi Hong

Received: 26 June 2023
Revised: 13 July 2023
Accepted: 16 July 2023
Published: 18 July 2023

1. Introduction

VSC-HVDC systems are emerging as a future transmission energy technology [1]. Near-synchronous [1–3] and harmonic [4,5] oscillatory instabilities due to low-damped resonances are a usual concern in these systems, and their accurate assessment is challenging for researchers. The state-space and frequency-domain methods are the most common approaches used to analyze these oscillatory phenomena [1–10]. The state-space method [3] is a global stability approach (i.e., stability is determined regardless of location [2]) but uses high-order dynamic models and requires specific information about all system elements, which is not always accessible. On the other hand, the frequency-domain methods assess system stability with less computing effort than the state-space method and are used either with simulations or system measurements. Broadly speaking, frequency-domain methods are applied using impedance-based stability criteria, such as the Nyquist criterion [4–6], impedance-based analysis [1], and the positive-net-damping stability criterion [4–9] for load–source equivalent systems, and the generalized Nyquist criterion (GNC) [2,6,10] and impedance matrix determinant analysis [3] for interconnected systems. Most of these criteria are local stability approaches (i.e., they are locally applied) sensitive to system partitions, which can lead to inaccurate stability predictions.

This weakness has been discussed in the literature to understand the proper application of the different criteria [2].

Stability studies in HVDC systems require DC-side equivalent modeling of converters. Early studies on VSC-HVDC systems considered weak and simple *R–L* AC grids and a few VSC controls to obtain the DC-side model of VSCs [4,9]. This, in general, leads to inaccurate stability predictions because AC-side and VSC control dynamics are overlooked. Improved DC-side models of VSCs that consider AC-side dynamics have recently been proposed [11–15]. The first DC-side equivalent model for VSC rectifier and inverter stations of VSC-HVDC systems with active power and DC voltage controls, respectively, and grid alternating voltage control is derived from *dq*-real space vectors in [11]. The model disregards the VSC delay and phase-locked loop (PLL) control. Moreover, the study considers a simple *R–L* impedance on the AC-side and wrongly concludes that AC-side instabilities are not detected on the DC side. A DC-side equivalent admittance model of a VSC-HVDC station in an offshore wind farm with DC voltage and PLL controls using *dq*-real space vectors is presented in [12]. Additionally, the modified sequence and phasor-domain models are derived from the previous model. The model disregards the VSC delay, grid voltage feedforward filter, and *q*-reference current controls. A general AC grid characterization to correctly analyze the influence of AC-side dynamics on DC-side instabilities (e.g., the mirror frequency effect [6,16]) is also considered in [12]. A DC impedance model of VSC rectifier and inverter stations of VSC-HVDC systems with active power and DC voltage controls, respectively, and PLL control is derived by using *dq*-real space vectors in [13–15]. The VSC delay and a general AC grid are also considered, but the grid voltage feedforward filter and the *q*-reference current controls are disregarded.

The above comments are summarized in Table 1. The first row (labeled "Phase-locked loop control") shows the models in which the PLL control is included. The second row (labeled "VSC delay") shows the models in which the VSC time delay transfer function is included. The third and fourth rows show the models which use voltage feedforward without and with a low-pass filter. The fifth and sixth rows show the models which include the outer control loops on channels *d* and *q*.

Table 1. Comparison of models.

	References					
Model Feature	**[11]**	**[12]**	**[13]**	**[14]**	**[15]**	**Proposed Model**
Phase-locked loop control	✗	✓	✓	✓	✗	✓
VSC delay	✗	✗	✓	✓	✗	✓
Grid voltage feedforward	✗	✗	✓	✓	✓	✓
Grid voltage feedforward filter	✗	✗	✗	✗	✗	✓
D-outer control loop	✓	✓	✓	✓	✓	✓
Q-outer control loop	✗	✗	✗	✗	✓	✓

The influence of AC-side dynamics on stability and the stability assessment using the impedance-based stability criteria have been discussed. To better explain frequency coupling dynamics between the AC and DC sides (i.e., the mirror frequency effect), modified sequence [16] and *dq*-complex-domain [17] approaches have been proposed from the *dq*-real-domain model based on the Park (*dq*-real) and Ku (*dq*-complex) transformations [18]. Both *dq*-real- and *dq*-complex-domain approaches can be used for VSC modeling, but they have advantages and disadvantages for Laplace and frequency studies of grid-connected VSC stability:

- *dq*-real-domain approach:
  - Laplace stability studies: It allows system poles and participation factors to be easily obtained but is not appropriate for integrating components of large power

systems. Moreover, frequency black-box models obtained from measurements cannot be used with this approach.
  - Frequency stability studies: It is the most used in the literature [11–15] but reducing AC networks to their equivalent Thevenin circuit in *dq*-real variables is more laborious than in *dq*-complex variables. This is because AC-side impedances are 2 $\times$ 2 matrices with *dq*-real variables.
- *dq*-complex-domain approach:
  - Laplace stability studies: The characterization of system poles poses numerical problems because of rounding errors produced by the complex coefficients of *dq*-complex-domain transfer functions. Polynomials in the *s*-domain with only real coefficients are required in the DC impedance to correctly obtain the system poles.
  - Frequency stability studies: It has recently been used in the literature [19,20]. In the *dq*-complex domain, the AC-side elements use the impedances $Z_{Th}(s + j\omega_1)$ and Z_{Th} $(s - j\omega_1)$ obtained by shifting the operational Thevenin impedance $Z_{Th}(s)$. The *dq*-complex domain is more appropriate for integrating VSC models in large power systems.

According to the above comments, the benefits of the *dq*-complex-domain approach over the *dq*-real-domain approach for grid impedance modeling and stability studies are greater.

This paper contributes to filling the above gaps in the characterization of the DC admittance model of VSCs. Moreover, it studies the inaccuracies of impedance-based stability criteria due to partitions in VSC-HVDC systems in two ways. The first contribution is a novel DC admittance model for VSCs based on *dq*-complex space vectors. The model extends existing models by considering the VSC delay, grid voltage feedforward filter, PLL, and main *d*-reference (active power and DC voltage controls) and *q*-reference (reactive power and grid alternating voltage controls) current outer loops, and also a general AC grid. The second contribution involves the study of VSC-HVDC system stability. The sensitivity of impedance-based stability criteria to VSC-HVDC system partitions is discussed, and the positive-net-damping stability criterion is proposed to overcome this issue. Finally, the influence of AC-side dynamics on the DC-side stability of VSC-HVDC systems was analyzed. The stability issues in a VSC-HVDC system were studied with the proposed DC admittance model to show the paper's contribution. PSCAD/EMTDC simulations were performed to validate the study.

2. Model of VSCs in DQ-Complex Domain

Complex transfer functions are used in this section to model VSCs with *dq*-complex space vectors [6,16,17]. This approach is used in the next section to obtain the DC admittance of the VSCs.

The Laplace variable *s* will be omitted in the *dq*-complex transfer functions hereinafter to simplify notation, and the *dq*-complex space vectors are defined as $\mathbf{U_{dq}} = (U_d + jU_q)/\sqrt{2}$ and $\mathbf{I_{dq}} = (I_d + jI_q)/\sqrt{2}$.

2.1. AC Grid Relations

The relation between voltage and current at the point of common coupling in Figure 1 is expressed as

$$\begin{bmatrix} \Delta\mathbf{E_{dq}} \\ \Delta\mathbf{E^*_{dq}} \end{bmatrix} = -\begin{bmatrix} \mathbf{Z^+_{Th}} & \mathbf{Z^-_{Th}} \\ (\mathbf{Z^-_{Th}})^* & (\mathbf{Z^+_{Th}})^* \end{bmatrix} \begin{bmatrix} \Delta\mathbf{I_{dq}} \\ \Delta\mathbf{I^*_{dq}} \end{bmatrix} \quad \Rightarrow \quad \Delta\mathbf{E^m_{dq}} = -\mathbf{Z^\pm_{Th}}\Delta\mathbf{I^m_{dq}}, \tag{1}$$

where $\mathbf{Z^+_{Th}} = Z_{Th}(s + j\omega_1)$, $(\mathbf{Z^+_{Th}})^* = Z_{Th}(s - j\omega_1)$ and $\mathbf{Z^-_{Th}} = 0$. The voltage balance across the converter $R_c - L_c$ filter [20] is expressed as

$$\begin{bmatrix}\Delta\mathbf{V_{dq}}\\ \Delta\mathbf{V^*_{dq}}\end{bmatrix} = -\begin{bmatrix}\mathbf{Z_c^+} & \mathbf{Z_c^-}\\ (\mathbf{Z_c^-})^* & (\mathbf{Z_c^+})^*\end{bmatrix}\begin{bmatrix}\Delta\mathbf{I_{dq}}\\ \Delta\mathbf{I^*_{dq}}\end{bmatrix} + \begin{bmatrix}\Delta\mathbf{E_{dq}}\\ \Delta\mathbf{E^*_{dq}}\end{bmatrix} \Rightarrow \Delta\mathbf{V^m_{dq}} = -\mathbf{Z_c^\pm}\Delta\mathbf{I^m_{dq}} + \Delta\mathbf{E^m_{dq}}, \quad (2)$$

where $\mathbf{Z_c^+} = R_c + L_c(s + j\omega_1)$, $(\mathbf{Z_c^+})^* = R_c + L_c(s - j\omega_1)$, and $\mathbf{Z_c^-} = 0$.

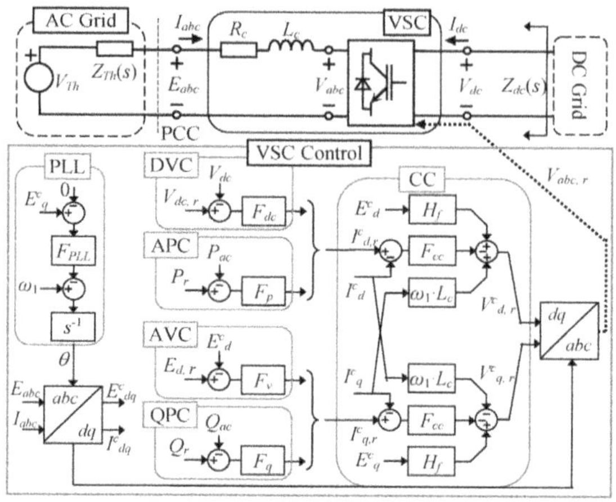

Figure 1. Grid-connected VSC system for DC impedance characterization.

Note that $\mathbf{Z^-_{Th}} = \mathbf{Z_c^-} = 0$ because the grid and the $R_c - L_c$ filter are three-phase symmetric systems whose impedance matrices in the dq-domain verify the relations $Z_{dd}(s) = Z_{qq}(s)$ and $Z_{dq}(s) = -Z_{qd}(s)$. Considering the above, the terms $\mathbf{Z^+_{Th}}$ and $\mathbf{Z^+_c}$ can also be expressed by using the coordinate transformation properties $\mathbf{Z^+_{Th}} = Z_{Th}(s + j\omega_1)$ and $\mathbf{Z^+_c} = Z_c(s + j\omega_1) = R_c + L_c(s + j\omega_1)$ [6].

The relation between VSC output voltage and AC current is derived from (1) and (2):

$$\Delta\mathbf{V^m_{dq}} = -\mathbf{Z_t^\pm}\Delta\mathbf{I^m_{dq}} \;\; \mathbf{Z_t^\pm} = (\mathbf{Z_c^\pm} + \mathbf{Z^\pm_{Th}}), \quad (3)$$

where $\mathbf{Z^+_t} = Z_t(s + j\omega_1)$ and $Z_t(s) = Z_{Th}(s) + R_c + L_c s$ is the AC impedance.

2.2. DC-AC Converter Relations

The relation between the VSC output voltage and the DC voltage and the relation between the DC and AC currents are expressed [20] as

$$\begin{bmatrix}\Delta\mathbf{V_{dq}}\\ \Delta\mathbf{V^*_{dq}}\end{bmatrix} = \begin{bmatrix}\Delta\mathbf{m_{dq}}\\ \Delta\mathbf{m^*_{dq}}\end{bmatrix} V_{dc0} + \begin{bmatrix}\mathbf{m_{dq0}}\\ \mathbf{m^*_{dq0}}\end{bmatrix}\Delta V_{dc} \Rightarrow \Delta\mathbf{V^m_{dq}} = \Delta\mathbf{m^m_{dq}} V_{dc0} + \mathbf{m^m_{dq0}}\Delta V_{dc}, \quad (4)$$

$$\begin{aligned}\Delta I_{dc} &= -\begin{bmatrix}\mathbf{m^*_{dq0}} & \mathbf{m_{dq0}}\end{bmatrix}\begin{bmatrix}\Delta\mathbf{I_{dq}}\\ \Delta\mathbf{I^*_{dq}}\end{bmatrix} - \begin{bmatrix}\mathbf{I^*_{dq0}} & \mathbf{I_{dq0}}\end{bmatrix}\begin{bmatrix}\Delta\mathbf{m_{dq}}\\ \Delta\mathbf{m^*_{dq}}\end{bmatrix}\\ \Rightarrow \Delta I_{dc} &= -(\mathbf{m^m_{dq0}})^H\Delta\mathbf{I^m_{dq}} - (\mathbf{I^m_{dq0}})^H\Delta\mathbf{m^m_{dq}},\end{aligned} \quad (5)$$

where $\mathbf{m_{dq}} = (m_d + jm_q)/\sqrt{2}$ and $\mathbf{I_{dq0}} = (I_{d0} + jI_{q0})/\sqrt{2}$, and the superscript H indicates the transpose and complex conjugate (Hermitian conjugate). Note that the three-phase set of modulation functions $\{m_a, m_b, m_c\} = M\{\cos(\omega_1 t + \phi_m), \cos(\omega_1 t + \phi_m - 2\pi/3), \cos(\omega_1 t + \phi_m + 2\pi/3)\}$ is transformed into the dq-complex space vector $\mathbf{m_{dq}} = (\sqrt{3}/2)M\angle\phi_m$ by applying the normalized Ku (dq-complex) transformation [6,18].

2.3. Simplified Model #1: Case without Current Control

In the case without current control, the modulation function is constant, $\Delta\mathbf{m}^*_{\mathbf{dq}} = 0$. Then, (4) and (5) result in

$$\Delta\mathbf{V}^{\mathbf{m}}_{\mathbf{dq}} = \mathbf{m}^{\mathbf{m}}_{\mathbf{dq0}}\Delta V_{dc} \;\; \Delta I_{dc} = -(\mathbf{m}^{\mathbf{m}}_{\mathbf{dq0}})^H \Delta\mathbf{I}^{\mathbf{m}}_{\mathbf{dq}}. \tag{6}$$

By considering (3) and (6), the DC admittance is

$$\Delta I_{dc} = Y_{dc}\Delta V_{dc} \;\; Y_{dc} = (\mathbf{m}^{\mathbf{m}}_{\mathbf{dq0}})^H (\mathbf{Z}^{\pm}_{\mathbf{t}})^{-1}\mathbf{m}^{\mathbf{m}}_{\mathbf{dq0}}, \tag{7}$$

where

$$\begin{aligned} Y_{dc} &= \begin{bmatrix} \mathbf{m}^*_{\mathbf{dq0}} & \mathbf{m}_{\mathbf{dq0}} \end{bmatrix} \begin{bmatrix} \mathbf{Z}^+_{\mathbf{t}} & 0 \\ 0 & (\mathbf{Z}^+_{\mathbf{t}})^* \end{bmatrix}^{-1} \begin{bmatrix} \mathbf{m}_{\mathbf{dq0}} \\ \mathbf{m}^*_{\mathbf{dq0}} \end{bmatrix} = \frac{3M^2}{4}\left(\frac{1}{\mathbf{Z}^+_{\mathbf{t}}} + \frac{1}{(\mathbf{Z}^+_{\mathbf{t}})^*}\right) \\ \mathbf{Z}^+_{\mathbf{t}} &= Z_t(s+j\omega_1) = Z_{Th}(s+j\omega_1) + R_c + L_c(s+j\omega_1) \\ (\mathbf{Z}^+_{\mathbf{t}})^* &= Z^*_t(s+j\omega_1) = Z_{Th}(s-j\omega_1) + R_c + L_c(s-j\omega_1), \end{aligned} \tag{8}$$

and the DC impedance is the inverse of the DC admittance, $Z_{dc} = 1/Y_{dc}$.

The AC grid is represented by the circuit in Figure 2 [12], with the Thevenin transfer function written as

$$Z_{Th}(s) = \frac{1}{\frac{1}{R_g+L_g s} + \frac{1}{Z_f(s)}} \;;\; Z_f(s) = \frac{1}{C_f s} + \frac{1}{\frac{1}{L_f s} + \frac{1}{R_f}}. \tag{9}$$

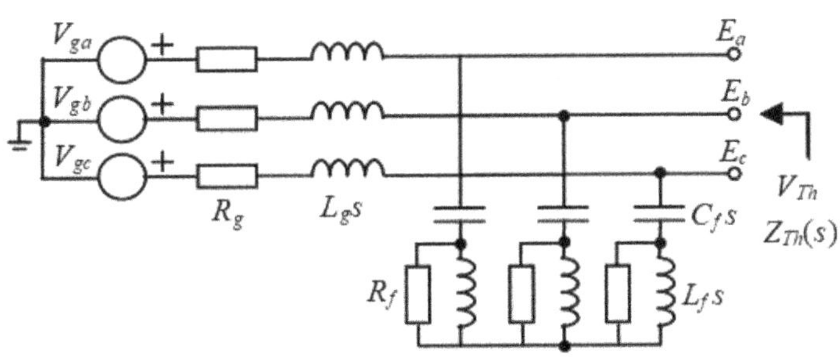

Figure 2. AC grid connected at the VSC terminals [12].

Note that the frequency response of a conjugate transfer function such as $(\mathbf{Z}^+{}_t)^*(s)$ in (8) can be calculated as [19].

$$Z^*(s) = [Z(s^*)]^* \;\Rightarrow\; Z^*(j\omega) = [Z(-j\omega)]^*. \tag{10}$$

2.4. Simplified Model #2: Case with Current Control

Only the modulation function and current control of the VSC with delay are considered. Then, the current control loop [20] is expressed as

$$\Delta\mathbf{V}^{\mathbf{m}}_{\mathbf{dq,\,r}} = \mathbf{F}^{\pm}_{\mathbf{cc}}\Delta\mathbf{I}^{\mathbf{m}}_{\mathbf{dq}} - \mathbf{Z}^{\pm}_{\omega}\Delta\mathbf{I}^{\mathbf{m}}_{\mathbf{dq}} + \mathbf{H}^{\pm}_{\mathbf{f}}\Delta\mathbf{E}^{\mathbf{m}}_{\mathbf{dq}}, \tag{11}$$

where $\mathbf{F}^+_{\mathbf{cc}} = D(s)F_{cc}(s)$, $\mathbf{Z}^+_{\omega} = D(s)\cdot jL_c\omega_1$, $\mathbf{H}^+_{\mathbf{f}} = D(s)H_f(s)$, and $\mathbf{F}^-_{\mathbf{cc}} = \mathbf{Z}^-_{\omega} = \mathbf{H}^-_{\mathbf{f}} = 0$, with $D(s) = e^{-sTd}$ being the VSC time delay transfer function, $F_{scc}(s) = k_{sp,\,cc} + k_{si,\,cc}/s$, and $H_f(s) = \alpha_f/(s + \alpha_f)$, with $k_{sp,\,cc,}$ and $k_{si,\,cc}$ being the proportional and integral gains of the current control, α_f being the bandwidth of the voltage feedforward low-pass filter, and T_d being the VSC time delay. The VSC reference voltage, $\Delta\mathbf{V}^{\mathbf{m}}{}_{\mathbf{dq,\,r}} = V_{dc0}\Delta\mathbf{m}^{\mathbf{m}}{}_{\mathbf{dq}}$ (11), is the *dq*-complex space vector determining the modulation function:

$$\Delta\mathbf{m}^{\mathbf{m}}_{\mathbf{dq}} = \frac{1}{V_{dc0}}\left(\mathbf{Z}^{\pm}_{\mathbf{pi}}\Delta\mathbf{I}^{\mathbf{m}}_{\mathbf{dq}} + \mathbf{H}^{\pm}_{\mathbf{f}}\Delta\mathbf{E}^{\mathbf{m}}_{\mathbf{dq}}\right);\ \mathbf{Z}^{\pm}_{\mathbf{pi}} = \mathbf{F}^{\pm}_{\mathbf{cc}} - \mathbf{Z}^{\pm}_{\omega}. \tag{12}$$

By considering (1) and (12), (5) becomes

$$\Delta I_{dc} = \left[-\left(\mathbf{m}^{\mathbf{m}}_{\mathbf{dq0}}\right)^{H} - \frac{\left(\mathbf{I}^{\mathbf{m}}_{\mathbf{dq0}}\right)^{H}}{V_{dc0}}\left(\mathbf{Z}^{\pm}_{\mathbf{pi}} - \mathbf{H}^{\pm}_{\mathbf{f}}\mathbf{Z}^{\pm}_{\mathbf{Th}}\right)\right]\Delta\mathbf{I}^{\mathbf{m}}_{\mathbf{dq}}, \tag{13}$$

and (4) can be expressed from (1), (2), and (12) as

$$\Delta\mathbf{V}^{\mathbf{m}}_{\mathbf{dq}} = -\mathbf{Z}^{\pm}_{\mathbf{c}}\Delta\mathbf{I}^{\mathbf{m}}_{\mathbf{dq}} + \Delta\mathbf{E}^{\mathbf{m}}_{\mathbf{dq}} = -\left(\mathbf{Z}^{\pm}_{\mathbf{c}} + \mathbf{Z}^{\pm}_{\mathbf{Th}}\right)\Delta\mathbf{I}^{\mathbf{m}}_{\mathbf{dq}} = \left(\mathbf{Z}^{\pm}_{\mathbf{pi}} - \mathbf{H}^{\pm}_{\mathbf{f}}\mathbf{Z}^{\pm}_{\mathbf{Th}}\right)\Delta\mathbf{I}^{\mathbf{m}}_{\mathbf{dq}} + \mathbf{m}^{\mathbf{m}}_{\mathbf{dq0}}\Delta V_{dc}, \tag{14}$$

which can be rewritten as

$$\Delta\mathbf{I}^{\mathbf{m}}_{\mathbf{dq}} = \left[-\mathbf{Z}^{\pm}_{\mathbf{c}} - \mathbf{Z}^{\pm}_{\mathbf{Th}} - \mathbf{Z}^{\pm}_{\mathbf{pi}} + \mathbf{H}^{\pm}_{\mathbf{f}}\mathbf{Z}^{\pm}_{\mathbf{Th}}\right]^{-1}\mathbf{m}^{\mathbf{m}}_{\mathbf{dq0}}\Delta V_{dc}. \tag{15}$$

By replacing (15) in (13), the DC admittance function is

$$\begin{aligned}&\Delta I_{dc} = Y_{dc}\Delta V_{dc}\quad \mathbf{Z}^{\pm}_{\mathbf{cf}} = \mathbf{Z}^{\pm}_{\mathbf{pi}} - \mathbf{H}^{\pm}_{\mathbf{f}}\mathbf{Z}^{\pm}_{\mathbf{Th}}\\ &Y_{dc} = \left(\left(\mathbf{m}^{\mathbf{m}}_{\mathbf{dq0}}{}^{T}\right)^{*} + \frac{\left(\mathbf{I}^{\mathbf{m}}_{\mathbf{dq0}}\right)^{H}}{V_{dc0}}\mathbf{Z}^{\pm}_{\mathbf{cf}}\right)\left(\mathbf{Z}^{\pm}_{\mathbf{t}} + \mathbf{Z}^{\pm}_{\mathbf{cf}}\right)^{-1}\mathbf{m}^{\mathbf{m}}_{\mathbf{dq0}}.\end{aligned} \tag{16}$$

The DC admittance in (16) can finally be expressed as

$$\begin{aligned}Y_{dc} &= \frac{3M^2}{4}\left(\frac{1}{\mathbf{Z}^{+}_{\mathbf{t}}+\mathbf{Z}^{+}_{\mathbf{cf}}} + \frac{1}{\left(\mathbf{Z}^{+}_{\mathbf{t}}+\mathbf{Z}^{+}_{\mathbf{cf}}\right)^{*}}\right)\\ &+ \frac{1}{2V_{dc0}}\left[\frac{(I_{d0}-jI_{q0})(m_{d0}+jm_{q0})\mathbf{Z}^{+}_{\mathbf{cf}}}{\mathbf{Z}^{+}_{\mathbf{t}}+\mathbf{Z}^{+}_{\mathbf{cf}}} + \frac{(I_{d0}+jI_{q0})(m_{d0}-jm_{q0})(\mathbf{Z}^{+}_{\mathbf{cf}})^{*}}{(\mathbf{Z}^{+}_{\mathbf{t}})^{*}+\left(\mathbf{Z}^{+}_{\mathbf{cf}}\right)^{*}}\right],\end{aligned} \tag{17}$$

where

$$\begin{aligned}\mathbf{Z}^{+}_{\mathbf{cf}} &= D(s)\left[F_{cc}(s) - jL_c\omega_1 - H_f(s)Z_{Th}(s+j\omega_1)\right]\\ \left(\mathbf{Z}^{+}_{\mathbf{cf}}\right)^{*} &= D(s)\left[F_{cc}(s) + jL_c\omega_1 - H_f(s)Z_{Th}(s-j\omega_1)\right],\end{aligned} \tag{18}$$

and the DC impedance is calculated with the inverse of the DC admittance, $\mathbf{Z_{DC}} = 1/\mathbf{Y_{DC}}$. The DC impedance in the previous simplified models, (8) and (17), is numerically validated by PSCAD/EMTDC simulation.

Figure 3 shows these DC impedances (see data in Table 2). The case without current control, $\mathbf{Z_{WC}}$, calculated with (8), is plotted with a solid line whereas the frequency response obtained by PSCAD/EMTDC simulation is plotted with crosses. The DC impedance case with current control, $\mathbf{Z_{CC}}$, calculated with (17), is represented in the same way. The accurate results of the DC-side equivalent admittance obtained with the proposed model are worth noting.

In the case without current control, the coupling between the DC and AC sides is revealed by the two peak values of $\mathbf{Z_{WC}}$. The DC-impedance $\mathbf{Z_{WC}}$ is affected by the mirror frequency effect. This is reflected in the expression of $\mathbf{Y_{DC}}$ (8) (see (1) in [12]), which depends on the impedances $Z_t(s + j\omega_1)$ and $Z_t(s - j\omega_1)$. As can be seen in the DC-impedance $\mathbf{Z_{CC}}$ in Figure 3, the control damps this coupling effect between the DC and AC sides [13].

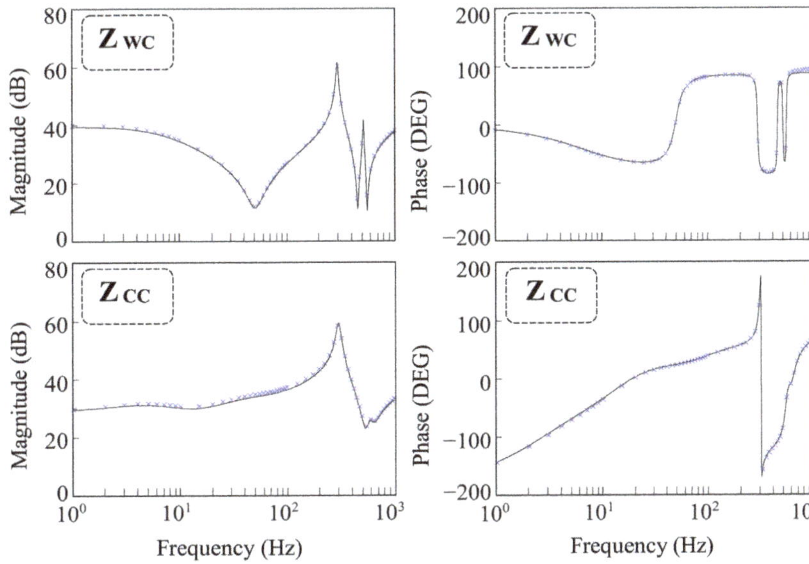

Figure 3. DC-impedance in the case without control, $\mathbf{Z_{WC}}$, and with current control, $\mathbf{Z_{CC}}$. (Lines: inverse of (8) and (17). Cross: PSCAD/EMTDC simulations.)

Table 2. AC Grid and VSC Parameters.

V_{ga}	$12.25/\sqrt{2}$ kV	P_{vsc}	30 MW	R_f, L_f	2.3 Ω, 0.18 mH
V_{dc0}	30 kV	R_g, L_g	0.52 Ω, 7.8 mH	R_c, L_c	0, 4 mH
C_f	35 μF	f_{sw}, T_d	10 kHz, 0.3 ms	I_{d0}, I_{q0}	−1996.3 A, 0 A
α_f	106.18 s^{-1}	m_{d0}	0.5168 pu	m_{q0}	0.0836 pu
$k_{p,cc}$	2 pu	$k_{p,dc}$	0.8662 pu	$k_{p,pll}$	0.2757 pu
$k_{i,cc}$	0.096 pu	$k_{i,dc}$	0.0334 pu	$k_{i,pll}$	0.2203 pu
$k_{p,ac}$	0.9 pu	$k_{p,P}$	0.13 pu	$k_{p,Q}$	0.13 pu
$k_{i,ac}$	0.008 pu	$k_{i,P}$	0.0083 pu	$k_{i,Q}$	0.0083 pu

3. Detailed Model of DC Admittance of VSCs

The PLL and the outer loop relations are included in the detailed model, where the current control loop is modified to account for the PLL effect. Now, the current control loop relations must consider the relationship between the variables in the converter *dq*-domain, which has the superscript "c", and the variables in the grid *dq*-domain, where no superscript is used. The current control loop [20] is expressed as

$$\Delta\mathbf{V}^{\mathbf{cm}}_{\mathbf{dq,\,r}} = -\mathbf{F}^{\pm}_{\mathbf{cc}}\left(\Delta\mathbf{I}^{\mathbf{cm}}_{\mathbf{dq,\,r}} - \Delta\mathbf{I}^{\mathbf{cm}}_{\mathbf{dq}}\right) - \mathbf{Z}^{\pm}_{\boldsymbol{\omega}}\Delta\mathbf{I}^{\mathbf{cm}}_{\mathbf{dq}} + \mathbf{H}^{\pm}_{\mathbf{f}}\Delta\mathbf{E}^{\mathbf{cm}}_{\mathbf{dq}}. \tag{19}$$

The *dq*-complex space vector of the VSC reference voltage is written as $\mathbf{\Delta V^{cm}}_{\mathbf{dq,\,r}} = V_{dc0}\mathbf{\Delta m^{cm}}_{\mathbf{dq}}$. The PLL relations [9,20] are expressed as

$$\begin{aligned} \Delta\mathbf{I}^{\mathbf{cm}}_{\mathbf{dq}} &= \Delta\mathbf{I}^{\mathbf{m}}_{\mathbf{dq}} + \mathbf{G}^{\pm}_{\mathbf{PLLi}}\Delta\mathbf{E}^{\mathbf{m}}_{\mathbf{dq}} \\ \mathbf{G}^{+}_{\mathbf{PLLi}} &= -\mathbf{G}^{-}_{\mathbf{PLLi}} = -G_{PLL}(s)(I_{d0} + jI_{q0})/2, \end{aligned} \tag{20}$$

$$\begin{aligned} \Delta\mathbf{E}^{\mathbf{cm}}_{\mathbf{dq}} &= \mathbf{G}^{\pm}_{\mathbf{PLLv}}\Delta\mathbf{E}^{\mathbf{m}}_{\mathbf{dq}} \\ \mathbf{G}^{+}_{\mathbf{PLLv}} &= 1 - G_{PLL}(s)(E_{d0} + jE_{q0})/2 \\ \mathbf{G}^{-}_{\mathbf{PLLv}} &= 1 - \mathbf{G}^{+}_{\mathbf{PLLv}} = G_{PLL}(s)(E_{d0} + jE_{q0})/2, \end{aligned} \tag{21}$$

$$\begin{array}{l}\Delta\mathbf{m}_{\mathbf{dq}}^{\mathbf{cm}} = \Delta\mathbf{m}_{\mathbf{dq}}^{\mathbf{m}} + \mathbf{G}_{\mathbf{PLLm}}^{\pm}\Delta\mathbf{E}_{\mathbf{dq}}^{\mathbf{m}} \\ \mathbf{G}_{\mathbf{PLLm}}^{+} = -\mathbf{G}_{\mathbf{PLLm}}^{-} = -G_{PLL}(s)(m_{d0} + jm_{q0})/2,\end{array} \tag{22}$$

where $G_{pll}(s) = F_{pll}(s)/(s + V_{d0}F_{pll}(s))$, $F_{pll}(s) = k_{p,\,pll} + k_{i,\,pll}/s$, and $k_{p,\,pll}$ and $k_{i,\,pll}$ are the proportional and integral gains of the PLL control.

The reference currents generated by the outer control loops are related to the DC voltage and the input AC voltage and current of the VSC with dq-complex space vectors as the following general expression:

$$\Delta\mathbf{I}_{\mathbf{dq,\,r}}^{\mathbf{cm}} = -\mathbf{F}_{\mathbf{o}}^{\mathbf{m}}\Delta V_{dc} - \mathbf{G}_{\mathbf{o}}^{\pm}\Delta\mathbf{I}_{\mathbf{dq}}^{\mathbf{cm}} - \mathbf{Y}_{\mathbf{o}}^{\pm}\Delta\mathbf{E}_{\mathbf{dq}}^{\mathbf{cm}}, \tag{23}$$

where the complex transfer matrices $\mathbf{F^m}_\mathbf{o}$, $\mathbf{G^\pm}_\mathbf{o}$, and $\mathbf{Y^\pm}_\mathbf{o}$ are the complex transfer matrices depending on the VSC operation mode (see Appendix A).

The outer control loop (23) can be introduced into the inner current control loop (19) as

$$V_{dc0}\Delta\mathbf{m}_{\mathbf{dq}}^{\mathbf{cm}} = \Delta\mathbf{V}_{\mathbf{dq,\,r}}^{\mathbf{cm}} = \mathbf{F}_{\mathbf{cc}}^{\pm}\mathbf{F}_{\mathbf{o}}^{\mathbf{m}}\Delta V_{dc} + \left(\mathbf{Z}_{\mathbf{pi}}^{\pm} + \mathbf{F}_{\mathbf{cc}}^{\pm}\mathbf{G}_{\mathbf{o}}^{\pm}\right)\Delta\mathbf{I}_{\mathbf{dq}}^{\mathbf{cm}} + \left(\mathbf{H}_{\mathbf{f}}^{\pm} + \mathbf{F}_{\mathbf{cc}}^{\pm}\mathbf{Y}_{\mathbf{o}}^{\pm}\right)\Delta\mathbf{E}_{\mathbf{dq}}^{\mathbf{cm}}, \tag{24}$$

where (24) is expressed in the converter dq-complex domain. By considering the PLL relations (20), (21), and (22) in the grid dq-domain (where no superscript is used), (24) becomes

$$\begin{array}{l}\Delta\mathbf{m}_{\mathbf{dq}}^{\mathbf{m}} = \frac{1}{V_{dc0}}\left\{\mathbf{F}_{\mathbf{cc}}^{\pm}\mathbf{F}_{\mathbf{o}}^{\mathbf{m}}\Delta V_{dc} + (\mathbf{Z}_{\mathbf{pi}}^{\pm} + \mathbf{F}_{\mathbf{cc}}^{\pm}\mathbf{G}_{\mathbf{o}}^{\pm})\Delta\mathbf{I}_{\mathbf{dq}}^{\mathbf{m}} + \mathbf{H}_{\mathbf{o}}^{\pm}\Delta\mathbf{E}_{\mathbf{dq}}^{\mathbf{m}}\right\} \\ \mathbf{H}_{\mathbf{o}}^{\pm} = (\mathbf{Z}_{\mathbf{pi}}^{\pm} + \mathbf{F}_{\mathbf{cc}}^{\pm}\mathbf{G}_{\mathbf{o}}^{\pm})\mathbf{G}_{\mathbf{plli}}^{\pm} + (\mathbf{H}_{\mathbf{f}}^{\pm} + \mathbf{F}_{\mathbf{cc}}^{\pm}\mathbf{Y}_{\mathbf{o}}^{\pm})\mathbf{G}_{\mathbf{pllv}}^{\pm} - V_{dc0}\mathbf{G}_{\mathbf{pllm}}^{\pm},\end{array} \tag{25}$$

and introducing (1) into (25) results in

$$\Delta\mathbf{m}_{\mathbf{dq}}^{\mathbf{m}} = \frac{1}{V_{dc0}}\left\{\mathbf{F}_{\mathbf{cc}}^{\pm}\mathbf{F}_{\mathbf{o}}^{\mathbf{m}}\Delta V_{dc} + (\mathbf{Z}_{\mathbf{pi}}^{\pm} + \mathbf{F}_{\mathbf{cc}}^{\pm}\mathbf{G}_{\mathbf{o}}^{\pm} - \mathbf{H}_{\mathbf{o}}^{\pm}\mathbf{Z}_{\mathbf{Th}}^{\pm})\Delta\mathbf{I}_{\mathbf{dq}}^{\mathbf{m}}\right\}. \tag{26}$$

The DC voltage in (4) can be related to the AC current from (26) and the AC grid relations (1) and (2) as follows:

$$\Delta\mathbf{V}_{\mathbf{dq}}^{\mathbf{m}} = -(\mathbf{Z}_{\mathbf{c}}^{\pm} + \mathbf{Z}_{\mathbf{Th}}^{\pm})\Delta\mathbf{I}_{\mathbf{dq}}^{\mathbf{m}} = (\mathbf{m}_{\mathbf{dq0}}^{\mathbf{m}} + \mathbf{F}_{\mathbf{cc}}^{\pm}\mathbf{F}_{\mathbf{o}}^{\mathbf{m}})\Delta V_{dc} + (\mathbf{Z}_{\mathbf{pi}}^{\pm} + \mathbf{F}_{\mathbf{cc}}^{\pm}\mathbf{G}_{\mathbf{o}}^{\pm} - \mathbf{H}_{\mathbf{o}}^{\pm}\mathbf{Z}_{\mathbf{Th}}^{\pm})\Delta\mathbf{I}_{\mathbf{dq}}^{\mathbf{m}}. \tag{27}$$

By using (1), (25) is rewritten as

$$\begin{array}{ll}\mathbf{G}_{\mathbf{V}}\Delta V_{dc} = -\mathbf{Z}_{\mathbf{V}}\Delta\mathbf{I}_{\mathbf{dq}}^{\mathbf{m}} & \mathbf{G}_{\mathbf{V}} = \mathbf{m}_{\mathbf{dq0}}^{\mathbf{m}} + \mathbf{F}_{\mathbf{cc}}^{\pm}\mathbf{F}_{\mathbf{o}}^{\mathbf{m}} \\ \mathbf{Z}_{\mathbf{V}} = \mathbf{Z}_{\mathbf{t}}^{\pm} + \mathbf{Z}_{\mathbf{a}}^{\pm} & \mathbf{Z}_{\mathbf{a}}^{\pm} = \mathbf{Z}_{\mathbf{pi}}^{\pm} + \mathbf{F}_{\mathbf{cc}}^{\pm}\mathbf{G}_{\mathbf{o}}^{\pm} - \mathbf{H}_{\mathbf{o}}^{\pm}\mathbf{Z}_{\mathbf{Th}}^{\pm}.\end{array} \tag{28}$$

The DC current in (5) can be related to the AC current using (25) and the AC grid relations (1) and (2) as follows:

$$\Delta I_{dc} = -\left(\mathbf{m}_{\mathbf{dq0}}^{\mathbf{m}\,T}\right)^{*}\Delta\mathbf{I}_{\mathbf{dq}}^{\mathbf{m}} - \frac{\left(\mathbf{I}_{\mathbf{dq0}}^{\mathbf{m}}\right)^{H}}{V_{dc0}}\left\{\mathbf{F}_{\mathbf{cc}}^{\pm}\mathbf{F}_{\mathbf{o}}^{\mathbf{m}}\Delta V_{dc} + \mathbf{Z}_{\mathbf{a}}^{\pm}\Delta\mathbf{I}_{\mathbf{dq}}^{\mathbf{m}}\right\}, \tag{29}$$

which can be rewritten as

$$\Delta I_{dc} = -\mathbf{Y}_{\mathbf{I}}\Delta V_{dc} - \mathbf{G}_{\mathbf{I}}\Delta\mathbf{I}_{\mathbf{dq}}^{\mathbf{m}} \quad \mathbf{Y}_{\mathbf{I}} = \frac{\left(\mathbf{I}_{\mathbf{dq0}}^{\mathbf{m}}\right)^{H}}{V_{dc0}}\mathbf{F}_{\mathbf{cc}}^{\pm}\mathbf{F}_{\mathbf{o}}^{\mathbf{m}} \quad \mathbf{G}_{\mathbf{I}} = \left(\mathbf{m}_{\mathbf{dq0}}^{\mathbf{m}\,T}\right)^{*} + \frac{\left(\mathbf{I}_{\mathbf{dq0}}^{\mathbf{m}}\right)^{H}}{V_{dc0}}\mathbf{Z}_{\mathbf{a}}^{\pm}. \tag{30}$$

The DC-admittance $Y_{dc} = \Delta I_{dc}/\Delta V_{dc}$ is obtained from the DC voltage and current relations (28) and (30), respectively, resulting in

$$Y_{dc} = \frac{\Delta I_{dc}}{\Delta V_{dc}} = -\mathbf{Y}_{\mathbf{I}} + \mathbf{G}_{\mathbf{I}}\mathbf{Z}_{\mathbf{V}}^{-1}\mathbf{G}_{\mathbf{V}} = -\frac{\left(\mathbf{I}_{\mathbf{dq0}}^{\mathbf{m}}\right)^{H}}{V_{dc0}}\mathbf{F}_{\mathbf{cc}}^{\pm}\mathbf{F}_{\mathbf{0}}^{\mathbf{m}} + \left(\left(\mathbf{m}_{\mathbf{dq0}}^{\mathbf{m}}\right)^{H} + \frac{\left(\mathbf{I}_{\mathbf{dq0}}^{\mathbf{m}}\right)^{H}}{V_{dc0}}\mathbf{Z}_{\mathbf{a}}^{\pm}\right)\left(\mathbf{Z}_{\mathbf{t}}^{\pm} + \mathbf{Z}_{\mathbf{a}}^{\pm}\right)^{-1}\left(\mathbf{m}_{\mathbf{dq0}}^{\mathbf{m}} + \mathbf{F}_{\mathbf{cc}}^{\pm}\mathbf{F}_{\mathbf{0}}^{\mathbf{m}}\right). \tag{31}$$

PSCAD/EMTDC simulations were performed to study the DC impedance calculation accuracy of different control modes. The parameters of each case are listed in Table 2. Figure 4 shows the DC impedance in the case of the outer control mode $P-Q$, Z_{PQ} and $P-E_{ac}$, Z_{PE}. Figure 5 shows the DC impedance in the case of the outer control mode $V_{dc}-Q$, Z_{UQ} and $V_{dc}-E_{ac}$, Z_{UE}. The simulation data are shown with x crosses in Figures 4 and 5. The continuous curves are calculated from the inverse of (31) and they fit well with the simulation results.

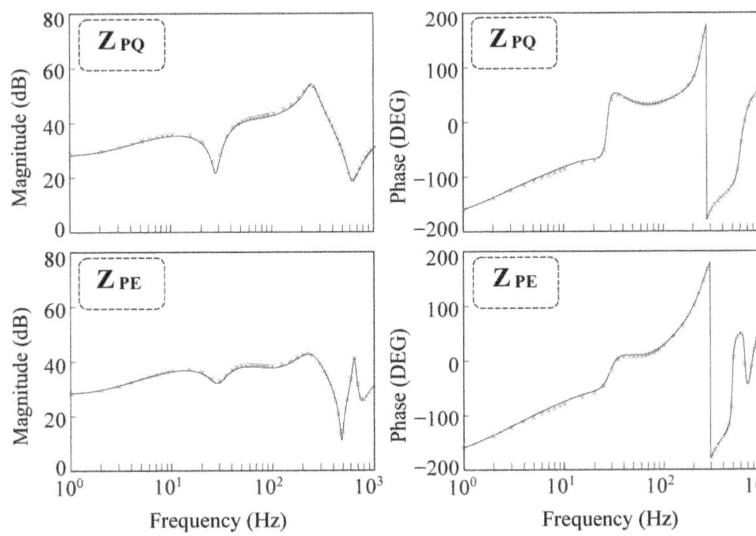

Figure 4. DC-impedance of P-Q and P-E control modes. (Lines: inverse of (31). Cross: PSCAD/EMTDC simulations.)

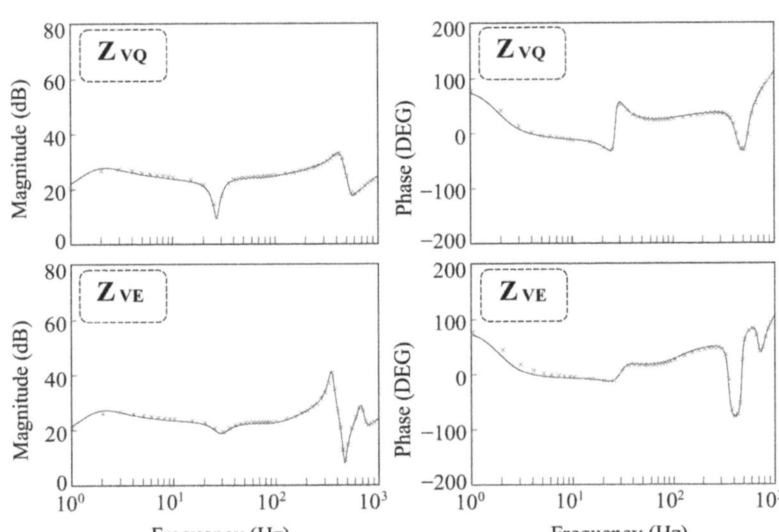

Figure 5. DC-impedance of V_{dc}-Q and V_{dc}-E control modes. (Lines: inverse of (31). Cross: PSCAD/EMTDC simulations.)

4. Impedance-Based Stability Criteria

Once the DC-side equivalent admittance of VSCs is obtained, the general HVDC interconnected system in Figure 6a can be considered for stability studies. The incremental symbol Δ of the small-signal variables is omitted for the sake of simplicity. The external components connected to the DC grid buses are represented by their equivalent circuits as follows:

- Grid-connected VSCs are characterized by the current balance at their Norton equivalent circuit buses:

$$I_i = I_{dc,\,i} - Y_{dc,\,i}(s)V_i \; (i = 1, \ldots a), \tag{32}$$

where the expression of $Y_{dc,\,i}(s)$ is the DC admittance in (31).

- DC loads are characterized by their admittance transfer function $Y_i(s)$ ($i = 1, \ldots b$).
- Terminals without any connected external component are represented as open-circuit buses characterized by zero values of the current source and admittance of the Norton equivalent circuit, $I_i = Y_i(s) = 0$ ($i = 1, \ldots c$).

Several impedance-based stability criteria are proposed in the literature [1–10]. Some of the differences between these criteria are discussed in [2]. This discussion is extended in the sections below considering the general DC system in Figure 6a where $Y_{dc,\,i}(s)$ (31) is used to assess DC interconnected system stability. The discussion is also supported by Section 5.

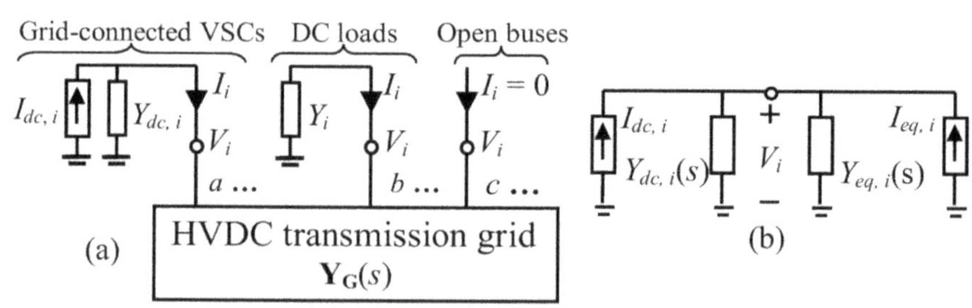

Figure 6. Stability assessment in HVDC transmission grids: (**a**) HVDC transmission grid. (**b**) SISO equivalent system partitioned at VSC terminals.

4.1. Norton Admittance Matrix-Based Stability Assessment

The relations between voltages and currents at the HVDC transmission grid terminals in Figure 6a are expressed as

$$\left.\begin{array}{l} \mathbf{i} = \mathbf{Y}_G(s)\mathbf{v} \\ \mathbf{i} = \mathbf{i}_E - \mathbf{Y}_E(s)\mathbf{v} \end{array}\right\} \Rightarrow \mathbf{v} = (\mathbf{Y}_G(s) + \mathbf{Y}_E(s))^{-1}\mathbf{i}_E, \tag{33}$$

where $\mathbf{Y}_G(s)$ is the admittance matrix of the HVDC transmission grid; $\mathbf{i}_E$ and $\mathbf{Y}_E(s)$ are the current source and the diagonal admittance matrix of the external components, respectively; and $\mathbf{v}$ and $\mathbf{i}$ are the voltage and the injected current at the grid terminals, respectively.

Stability can be analyzed in the frequency domain by rewriting (33) as

$$\mathbf{v} = (\mathbf{I} + \mathbf{Z}_G(s)\mathbf{Y}_E(s))^{-1}\mathbf{Z}_G(s)\mathbf{i}_E, \tag{34}$$

where $\mathbf{I}$ is the identity matrix, and $\mathbf{Z}_G(s) = \mathbf{Y}_G^{-1}(s)$ is the grid impedance matrix. The stability of the closed-loop system in (34) can be assessed by the GNC if and only if $\mathbf{L}(s)$ does not have a right-half-plane (RHP) pole. This criterion (called SC_1 in this paper) extends the traditional Nyquist criterion to the Nyquist curves of the eigenvalues of the loop transfer function $\mathbf{L}(s) = \mathbf{Z}_G(s)\mathbf{Y}_E(s)$ [2,6,10,12,13]. Note that the RHP pole condition fails when system instability comes from the AC side of any VSC because the corresponding VSC subsystem in the diagonal of $\mathbf{Y}_E(s)$ is unstable. This instability was not detected in this case and, obviously, was also not detected when AC grid dynamics were disregarded in the DC-side equivalent admittance modeling of the VSCs.

4.2. SISO-Based Stability Assessment

The SISO system in Figure 6b can be derived at the VSC terminals, where $I_{eq,\,i}$ and $Y_{eq,\,i}(s)$ represent the current source and the admittance of the DC grid equivalent circuit observed from any VSC. The relation between the voltage V_i and the current sources is expressed as

$$V_i = Z_t(s)(I_{dc,\,i} + I_{eq,\,i}) \;\; ; \; Z_t(s) = \frac{1}{Y_{eq,\,i}(s) + Y_{dc,\,i}(s)}. \tag{35}$$

Two approaches are used to assess the stability of the SISO closed-loop system in (35):

- **Impedance-ratio stability criterion**

  Stability can be analyzed in the frequency domain by rewriting (35) as

$$V_i = \frac{Z_{eq,\,i}(s)}{1 + Z_{eq,\,i}(s)Y_{dc,\,i}(s)}(I_{dc,\,i} + I_{eq,\,i}) = Z_{T,\,i}(s)I_{T,\,i}. \tag{36}$$

If and only if the impedance ratio $Z_{eq,\,i}(s)\cdot Y_{dc,\,i}(s)$ does not have a right-half-plane (RHP) pole, the stability of the closed-loop system in (36) can be assessed by the Nyquist stability criterion (denoted as SC in this paper), which evaluates the Nyquist curve of the impedance ratio [4–6,9,11,15]. This criterion can only be applied if instability comes from the interaction between the VSC and the DC grid equivalent circuit because the RHP pole condition fails when system instability comes from the AC side of the VSC or the DC grid equivalent circuit observed from the VSC.

- **Positive-net-damping stability criterion**

The positive-net-damping stability criterion (denoted as SC_3 in this paper) evaluates system transfer impedance damping in (36) at resonance frequencies, i.e., the value of $\mathrm{Re}\{Z_{T,\,i}(j\omega)\}$ at the frequencies of the peak values of $Z_{T,\,i}(j\omega)$ [4–9]. This criterion assesses the closed-loop transfer function of the system, which avoids inaccuracies derived from local instabilities in the stability conclusion.

5. Application

The stability in the VSC-HVDC system of Figure 7a was studied. Two converter stations, VSC_1 and VSC_2, were connected through a 1000 MW 440 kV HVDC transmission grid. VSC_1 operated in mode $V_{dc} - Q$, controlling the DC voltage of the HVDC transmission grid, whereas VSC_2 operated in mode $P - Q$, controlling the transmitted active power. Both converters worked at unity power factor, typical of normal operating conditions [8,12–15]. According to Figure 7b, VSC_1 is characterized by the DC-side equivalent admittance in (31), while VSC_2 is considered as constant power consumption P_{20} characterized by the fictive resistance $R_{20} = -(V_{dc2,\,0})^2/P_{20}$ [4,9]. This converter could also be characterized with the DC-side equivalent admittance in (31), but a simpler model was used in the stability study to focus the attention on VSC_1. VSC-HVDC system data are in Table 3. Three cases were studied in the application:

- Case #1 (stable reference case): this corresponds to the data in Table 3, with the power consumed by VSC_2 being half the nominal power (i.e., P_{20} = 500 MW).
- Cases #2 and #3: the influence of the short-circuit ratio SCR and the VSC_2-consumed power P_{20}, respectively, on system stability were analyzed.

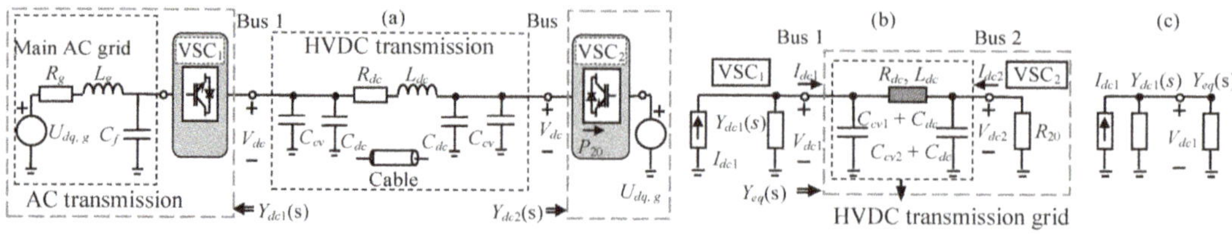

Figure 7. HVDC application: (**a**) HVDC transmission system. (**b**) HVDC transmission grid with VSC equivalent admittances. (**c**) Equivalent HVDC circuit.

Table 3. VSC-HVDC System Data ($U_{B,\,AC}$ = 220 kV, $U_{B,\,DC}$ = 440 kV, S_B = 1000 MVA).

AC main grid	U_0 (f_1), X_g/R_g $SCR = S_{cc}/(P_{vsc1,\,N})$	220 kV (50 Hz), 20 pu 15 pu
VSC AC filter	C_f	6 MF
DC cable	R_{dc}, L_{dc} C_{dc}, Length	10.8 mΩ/km, 0.149 mH/km 0.145 MF/km, 25 km
VSC DC filters	C_{cvi} (i = 1 to 2)	30 MF
VSC_1	$P_{vsc1,\,N}$, $V_{dc1,\,0}$	1000 MW, 440 kV
VSC_1 control	f_s, T_d R_c, L_c $k_{p,cc}$, $k_{i,cc}$ $k_{p,pll}$, $k_{i,pll}$ $k_{p,dc}$, $k_{i,dc}$ $k_{p,p}$, $k_{i,p}$	2 kHz, 0.3 ms 0.025 pu, 0.25 pu 1.21 pu, 0.121 pu 0.48 pu, 0.031pu 0.15 pu, 0.058 pu 0.15 pu, 0.058 pu

The system stability was verified by PSCAD/EMTDC time-domain simulations, and the predictions obtained by the stability criteria SC_1, SC_2, and SC_3 in Section 4 were investigated. The results are shown in Figures 8 and 9.

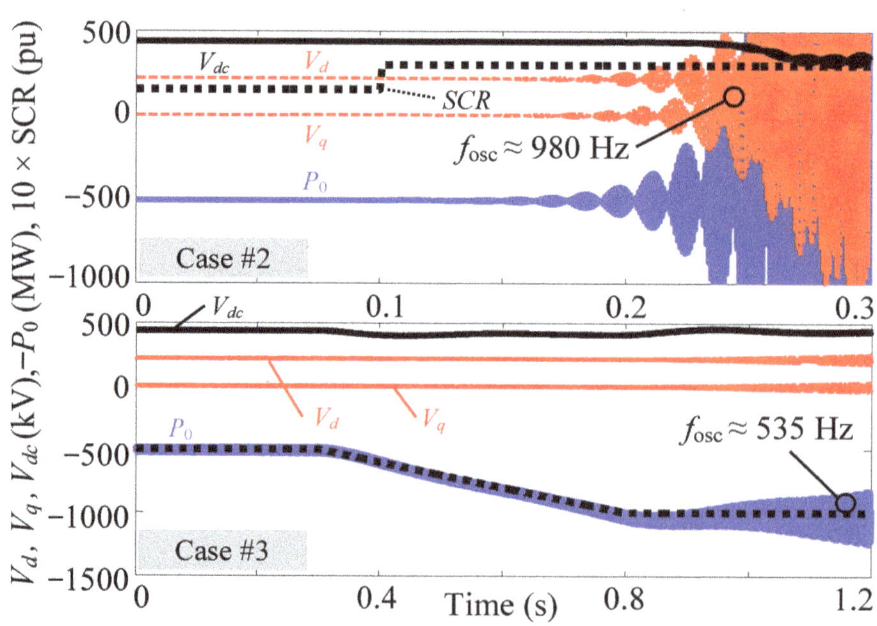

Figure 8. Stability study: Case #2 (top) and Case #3 (bottom).

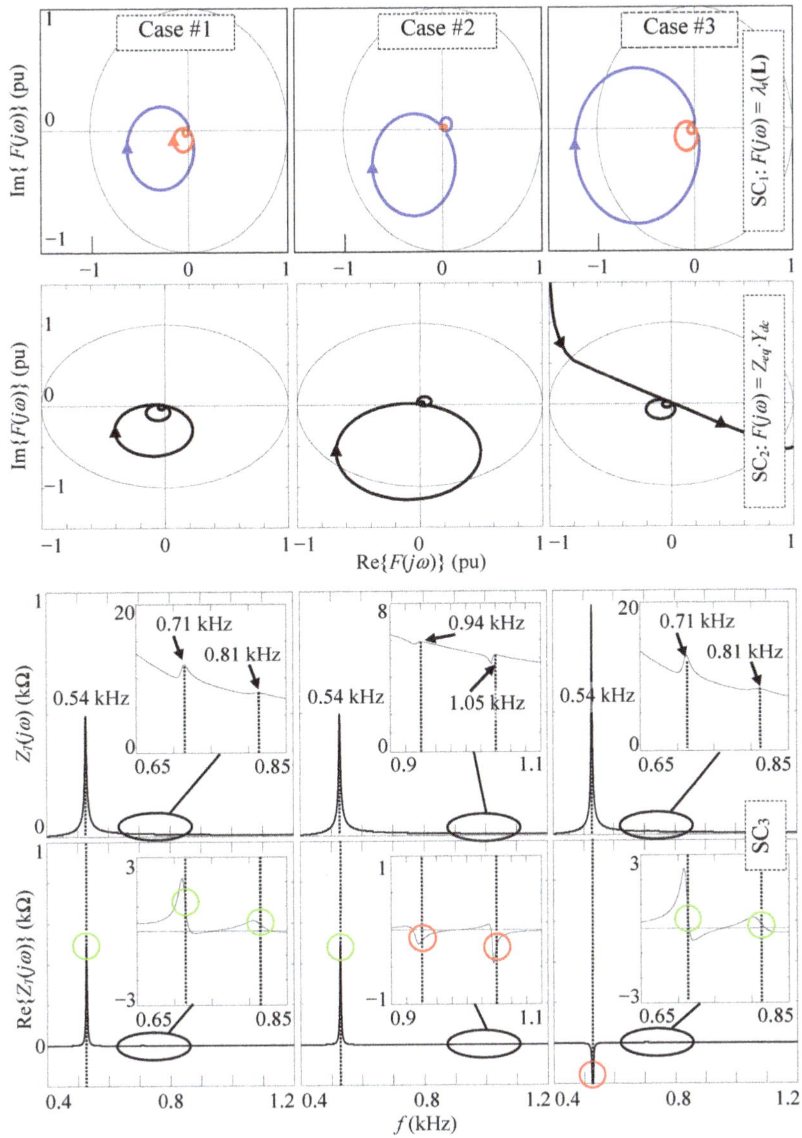

Figure 9. Impedance-based stability study.

Case #1: The PSCAD/EMTDC simulations before *SCR* and P_{20} changes in Figure 8 show the system stability. The results obtained from the stability criteria are as follows (see Figure 9, left):

- SC$_1$: the GNC verifies system stability because the curves of the eigenvalues of $\mathbf{L}(s)$ do not encircle the -1 point.
- SC$_2$: the Nyquist criterion verifies system stability because the Nyquist curve of $Z_{eq,i}Y_{dc,i}$ does not encircle the -1 point.
- SC$_3$: The positive-net-damping stability criterion verifies system instability because the damping Re{$Z_T(j\omega)$} at resonance frequencies $f_r \approx 537.5$, 712.3 and 810.3 Hz is positive. The first resonance is due to the HVDC grid capacitors and inductors whereas the second and third resonances are at the DC mirror resonance frequencies of the

AC-side resonance between the AC grid inductance L_g and the VSC_1 filter capacitor C_f at f_r = 760 Hz. It is observed that the VSC control damps (but does not eliminate) the AC-side resonance at the DC side. This means that AC-side dynamics are only damped at the DC side by the VSC control and should therefore be considered in the DC-side equivalent admittance modeling of the VSCs [12,13].

Case #2: The short-circuit ratio *SCR* increases from 15 pu to 30 pu at 0.1 s. The PSCAD/EMTDC simulation in Figure 8 shows that the system becomes unstable after the *SCR* value steps up. It was numerically verified that this is because the AC system resonance frequency is shifted to a negative damping frequency range and the AC-side voltage oscillations at $f \approx 980$ Hz are amplified. The results obtained by the stability criteria are as follows (see Figure 9, middle):

- SC_1: the GNC leads to an inaccurate stability assessment (i.e., the curves of the eigenvalues of $\mathbf{L}(s)$ do not encircle the -1 point) because the instability comes from the AC system resonance, causing the DC-admittance $Y_{dc,\,1}(s)$ in the diagonal of $\mathbf{Y_E}\,(s)$ to be unstable.
- SC_2: the Nyquist criterion also leads to an inaccurate stability assessment because $Y_{dc,\,1}(s)$ is unstable [13].
- SC_3: The positive-net-damping stability criterion verifies system stability (i.e., $\mathrm{Re}\{Z_T(j\omega)\}$ is negative at mirror resonance frequencies $f_r \approx 935.5$ and 1045 Hz). Closed-loop transfer function assessment by the positive-net-damping stability criterion avoids the SC_1 and SC_2 stability inaccuracy derived from the instability of $Y_{dc,\,1}(s)$. If AC-side dynamics are disregarded in the DC-side equivalent admittance model, AC-side resonances are overlooked, which leads to inaccurate stability assessments [12,13].

Case #3: The VSC_2-consumed active power increases from P_{20} = 500 MW to the nominal value P_{20} = 1000 MW at 0.3 s. The PSCAD/EMTDC simulation in Figure 8 shows that the system becomes unstable after the P_{20} value ramps up. It was numerically verified that this is because the damping becomes negative at the DC resonance frequency, and the DC-side voltage oscillations at $f \approx 535$ Hz are amplified. The results obtained by the stability criteria are as follows (see Figure 9, right):

- SC_1: the GNC verifies system instability because the curve of one eigenvalue intersects the unit circle at approximately 527 Hz, enclosing the -1 point in a clockwise direction.
- SC_2: the Nyquist criterion leads to an inaccurate stability assessment because of instability in $Y_{eq,\,1}(s)$.
- SC_3: The positive-net-damping stability criterion verifies system instability (i.e., $\mathrm{Re}\{Z_T(j\omega)\}$ is negative at $f_r \approx 537$ Hz). The closed-loop transfer function assessment in the positive-net-damping stability criterion avoids the SC_2 stability inaccuracy derived from the instability of $Y_{eq,\,1}(s)$.

6. Comparison of DC Admittance Models

The influence of the different simplifications that are usually carried out in DC admittance models is shown in Table 4. The results obtained with the full model in Section 3 for the three cases in Section 4 were compared with those obtained considering five different simplifications. It can be observed that the simplification of the *d*-outer control loop and the PLL do not affect the results of the full model, while this is not true for the simplifications related to the inner control loop (i.e., the VSC delay and the bandwidth of the grid voltage feedforward simplifications). Nevertheless, it must be noted that the above conclusions correspond to the harmonic instabilities of the studied cases, and they cannot be extrapolated to other types of instabilities. In general, harmonic instabilities are mainly affected by the inner control loop parameters, while synchronous and subsynchronous instabilities are affected by the outer control loops and the PLL.

Table 4. Comparison of simplifications.

Model Feature	Cases		
	Case #1	**Case #2**	**Case #3**
Full model	Stable	Instable	Instable
No phase-locked loop control	Stable	Instable	Instable
No VSC delay	Stable	Stable	Stable
No grid voltage feedforward ($\alpha_f = 0$)	Stable	Stable	Instable
No grid voltage feedforward filter ($\alpha_f = \infty$)	Stable	Instable	Instable
No d-outer control loop	Stable	Instable	Instable

7. Conclusions

This paper presents a complete and detailed DC admittance model for VSC rectifier and inverter stations of VSC-HVDC systems with active power, DC voltage, grid alternating voltage, and reactive power controls. The model also considers the VSC delay, grid voltage feedforward filter, PLL, and several outer control loops. It is derived from dq-complex space vectors commonly used in circuit theory. A comparison between the different models in the literature and a study of the impact on the instability of the different model simplifications were carried out to illustrate the contribution of the paper. The proposed model was applied to discuss VSC-HVDC system stability and the inaccuracies of impedance-based stability criteria due to system partitions. It is concluded that the stability assessment of VSC-HVDC systems from VSC and DC grid equivalent circuits could lead to inaccurate conclusions by the Nyquist criterion (e.g., when instability comes from AC-side dynamics) whereas the positive-net-damping criterion overcomes this issue because it is a global stability criterion which evaluates system transfer impedance damping at resonance frequencies.

Author Contributions: Conceptualization and methodology, J.P. and L.S.; software development and validation, L.M. and J.P.; writing—original draft preparation, J.P. and L.S.; visualization, L.M. All authors have read and agreed to the published version of the manuscript.

Funding: This research was funded by the MCIN (Proyectos de Generación de Conocimiento) under grant number PID2021-123633OB-C33 (J-02895).

Data Availability Statement: Data sharing not applicable.

Conflicts of Interest: The authors declare no conflict of interest.

Appendix A. Outer Loop Controls

The VSC control is characterized by the current controller (CC) and the outer control loops, which make it possible to characterize the main VSC operation modes (i.e., mode $P - Q$, mode $V_{dc} - Q$, mode $V_{dc} - E_{ac}$, and mode $P - E_{ac}$) [8].

$\underline{V_{dc} - Q \text{ control:}}$ the direct voltage control (DVC) and reactive power control (QPC) of Figure 1 are expressed with dq-complex space vectors [20] using the notation of (23) as

$$\mathbf{F}^{\mathbf{m}}_{\mathbf{o},\,\mathbf{UQ}} = \frac{1}{\sqrt{2}}\begin{bmatrix} F_{dc}(s) \\ F_{dc}(s) \end{bmatrix} \; ; \; \mathbf{G}^{\pm}_{\mathbf{o},\,\mathbf{UQ}} = \frac{F_q(s)E_{d0}}{2}\begin{bmatrix} 1 & -1 \\ -1 & 1 \end{bmatrix} \; ; \; \mathbf{Y}^{\pm}_{\mathbf{o},\,\mathbf{UQ}} = \frac{F_q(s)}{\sqrt{2}}\begin{bmatrix} -\mathbf{I}^{*}_{\mathbf{dq0}} & \mathbf{I}_{\mathbf{dq0}} \\ \mathbf{I}^{*}_{\mathbf{dq0}} & -\mathbf{I}_{\mathbf{dq0}} \end{bmatrix}, \tag{A1}$$

where $F_{dc}(s) = k_{p,\,dc} + k_{i,\,dc}/s$, with $k_{p,\,dc}$ and $k_{i,\,dc}$ being the proportional and integral gains of the DVC, and $F_p(s) = k_{p,\,q} + k_{i,\,q}/s$, with $k_{p,\,q}$, and $k_{i,\,q}$ being the proportional and integral gains of the QPC.

$\underline{P - Q \text{ control:}}$ the alternating power control (APC) and reactive power control of Figure 1 are expressed with dq-complex space vectors [20] using the notation of (23) as

$$\mathbf{G}^{\pm}_{\mathbf{o},\,\mathbf{PQ}} = F_p(s)E_{d0}\begin{bmatrix} 1 & 0 \\ 0 & 1 \end{bmatrix} \; ; \; \mathbf{Y}^{\pm}_{\mathbf{o},\,\mathbf{PQ}} = \sqrt{2}F_p(s)\begin{bmatrix} 0 & \mathbf{I}_{\mathbf{dq0}} \\ \mathbf{I}^{*}_{\mathbf{dq0}} & 0 \end{bmatrix}, \tag{A2}$$

where $F_p(s) = k_{p,p} + k_{i,p}/s$, with $k_{p,p} = k_{p,q}$ and $k_{i,p} = k_{i,q}$ being the proportional and integral gains of the APC.

$\underline{V_{dc} - E_{ac}}$ control: the direct voltage control (DVC) and alternating voltage control (AVC) control law of Figure 1 are expressed with dq-complex space vectors using the notation of (23) as

$$\mathbf{F}^{\mathbf{m}}_{\mathbf{o},\,\mathrm{UE}} = \frac{1}{\sqrt{2}}\begin{bmatrix} F_{dc}(s) \\ F_{dc}(s) \end{bmatrix} \; ; \; \mathbf{Y}^{\pm}_{\mathbf{o},\,\mathrm{UE}} = \frac{F_v(s)}{2}\begin{bmatrix} j & j \\ -j & -j \end{bmatrix}, \tag{A3}$$

where $F_v(s) = k_{p,v} + k_{i,v}/s$, with $k_{p,v}$ and $k_{i,v}$ being the proportional and integral gains of the AVC.

$\underline{P - E_{ac}}$ control: the alternating power control (APC) and alternating voltage control (AVC) control law of Figure 1 are expressed with dq-complex space vectors using the notation of (23) as

$$\mathbf{G}^{\pm}_{\mathbf{o},\,\mathrm{PE}} = \frac{F_p(s)E_{d0}}{2}\begin{bmatrix} 1 & 1 \\ 1 & 1 \end{bmatrix} \; ; \; \mathbf{Y}^{\pm}_{\mathbf{o},\,\mathrm{PE}} = \frac{F_p(s)}{\sqrt{2}}\begin{bmatrix} \mathbf{I}^{*}_{\mathbf{dq0}} & \mathbf{I}_{\mathbf{dq0}} \\ \mathbf{I}^{*}_{\mathbf{dq0}} & \mathbf{I}_{\mathbf{dq0}} \end{bmatrix} + \frac{F_v(s)}{2}\begin{bmatrix} j & j \\ -j & -j \end{bmatrix}. \tag{A4}$$

References

1. Sun, J.; Li, M.; Zhang, Z.; Xu, T.; He, J.; Wang, H.; Li, G. Renewable energy transmission by HVDC across the continent: System challenges and opportunities. *CSEE J. Power Energy Syst.* **2017**, *3*, 353–364. [CrossRef]
2. Zhang, C.; Molinas, M.; Rygg, A.; Cai, X. Impedance-Based Analysis of Interconnected Power Electronics Systems: Impedance Network Modeling and Comparative Studies of Stability Criteria. *IEEE J. Emerg. Sel. Top. Power Electron.* **2020**, *8*, 2520–2533. [CrossRef]
3. Liu, H.; Xie, X.; Liu, W. An Oscillatory Stability Criterion Based on the Unified *dq*-Frame Impedance Network Model for Power Systems with High-Penetration Renewables. *IEEE Trans. Power Syst.* **2018**, *33*, 3472–3485. [CrossRef]
4. Stamatiou, G.; Bongiorno, M. Stability Analysis of Two-Terminal VSC-HVDC Systems Using Net-Damping Criterion. *IEEE Trans. Power Deliv.* **2016**, *31*, 1748–1756. [CrossRef]
5. Sainz, L.; Cheah-Mane, M.; Monjo, L.; Liang, J.; Gomis-Bellmunt, O. Positive-net-damping Stability Criterion in Grid-connected VSC Systems. *IEEE J. Emerg. Sel. Top. Power Electron.* **2017**, *4*, 1499–1512. [CrossRef]
6. Harnefors, L. Modeling of Three-Phase Dynamic Systems Using Complex Transfer Functions and Transfer Matrices. *IEEE Trans. Ind. Electron.* **2007**, *54*, 2239–2248. [CrossRef]
7. Harnefors, L. Proof and Application of the Positive-net-damping Stability Criterion. *IEEE Trans. Power Syst.* **2011**, *26*, 481–482. [CrossRef]
8. Harnefors, L.; Bongiorno, M.; Lundberg, S. Input-admittance Calculation and Shaping for Controlled Voltage-source converters. *IEEE Trans. Ind. Electron.* **2007**, *54*, 3323–3334. [CrossRef]
9. Wen, B.; Boroyevich, D.; Burgos, R.; Mattavelli, P.; Shen, Z. Analysis of D-Q Small-Signal Impedance of Grid-Tied Inverters. *IEEE Trans. Power Electron.* **2016**, *31*, 675–687. [CrossRef]
10. Pinares, G.; Bongiorno, M. Analysis and Mitigation of Instabilities Originated from DC-side Resonances in VSC-HVDC Systems. *IEEE Trans. Ind. Appl.* **2016**, *52*, 2807–2815. [CrossRef]
11. Xu, L.; Fan, L.; Miao, Z. DC Impedance-model Based Resonance Analysis of a VSC-HVDC System. *IEEE Trans. Power Deliv.* **2015**, *30*, 1221–1230. [CrossRef]
12. Shah, S.; Parsa, L. Impedance Modeling of Three-phase Voltage Source Converters in dq, Sequence and Phasor Domains. *IEEE Trans. Energy Convers.* **2017**, *32*, 1139–1150. [CrossRef]
13. Amin, M.; Molinas, M. Small-signal Stability Assessment of Power Electronics Based Power Systems: A Discussion of Impedance- and Eigenvalue-based Methods. *IEEE Trans. Ind. Appl.* **2017**, *53*, 5014–5030. [CrossRef]
14. Amin, M.; Molinas, M.; Lyu, J.; Cai, X. Impact of Power Flow Direction on the Stability of VSC-HVDC Seen from the Impedance Nyquist Plot. *IEEE Trans. Power Electron.* **2017**, *32*, 8204–8217. [CrossRef]
15. Liu, J.; Tao, X.; Yu, M.; Xia, Y.; Wei, W. Impedance Modeling and Analysis of Three-Phase Voltage-Source Converters Viewing from DC Side. *IEEE J. Emerg. Sel. Top. Power Electron.* **2020**, *8*, 3906–3916. [CrossRef]
16. Rygg, A.; Molinas, M.; Zhang, C.; Cai, X. A Modified Sequence-domain Impedance Definition and its Equivalence to the dq-domain Impedance Definition for the Stability Analysis of AC Power Electronic Systems. *IEEE J. Emerg. Sel. Top. Power Electron.* **2016**, *4*, 1383–1396. [CrossRef]
17. Wang, X.; Harnefors, L.; Blaabjerg, F. Unified Impedance Model of Grid-connected Voltage-source Converters. *IEEE Trans. Power Electron.* **2018**, *33*, 1775–1785. [CrossRef]
18. Paap, G.C. Symmetrical Components in the Time Domain and their Application to Power Network Calculations. *IEEE Trans. Power Syst.* **2000**, *15*, 522–528. [CrossRef]

19. Harnefors, L.; Wang, X.; Chou, S.; Bongiorno, M.; Hinkkanen, M.; Routimo, M. Asymmetric Complex-Vector Models With Application to VSC–Grid Interaction. *IEEE J. Emerg. Sel. Top. Power Electron.* **2020**, *8*, 1911–1921. [CrossRef]
20. Pedra, J.; Sainz, L.; Monjo, L. Three-port small-signal admittance-based model of VSC for studies of multi-terminal HVDC hybrid AC/DC transmission grids. *IEEE Trans. Power Syst.* **2021**, *36*, 732–743. [CrossRef]

energies

MDPI

Article

Actions to Handle Increasing Distributed Generation and Rising Voltage during Low-Demand Periods: Tap Staggering Practices and Field Tests in the Italian Transmission Network

Stefano Quaia [1,*], Alessandro Mauri [1], Alessio Marchesin [2] and Paolo Manià [3]

[1] Department of Engineering and Architecture, University of Trieste, Via Valerio 10, 34127 Trieste, Italy; alessandro.mauri@phd.units.it
[2] Terna Rete Italia S.p.A., Via Botticelli 139, 10154 Torino, Italy; alessio.marchesin@terna.it
[3] AcegasApsAmga S.p.A., Via del Teatro 5, 34121 Trieste, Italy; pmania@acegasapsamga.it
* Correspondence: quaias@units.it

Abstract: This paper deals with the voltage regulation strategies implemented in the Italian transmission network to face the increasing problem of high voltages during time periods characterized by low demand. After an introduction in which this very actual problem is discussed, the focus is on tap staggering practices. Although tap staggering is not a new idea, it is gaining practical importance only in the very last few years as a means of enhancing the inductive power drawn from the grid and, therefore, limiting the voltage rise. Accordingly, tap staggering contributes to the mitigation of the problems caused by the increasing penetration of renewable energy sources and thus can allow an increase in the share of renewable energy sources. The paper presents the different tap staggering practices that are being defined by the Italian transmission system operator and reports some tap staggering tests recently performed on large autotransformers as well as phase-shifter transformers.

Keywords: voltage regulation; transformer tap staggering; autotransformers; phase-shifting transformers

Citation: Quaia, S.; Mauri, A.; Marchesin, A.; Manià, P. Actions to Handle Increasing Distributed Generation and Rising Voltage during Low-Demand Periods: Tap Staggering Practices and Field Tests in the Italian Transmission Network. *Energies* **2023**, *16*, 6258. https://doi.org/10.3390/en16176258

Academic Editor: Ying-Yi Hong

Received: 13 July 2023
Revised: 17 August 2023
Accepted: 24 August 2023
Published: 28 August 2023

1. Introduction

In the last decade, an increasing upward trend in voltages in some parts of the transmission systems has been taking place, mainly in low-demand periods like nights, weekends, and holidays [1–3]. Two important factors are behind this trend. First, in operating conditions characterized by low demand and high production from nonprogrammable renewable sources, the committed sets of production units (PUs) with regulation capability are particularly small. Second, the reactive energy drawn/introduced from/into the transmission systems is changing, with a clear progressive reduction in the total annual reactive energy drawn ("inductive energy") and a clear progressive increase in the total annual reactive energy introduced ("capacitive energy"). In Italy, the total amount of capacitive energy introduced in HV networks roughly doubled in the period 2014–2020, approaching the total amount of inductive energy drawn [3]. Table 1 shows that, in this period, the total amount of reactive energy exchanged between HV networks and the sum of distribution networks and HV users has remained approximately constant at nearly 50 TVArh/year, but there has been a clear progressive decrease in inductive energy and a clear progressive increase in capacitive energy [4]. In 2020, the latter almost matched the former. In the figures reported, distribution systems are by far the main part.

Transmission system operators (TSOs) control voltages and manage reactive energy in their HV networks by means of proper devices/techniques (which include synchronous compensators, capacitor banks, shunt reactors, STATCOMs, and transformer tap staggering) and stocking up resources in the dispatching services market (MSD). In order to meet voltage regulation requirements, the current upward trend in voltages in some parts of the transmission systems forces TSOs to commit additional PU compared to those committed as a result of the electricity markets. On the one hand, this leads to increased dispatching costs and, on the other hand, constitutes an obstacle to the increasing penetration of nonprogrammable renewable sources required by the ongoing energy transition process. This latter aspect is made even more crucial and urgent by the recent tensions in energy prices and dramatic geopolitical events.

Table 1. Total amounts of reactive energy exchanged between the HV grids and distribution systems/HV users.

Year	Inductive Energy [TVArh/year]	Capacitive Energy [TVArh/year]	Total Reactive Energy Exchanged [TVArh/year]
2014	36.1	13.4	49.5
2018	32.5	17.9	50.4
2019	29.9	20.3	50.4
2020	25.3	23.1	48.4

As for dispatching costs, in 2014, the cost incurred by Terna, the Italian TSO, on the MSD for reactive energy management was estimated at approximately 150 M€ [5]. But, in the two years 2018–2019, this cost rose to more than 1000 M€/year [4].

In addition, the increasing voltages in some parts of the transmission system have caused a significant increase in TSO investments for devices for voltage regulation and reactive energy control (in the period 2020–2026, the total investment costs, made and planned, are estimated at approximately 1 G€ [4]).

As for nonprogrammable renewable sources, their increasing penetration has important effects on the generators in service, as they lead to small sets of regulating PUs committed as a result of the electricity markets and, consequently, to the need to commit additional PUs in order to provide adequate system security and network services (like voltage regulation).

Aiming to allow an increasing share of nonprogrammable renewable generation, Terna has started, since the second half of 2021, a revision of its real-time management processes. In particular, special effort has been focused on reducing the commitment of PUs not resulting from the electricity markets but required to provide adequate system security and services.

The TSO has several means to counteract the rising trend in voltage values during low-demand periods. In detail, the actions based on already existing network components are listed below. Note that the actions performed on compensation devices generally have higher priority than configuration changes or power flow changes. For example, shunt reactor management has higher priority than PST control, tap change on autotransformers (ATRs), disconnection of ATRs, disconnection of lines, and so on.

- Synchronous generator excitation control on the primary transmission network (which includes the 380–220 kV voltage levels);
- Shunt reactor management;
- Active power flow control on the primary kV transmission network by means of phase-shifting transformers (PSTs) and set point changes in HVDC links;
- Changes in HVDC filter arrangement;
- Network configuration changes by acting on parallel busbar switch breakers;
- Control of ATRs on-load tap changers (OLTCs) and disconnection of ATRs;
- Distributed tap staggering (TS) of ATRs with limited tap unbalance (see Section 3) and/or localized TS of ATRs (and, when available, of PSTs).

In addition, the following actions can also be performed, but note that they are costly for the system:

- Active power flow changes by means of PU re-dispatching;
- Early PU commitment and/or delayed PU decommitment;

- Commitment of PUs not programmed as a result of the electricity markets.

In agreement with what is reported above, the target of Terna is to minimize the impact of the three latter actions while trying to maximize the benefit obtained through the implementation of the previous "not costly" actions. The benefit consists of a higher share of nonprogrammable renewable source generation allowed, reduced dispatching costs, or both.

From the viewpoint of organization, different actions have been put in place:

- A push/acceleration for the installation of reactive power compensators like shunt reactors, synchronous compensators, and STATCOMs;
- Close coordination between the National and Regional Control Centers for managing the primary transmission and subtransmission networks with a view to voltage regulation in the overall system;
- Creation of a list of maneuvers on the network intended to limit system voltages. For example, the list indicates the lines that can be disconnected, the effect of each disconnection, and information concerning the possibility of performing TS;
- Creation of teams of experts supporting 24 h the control rooms in the decisions concerning the acquisition of PUs not committed as a result of the electricity markets.

Although TS is a means for reactive power management that has been known for a long time, it has been gaining increasing importance only in the very last few years to help counteract the actual problem of high voltages during periods of low demand. As TS does not require the connection of new components to the power system, it is cheaper than other solutions like the installation of new shunt reactors (in Italy, the investment costs for the installation of new shunt reactors in the 4 years 2021–2024 will be approximately 48 M€ [4]). Moreover, operative field data suggest that the reliability of shunt reactors could be lower than that of TRSs, ATRs, and PSTs involved in TS. However, TS stresses the involved machines more, and it is still early to evaluate any reliability reductions.

The following sections specifically concern TS and describe the application rules that are being defined by the Italian TSO. Section 2 briefly reports the principles of TS performed on TRSs/ATRs and PSTs and includes a short comparison among different available techniques for reactive power management.

Section 3 analyzes the different modalities adopted by the Italian TSO for TS implementation/management. Section 4 reports the results of recent field tests performed on both ATRs and PSTs in the Italian primary transmission network. Sections 3 and 4 present some original rules/solutions for what concerns the TS performed using PSTs and distributed or localized TS.

2. Principles of Tap Staggering

The theory of TS is briefly described in this section. TS consists of operating in parallel two tap-change transformers (TRSs) or ATRs on a dedicated secondary busbar with different tap positions [6–8]. This configuration provides a means of absorbing reactive power and thus contributes to regulating voltage in the substation area in low-load periods. The reactive power absorbed at the input node and therefore the voltage variation on the primary side depends on its short-circuit power and steady-state operating conditions before TS implementation.

Consider two tap-change TRSs connected in parallel. The number of turns of the secondary windings depends on the selected taps. Assume the upper TRS has $(1 + n_a)$ turns, and the bottom TRS has $(1 + n_b)$ turns. By neglecting the no-load currents and the series (short-circuit) resistances, this arrangement is represented by the circuit reported in Figure 1 [9]. This circuit can be transformed in the "equivalent transformer" illustrated in Figure 2.

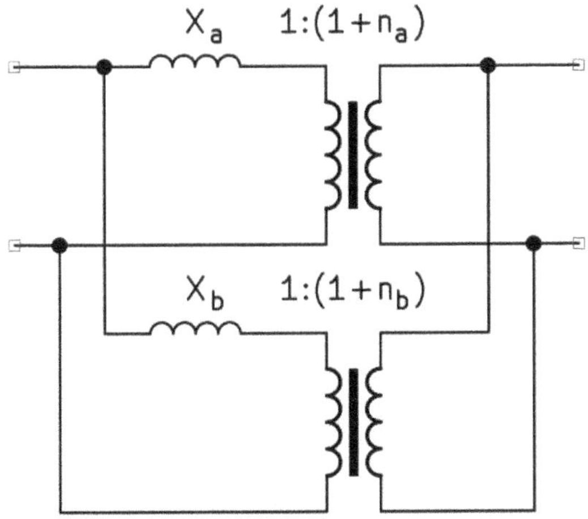

Figure 1. Electrical scheme of two tap-change transformers connected in parallel.

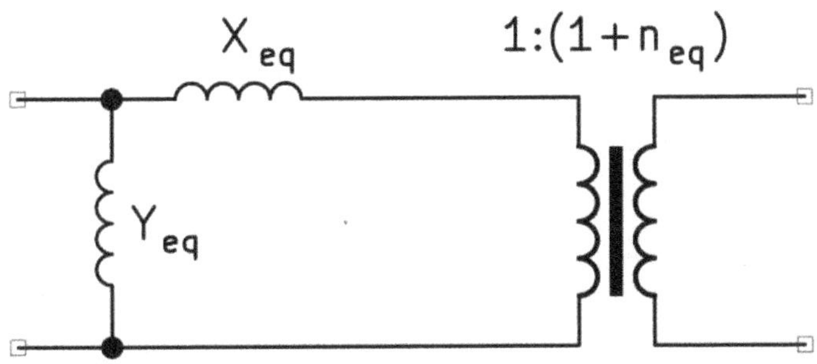

Figure 2. Equivalent transformer.

The parameters of the resulting equivalent transformer (i.e., series reactance X_{eq}, per unit tap-change setting neq and transversal admittance Y_{eq}) can be expressed as follows as a function of the per unit tap-change settings n_a and n_b and the series (or short circuit) reactances X_a and X_b:

$$Y_{eq} = -j\left[\frac{(n_b - n_a)^2}{X_a(1+n_a)^2 + X_b(1+n_b)^2}\right] \tag{1}$$

$$n_{eq} = n_a - \frac{X_a(1+n_a)^2(n_a - n_b)}{X_a(1+n_a)^2 + X_b(1+n_b)^2} \tag{2}$$

$$X_{eq} = \frac{X_a X_b\left[X_a(1+n_a)^2 + X_b(1+n_b)^2\right]}{\left[X_a(1+n_a)^2 + X_b(1+n_b)^2\right]^2} \tag{3}$$

Equation (1) shows that unequal tap-change settings ($n_a \neq n_b$) give rise to an equivalent transversal reactive admittance, with reactive power absorption from the upstream power system. Therefore, this arrangement can be used to improve voltage regulation during light-load conditions. If TS is performed with TRSs in no-load operation (see Section 3), according to Figure 2, the reactive power absorbed is only due to the resulting transversal admittance Y_{eq}.

Note that, in principle, this action has the same qualitative effect of other devices capable of absorbing reactive power, like shunt reactors, TCRs (i.e., continuously controlled reactors [10]), and synchronous compensators. However, the costs are very different. Unlike the existing TRSs or ATRs of an electrical substation, shunt reactors, TCRs, and synchronous compensators must be regarded as new devices to be installed by the TSO. Accordingly, if TRSs or ATRs can be operated, during light-load conditions, in TS configuration (this possibility must be verified case-by-case by the TSO), this practice can allow the installation of new devices to be avoided.

Aiming to compare these different techniques, first we note that the high and very high dynamic responses of respectively synchronous compensators and TCRs are not required to face the problem of rising voltages in low-demand periods. As is known, in most cases, these techniques are used in order to obtain other technical benefits: system inertia and short-circuit power increase (synchronous compensators), fast dynamic voltage regulation, and/or system stability increase (TCRs). It follows that, in practice, the most significant comparison can be performed between transformer TS and shunt reactors. Shunt reactors are relatively simple but allow just an on/off operation (reactive power cannot be regulated). On the other hand, their connection/disconnection is not subject to the network checks required by TS (see Section 3). Table 2 briefly summarizes the main differences concerning both costs and technical aspects of these four techniques.

Table 2. Comparison among different techniques capable of facing the problem of rising voltage during low-demand periods.

Technique	Cost	Main Advantages	Main Disadvantages
Transformer Tap Staggering	Low	Connection of new devices is not needed; several available machines can be used.	Operation must be verified case-by-case by the TSO.
Shunt reactors	High	Relatively simple; operation is largely independent of network conditions.	Only on/off operation.
TCR	Very high	Continuous reactive power control; high dynamics.	High dynamics are not needed for this voltage regulation problem.
Synchronous compensators	Very high	Continuous reactive power control; increase in system inertia and short-circuit power.	Inertia is not involved in this voltage regulation problem.

Aiming to increase the circulation current between the two TRSs and thus the reactive power absorbed from the upstream network, the taps selected on the two OLTCs must be changed in opposite directions. Assuming that:

- The same OLTC tap-change settings (with opposite sign) are used for the two TRSs: $n_a = \Delta n$ and $n_b = -\Delta n$;
- The series reactances X_a and X_b are equal (in practice, often the two involved TRSs/ATRs are twins and the limited difference between the two reactances due to the different number of turns n_a and n_b can be neglected).

From Equation (1), denoting by X the series reactances X_a and X_b, the resulting transversal admittance Y_{eq} is:

$$\dot{Y}_{eq} \cong -j\frac{4\Delta n^2}{2X(1+\Delta n^2)} \cong -j\frac{2\Delta n^2}{X} \tag{4}$$

More generally, given the nominal data of the involved machines and the tap position settings, the reactive power absorbed in TS operation can be estimated.

The typical values of the ATRs used in Italian HV substations are as follows: nominal power A is 250 MVA, short-circuit impedance X_{sc} goes from 11% to 12%, and OLTC ±10%

with ±5 taps and 11 total positions. According to Equation (4), the maximum reactive power that two typical ATRs can absorb in a no-load configuration is approximately 40 MVAr.

TS can also be performed using two TRSs under normal load. In this case, the currents caused by the TS add to the load currents, and therefore, the two TRSs carry different currents. It follows that the maximum allowed unbalance in the selected taps is generally lower than the maximum one. At the rated current, the reactive power that can be absorbed by two TRSs operated in TS in under-load mode depends on (1) the load level and (2) the power factor. Obviously, the reactive power that can be absorbed is lower than that in the no-load mode. In addition, the secondary voltage of the two TRSs in parallel is subject to more stringent constraints.

In principle, TS can also be performed by exploiting the normal layout of the primary substations (owned by the distribution system operators, DSO), which include two HV/MV TRSs in parallel [7,11,12]. However, in this case, the benefits mainly concern the upstream TSO HV network, whereas TS is performed by DSO machines.

In practice, until today, Italian DSOs have not performed TS on their TRSs. However, as already mentioned in Section 1, DSOs are responsible for most of the capacitive energy introduced in HV networks. Therefore, for DSOs (and HV users), recent tariff regulation concerning the reactive energy exchanged with HV networks, in force since 1 April 2023, states that capacitive energy cannot be introduced into HV networks in any time slot and sets the unit fees to be applied in case of noncompliance with this prescription [13]. In this regard, DSOs are developing network models to predict grid behavior. In addition, they are utilizing the provisions of the Italian compulsory standards for connections to improve their capability of managing reactive energy.

A general assumption for Terna is that the available assets already installed on the grid and able to provide multi-service performance should do so. This idea can also be applied to the case of PSTs. It is known that these machines are used to control the active power flow on the electric transmission lines they are connected to, by acting on the voltage phase angle difference between the two nodes at the extremities of the lines. However, exploiting a proper arrangement, a PST can also be used to absorb reactive power, and therefore, in "TS mode" [9].

The TS mode is obtained by closing the PST bypass circuit (see the field test described in Section 4.2 below). Accordingly, unlike TS performed with two TRSs/ATRs, one PST is sufficient. The current absorbed depends on the equivalent reactances of the PST, which depend in turn on the selected tap. However, PSTs are generally connected on interconnection lines between neighboring countries or on important transmission lines. Therefore, the possibility of operating a PST in TS mode must be verified case-by-case, since in this case, the machine is subtracted to its primary task of active power flow regulation.

For a PST with a symmetrical structure (as in the case of the field test illustrated in Section 4.2), the input and output voltages have equal magnitudes, and the voltage difference between the input and output, ΔV, depends on the selected tap (which determines the phase shift angle introduced, α) and on the input voltage magnitude V:

$$\Delta V = 2V \sin(\alpha/2) \tag{5}$$

The TS mode is obtained by closing the PST bypass circuit. The current absorbed depends on the equivalent reactances of the PST, which depend in turn on the selected tap.

For a two magnetic core machine (depicted in Figure 3), as the involved PST is in the field test reported below, reference [14] provides the equivalent shunt and series reactances as a function of the phase shift angle, α.

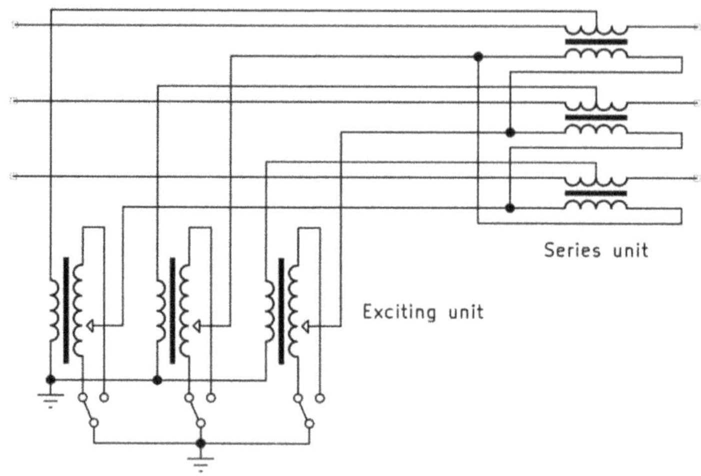

Figure 3. Circuit diagram of a two magnetic core symmetrical PST.

The shunt reactance $X_{shunt}(\alpha)$ is defined as the equivalent reactance, which crossed by the series input current I of the PST produces the reactive losses of the shunt (exciting) unit. The series reactance $X_{series}(\alpha)$ is defined as the equivalent reactance, which crossed by the same series input current I produces the reactive losses of the series unit. The equations are as follows:

$$X_{shunt}(\alpha) = 4\left(\sin\left(\frac{\alpha}{2}\right)\right)^2\left(X_3 + X_{4\max}\left(\frac{n_3}{n_{4\max}}\right)^2\right) \tag{6}$$

$$X_{series}(\alpha) = X_1 + X_2\left(\frac{n_1}{n_2}\right)^2 - X_2\left(\frac{n_1}{n_2}\right)^2\left(\sin\left(\frac{\alpha}{2}\right)\right)^2 \tag{7}$$

where:

- n_1 (fixed) is the number of turns of the series windings (series unit);
- n_2 (fixed) is the number of turns of the delta-connected "booster" windings (series unit);
- n_3 (fixed) is the number of turns of the star-connected exciting windings (exciting unit);
- n_4 (variable and selected by the OLTC) is the number of turns of the star-connected regulation windings (exciting unit).

The total equivalent reactance of the PST $X(\alpha)$, which crossed by the series input current I produces the total PST reactive power absorption, is then the sum of $X_{shunt}(\alpha)$ and $X_{series}(\alpha)$:

$$X(\alpha) = \left(X_1 + X_2\left(\frac{n_1}{n_2}\right)^2\right) + 4\left(\sin\left(\frac{\alpha}{2}\right)\right)^2\left(X_3 + X_{4\max}\left(\frac{n_3}{n_{4\max}}\right)^2 - X_2\left(\frac{n_1}{2n_2}\right)^2\right) \tag{8}$$

where each reactance X_i refers to the i-th winding made up of n_i turns.

Finally, the total reactive power absorbed Q is:

$$Q = 3\,X(\alpha)I^2 \tag{9}$$

Using these models, the reactive power that can be absorbed by any given PST can be theoretically calculated.

3. Tap Staggering Application Rules of the Italian TSO

As already reported, TS is one of the actions that can be used to counteract rising voltage values in low-demand time periods. TS can be either programmed to define a time for its actuation or implemented in real time. In the Italian primary transmission network, TS can be performed using the following:

- Two ATRs connected in the same electrical substation;
- One PST (using a PST in "TS mode" requires giving up the active power control of the PST: if this is acceptable, the machine can be used during a limited time interval for TS purposes. The TS test with PST described below exploits a (not frequent) situation where two PSTs are installed in the same substation).

From a power system perspective, TS can be performed involving a higher or lower number of nearby electrical substations. The former case is referred to as "distributed" TS. It is performed using ATRs with a limited tap unbalance, typically not exceeding 50% of the maximum unbalance allowed by OLTCs. The latter case is referred to as "localized" TS and can be performed using ATRs but also PSTs (when available), setting the tap unbalance to the maximum (provided that the rules listed below are satisfied). Of course, distributed TS causes less stress to the involved machines but, on the other hand, involves more machines than localized TS. In all cases, once the critical conditions time is over, best practice says to stop TS in order to limit wear on the involved machines.

As already reported in Section 2, TS performed with two ATRs of a substation can be performed either at no-load (more effective case) or under normal-load conditions. Performing TS at no-load requires the real-time control rooms to check the possibility of excluding two ATRs from their normal operation. If the network conditions do not allow this exclusion, TS can be performed with the machines under load.

3.1. Tap Staggering with ATRs in No-Load Operation

In this operative mode, the two ATRs run at no-load on a dedicated (insulated) busbar. This arrangement is possible when:

- The load supplied by the underlying local network allows the exclusion from service of two ATRs in the considered substation;
- The ATRs in normal operation (i.e., those not dedicated to TS) and/or the actual network conditions allow the safe operation of the underlying network;
- In TS operation, the following application rules are currently established:
- The real-time room controls that the currents circulating in the two machines remain within their limits;
- In the case that one ATR opens, the secondary of the other (still in service) ATR must be immediately opened as well.

As to the stress caused by TS in the involved machines, Terna is performing characterization campaigns to define the usage limits of OLTCs as a function of ATR secondary voltage. At present, for not-yet characterized families of devices, conservative limits are used (i.e., 50% of the OLTC regulation range).

3.2. Tap Staggering with Normally Loaded ATRs

When the operating conditions of the network and/or the number of ATRs available in the area do not allow for the dedication of two ATRs for TS in no-load operation, TS in the involved substation can be performed using two ATRs in normal under-load operation with the secondary sides connected to the same busbar. The two ATRs are connected in parallel, and TS is performed, changing in opposite directions the operating taps of the two OLTCs.

The current caused by the unbalance in OLTCs adds vectorially to the load currents and thus different currents are obtained in the two ATRs. As the taps are changed, one ATR takes more load, and the other ATR takes less load. This implies that the unbalance in OLTC positions must be limited and, consequently, a lower reactive power can be absorbed compared with the maximum reactive power absorbed in the no-load mode [12]. Also, in the under-load mode, the currents circulating in the two machines must be continuously controlled.

If one ATR is opened during the operation in TS, the OLTC of the other ATR must be adjusted in order to keep the secondary voltage at a proper value within normal operation limits. A correct secondary voltage will be restored in the times required by OLTC maneuvering.

As explained in the previous paragraph with regard to the no-load mode, also in the under-load mode, the control room controls that the currents circulating in the two machines remain within their limits.

4. Field Tests

4.1. *Tap Staggering with ATRs in No-Load Operation*

This paragraph briefly reports a TS field test with ATRs in no-load operation recently performed in the 380 kV substation Parma Vigheffio, in Northern Italy. The two ATRs involved are both rated 250 MVA and have the following data: short-circuit impedance $X_{sc} = 0.12$ and OLTC ±10% with ±5 taps.

As a first step (at approximately 11:30 AM in Figure 4), the two ATRs were set in parallel with the secondary side closed on the isolated 132 kV busbar dedicated to the test. Of course, this action resulted in the cancellation of the previous active power flow, as indicated by the dotted curve in Figure 4.

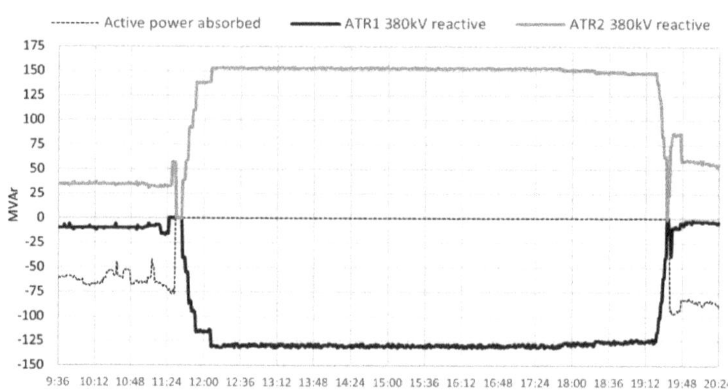

Figure 4. Active and reactive power of two ATRs in TS.

Next, the two transformers were operated at different (not symmetrical) tap positions: +5 taps for one ATR (which means $\Delta n = 0.1$ p.u.) and −2 taps for the second ATR (which means $\Delta n = -0.04$ p.u.). The reactive power measurements show opposite flows on the two transformers, with a total balance of approximately 22 MVAr absorbed at the network connection point. The reactive power absorbed is actually approximately 8% of the rated power of the machines, as expected on the basis of the theoretical models reported in Section 2. This result confirms experimentally very well what was foreseen analytically. Overall, the TS test lasted approximately 8 h.

4.2. *Tap Staggering with PST*

This paragraph reports a TS field test performed with a PST in the 380 kV substation Foggia, in Southern Italy. Under the supervision of the TSO National Control Center, the test started after checking the network operative conditions and power flows involved.

Figure 5 shows the operational configuration adopted for the test in the involved substation in which two PSTs are installed.

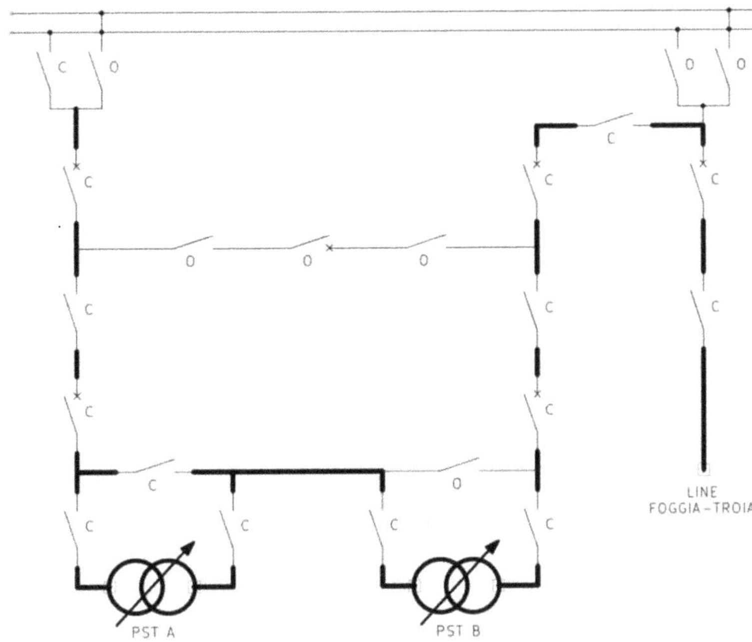

Figure 5. Substation layout, where PST A is in TS and PST B is in pushing mode (C and O stand, respectively, for close and open).

PST A is the machine under test. TS operation is obtained by closing the bypass disconnector and selecting a proper tap (angle shift) (in order to allow PST A operation with the bypass closed, some block signals of the PST automatic control must be turned off). Meanwhile, PST B is kept in normal operation, pushing active power on the 380 kV line Foggia-Troia, on which it is connected.

This configuration (evidenced by the thick lines in Figure 4) allowed the normal operation of the 380 kV line Foggia-Troia, on which the active power flow was approximately 400 MW.

The main technical data of the two identical PSTs are as follows: symmetrical structure with two magnetic cores (see Figure 3), buck and boost regulation, throughput rating P_T is 1800 MVA, nominal rating of the two units P_D is 543.2 MVA, 33 taps, maximum phase shift at no-load α_0 is 17.5°.

The theoretical analysis based on the considerations reported in Section 2 provides a maximum expected reactive power absorption of approximately 150 MVAr for the PST in TS operation. The test had the following main targets:

- To check the maximum reactive power absorption estimated (approximately 150 MVAr);
- To check the effect produced on the voltages in the area.

We briefly report the main results of the test. The first result, shown in Figure 6, regards the maximum reactive power absorbed, which was approximately 140 MVAr, broadly in line with the expected theoretical value.

As to the effect of TS on voltage regulation, the reactive power absorbed during the test led to a voltage reduction of approximately 2.5 kV at the Foggia 380 kV node, as Figure 7 illustrates. The other voltages shown in Figure 7 are those recorded at two neighboring nodes of the 380 kV network.

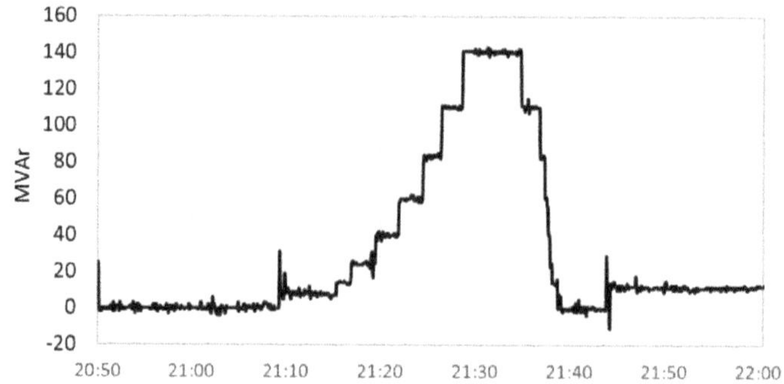

Figure 6. Reactive power absorbed during the TS test (changing with the selected taps).

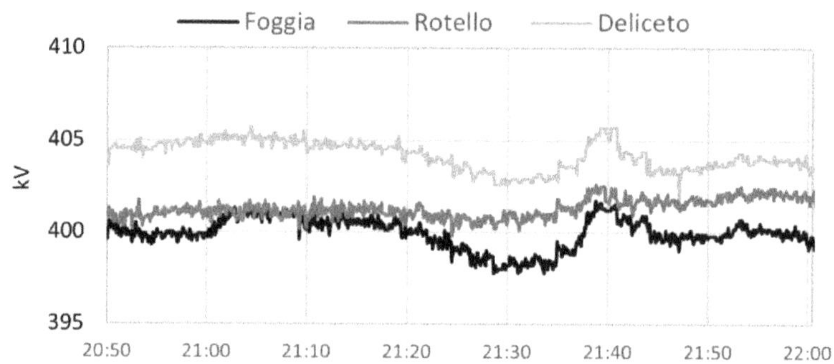

Figure 7. Voltages measured at the 380 kV nodes of Foggia, Deliceto, and Rotello.

For reasons of safe operation during the TS test, the automatic voltage regulations (AVRs) of the nearby rotating PUs were not suppressed/modified. The AVR operation reduces the TS effect on the voltages in the area (in other words, the TS effect is weakened). Measurements performed during the test showed that the voltage reduction caused an overall reactive power change of approximately 75 MVAr coming from the four synchronous generators in operation in the area (see Figure 8). This reactive power amount would have resulted in a further voltage reduction at the Foggia 380 kV substation that could be evaluated at roughly 1–1.5 kV, for a total of almost 4 kV (approximately 1% of the nominal voltage). Figure 8 clearly shows the reduction in the reactive power absorbed by the generators in correspondence with the peak of the reactive power absorbed by the PST under the test.

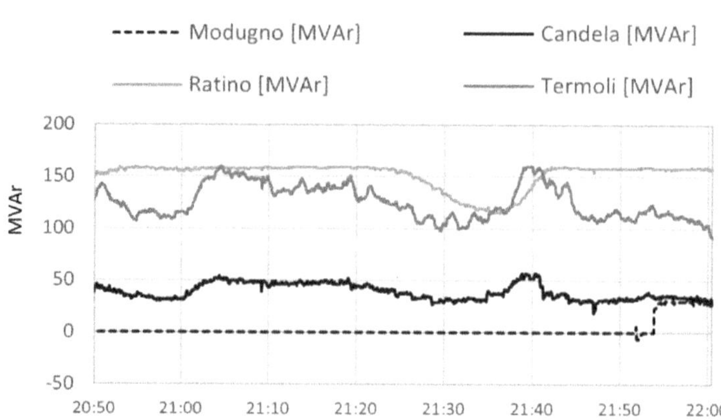

Figure 8. Reactive power absorbed by four rotating PUs in the area of the test.

5. Conclusions

In the very last few years, TS has been gaining practical importance as a possible action aimed at counteracting the rising voltage trend that occurs in low-demand time periods in some areas of HV networks.

This paper describes the TS application rules that are being defined by the Italian TSO and reports some recent field tests performed on large ATRs and PSTs. Considering the very limited literature available today on this subject, these contents can be considered, at least to a certain extent, original.

TS can be performed using couples of ATRs located in the same substation, either in no-load operation or in normal under-load operation. The former case provides the most effective results, but this operative mode is not always allowed, as it requires the exclusion of two ATRs from their normal operation.

A further possibility is given by TS performed with PSTs. Even though ATRs are far more widespread than PSTs, PSTs usually have a larger rated power and regulation range and can absorb large reactive power in TS operation.

Finally, this paper reports the main results of recent TS field tests performed by the Italian TSO using both ATRs and PSTs. The test results confirm the effectiveness of TS and the correctness of the theoretical models used.

From an overall power system perspective, an important technical advantage connected with TS implementation is an increase in the nonprogrammable renewable generation allowed without violation of voltage constraints. Of course, this advantage can also be obtained by installing various types of voltage-controlling devices (for example, shunt reactors). A first qualitative comparison between TS and shunt reactors can be performed, considering both the costs and reliability of the two solutions. As for the costs, the comparison is clearly in favor of TS. Concerning reliability, some faults that happened on shunt reactors in Italy could lead to the same conclusion, even though more time is still needed to obtain the necessary operative experience for TS. A further advantage of TS is the possibility of controlling, step-by-step, the reactive power absorbed, whereas shunt reactors usually allow just on–off operation. On the other hand, shunt reactors do not require careful network checks for the connection/disconnection required by transformer TS.

Author Contributions: All the authors contributed to design the research approach. S.Q. and A.M. (Alessio Marchesin) dealt with conceptualization and methodology; S.Q. and A.M. (Alessandro Mauri) wrote and edited the original draft, including formal analysis and graphing; P.M. and A.M. (Alessio Marchesin) took care of resources and data. All the authors revised the manuscript. All authors have read and agreed to the published version of the manuscript.

Funding: This research received no external funding.

Data Availability Statement: Additional data and measurements not explicitly provided in the manuscript may be subject to censorship by Terna S.p.A. as they are of a sensitive nature.

Acknowledgments: The authors acknowledge Davide Rampazzo for his precious technical support.

Conflicts of Interest: The authors declare no conflict of interest.

Nomenclature

PU	production unit
HV	high voltage
MV	medium voltage
TSO	transmission system operator
DSO	distribution system operator
MSD	dispatching services market
PST	phase-shifting transformer
ATR	autotransformer

TRS	transformer
OLTC	on-load tap changer
TS	tap staggering
TCR	thyristor controlled reactor
AVR	automatic voltage regulator

References

1. National Grid. *Electricity Ten Year Statement*; National Grid: London, UK, 2012; pp. 221–223.
2. Kaloudas, C.G.; Ochoa, L.F.; Fletcher, I.; Marshall, B.; Majithia, S. Investigating the declining reactive power demand of UK distribution networks. In Proceedings of the IEEE Power & Energy Society General Meeting (PESGM), Denver, CO, USA, 26–30 July 2015; pp. 1–5. [CrossRef]
3. ARERA. The Italian Regulatory Authority for Energy, Networks and Environment. DCO 515/21, Appendix 2, Analisi Dei Flussi Di Energia Reattiva Nelle Reti in Alta e Altissima Tensione (Analysis of Reactive Energy Flows in HV Networks). Available online: https://www.arera.it/allegati/docs/21/515-21appendice2.pdf (accessed on 17 August 2023).
4. ARERA. The Italian Regulatory Authority for Energy, Networks and Environment. DCO 515/21, Completamento Della Regolazione Tariffaria Dell'energia Reattiva (Completion of Reactive Energy Tariff Regulation). Available online: https://www.arera.it/allegati/docs/21/515-21.pdf (accessed on 17 August 2023).
5. ARERA. The Italian Regulatory Authority for Energy, Networks and Environment. DCO 420/16, Regolazione Tariffaria Dell'energia Reattiva per le Reti in Alta e Altissima Tensione e per le Reti di Distribuzione (Reactive Energy Tariff Regulation for HV Networks and Distribution Networks). Available online: https://www.autorita.energia.it/allegati/docs/dc/420-16.pdf (accessed on 17 August 2023).
6. Chen, L.; Li, H. Optimised Reactive Power Supports using Transformer Tap Stagger in Distribution Networks. *IEEE Trans. Smart Grid* **2017**, *8*, 1987–1996. [CrossRef]
7. Chen, L.; Li, H.; Cox, S.; Bailey, K. Ancillary Service for Transmission Systems by Tap Stagger Operation in Distribution Networks. *IEEE Trans. Power Deliv.* **2016**, *31*, 1701–1709. [CrossRef]
8. Weedy, B.M.; Cory, B.J.; Jenkins, N.; Ekanayake, J.B.; Strbac, G. *Electric Power Systems*, 5th ed.; John Wiley & Sons: Chichester, UK, 2012; pp. 180–181. ISBN 978-0-470-68268-5.
9. O'Kelly, D. The Representation of Tap-Change Transformers in Power System Networks. *Int. J. Electr. Eng. Educ.* **1970**, *8*, 187–192. [CrossRef]
10. Constantin, M.; Eremia, M.; Toma, L. Comparative analysis between conventional voltage control using reactors and continuous voltage control using TCR in the Romanian transmission grid. In Proceedings of the IEEE PowerTech, Grenoble, France, 16–20 June 2013; pp. 1–6. [CrossRef]
11. Chen, L.; Li, H.; Turnham, V.; Brooke, S. Distribution network supports for reactive power management in transmission systems. In Proceedings of the IEEE PES Innovative Smart Grid Technologies Conference Europe (ISGT Europe), Istanbul, Turkey, 15–16 October 2014; pp. 1–6. [CrossRef]
12. Wang, D.; Chen, L.; Li, H.; Wang, Z.; Turnham, V. Impacts of tap stagger on currents of power transformers. In Proceedings of the IEEE PES Innovative Smart Grid Technologies Conference Europe (ISGT-Europe), Turin, Italy, 26–29 September 2017. [CrossRef]
13. ARERA. The Italian Regulatory Authority for Energy, Networks and Environment. Testo Integrato Delle Disposizioni per L'erogazione Dei Servizi di Trasmissione e Distribuzione Dell'energia Elettrica. Available online: https://www.arera.it/allegati/docs/19/568-19TIT.pdf (accessed on 17 August 2023).
14. European Network of Transmission System Operators for Electricity—ENTSO-E. Phase Shift Transformers Modeling, Version 1.0.0. 2014. Available online: https://eepublicdownloads.entsoe.eu/clean-documents/CIM_documents/Grid_Model_CIM/ENTSOE_CGMES_v2.4_28May2014_PSTmodelling.pdf (accessed on 17 August 2023).

energies

MDPI

Article

Analysis of Secondary Controller on MTDC Link with Solar PV Integration for Inter-Area Power Oscillation Damping

Oluwafemi Emmanuel Oni * and Omowunmi Mary Longe

Department of Electrical and Electronic Engineering Science, University of Johannesburg, Johannesburg 2006, South Africa; omowunmil@uj.ac.za
* Correspondence: 223248712@student.uj.ac.za; Tel.: +27-11-678984406

Abstract: Integration of renewable energy sources is important in limiting the continuous environmental degradation and emissions caused by energy generation from fossil fuels and thus becoming a better alternative for a large-scale power mix. However, an adequate analysis of the interaction with the alternating current (AC) network during network disturbance, especially during inter-area power (IAP) oscillations is needed. Insufficient damping of oscillations can significantly impact the reliability and effective operation of a whole power system. Therefore, this paper focuses on the stability of the modified Kundur two-area four-machine (MKTAFM) system. A robust secondary controller is proposed and implemented on a line commutated converter (LCC)-based multi-terminal high voltage direct current (MTDC) system. The solution consists of a local generator controller and the LCC MTDC (LMTDC) system, voltage-dependent current order limiter, and extinction angle controller. The proposed robust controller is designed for the LMTDC systems to further dampen the inter-area power oscillations. Three operational scenarios were implemented in this study, which are the local generator controller and double circuits AC line, local generator controller with LMTDC controllers, and local generator controller with LMTDC controllers and secondary controller. The simulation result carried out on PSCAD/EMTDC recorded better damping of the inter-area power oscillation with LMTDC. A considerable improvement of 100% damping of the IAP oscillations was observed when a secondary controller was implemented on the LMTDC.

Keywords: inter-area power oscillation; two-area four-machine network; solar PV; high voltage direct current; PSCAD; secondary controller

Citation: Oni, O.E.; Longe, O.M. Analysis of Secondary Controller on MTDC Link with Solar PV Integration for Inter-Area Power Oscillation Damping. *Energies* **2023**, *16*, 6295. https://doi.org/10.3390/en16176295

Academic Editor: Ying-Yi Hong

Received: 22 July 2023
Revised: 26 August 2023
Accepted: 27 August 2023
Published: 29 August 2023

1. Introduction

An increase in the number of power system interconnections has resulted in inter-area power oscillations, a phenomenon involving groups of generators oscillating relatively to each other that are sometimes hard to control due to their scale and complexity [1]. Being often of low frequency, these oscillations are becoming a big challenge to the safe operation of modern power networks. When poorly damped, inter-area oscillation can further lead to voltage or rotor angle instability, or system collapse. Inter-area oscillations have long been a point of discussion among researchers, power utilities, and industries. The frequency is always in the range of 0.1 to 1 Hz [2]. These oscillations lack damping or are unstable, and when these disturbances take place due to crucial faults on the line, the network becomes less stable.

Engineers and researchers in the power and control systems have worked extremely hard over the past 20 years to increase the stability of power systems [3–6]. The standard controllers, which include the power system stabiliser (PSS), automatic voltage regulator (AVR), and speed governor control, are single-input single-output non-coordinated linear controllers that are mostly not capable of ensuring stability when significant perturbations or contingencies arise. In addition, the AVRs, which are generally employed to maintain the terminal voltage magnitude of the synchronous generator, introduce negative damping torques, which have a detrimental effect on stability. Because of operating point variations and short circuit disruptions, power systems exhibit electromechanical oscillations. These frequency oscillations must be suppressed to the amount that is acceptable; otherwise, instability might develop as their amplitude grows. In order to address these problems, power system stabilisers (PSSs) are further utilised to produce an extra stabilising signal to the excitation system to dampen these oscillations [7].

Other controllers like the Static Var Compensators (SVC) and flexible AC transmission systems (FACTS) have also been deployed for inter-area oscillations damping; however, under specific operating circumstances, an inter-area mode may be viewable from one area and controllable from another [8,9]. In this situation, a failure or system disturbance in one area will tend to have a greater effect than before in another.

Researchers have developed several efficient techniques to examine the low-frequency oscillation issue [10–12]. Among these, the linearised state matrix's modal analysis offers a fundamental technique for analysing low-frequency oscillations. This approach may be used to identify the system's weakly damped oscillation modes, whether they are local or inter-area. Using linear participation factors, it is possible to pinpoint the high association between a few chosen generators and the weak-damped mode. A fuzzy controller can also be used to provide a fast response and coordinated control of a wide area network for quick damping of inter-area oscillations [13]. In [14], a Factional Order Proportional Integral (FOPI) was used against the conventional PI controller with an adaptive differential evolution algorithm to further provide an optimal tunning of the FOPI for performance enhancements. Another work also carried out a data-enabled predictive control algorithm on a voltage source converter-based high-voltage DC (HVDC) for optimal damping of wide-area power oscillations [15]. A wide area damping controller (WADC) was proposed for energy storage systems such as capacitors for the damping of low-frequency oscillations [16]. The authors used an accelerated particle swarm optimization technique for the turning of the wide-area controllers. In the related research [17], the authors use dual power oscillation dampers on the active and reactive power of a double-fed induction generator for the damping of the inter-area power (IAP) oscillations.

However, the recent trend and transition towards renewable energy and the deregulation of the electricity market point towards the usage of different DC converter links, which has led to different changes in generation and power transmission patterns as well as grid topology [18]. Therefore, a change is required for a better controller to adequately dampen inter-area oscillations before the generator control goes out of step, especially when operating a thyristor-based multi-terminal HVDC system.

Therefore, this paper presents a novel method to model a multi-terminal line commutated converter (LCC) transmission link to provide a robust, flexible, and expandable oscillation damping for a modified Kundur two-area four-machine (MKTAFM) network. The method used in this paper is to build and validate a secondary controller for one of the rectifying stations of the multi-terminal direct current (MTDC) system. The topology of the secondary controller is first tuned and adjusted precisely for the MTDC network. The LCC-MTDC (LMTDC) grid is equipped with an overall power controller that optimises the power flow and current sharing between the two rectifier stations. Therefore, this paper focused on the control strategy of LMTDC grids in reducing the inter-area oscillation of two-area networks. This control aims to minimise the ripple effect of faults on the AC grid voltage and maintain a power balance in case of a fault in one of the areas.

In the first contribution of this paper, it shows how the stability of two asynchronous grids can be improved. The second contribution of the paper shows how a coordinated control of three-terminal LCC MTDC links can be used to avoid inter-area oscillations interaction between two inter-area modes. The investigation carried out in this study evaluates the performance of an MKTAFM and proffers a better means of inter-area oscillation reduction with the use of some multivariable as well as robust controller designs. Furthermore, this research design handles voltage control and power oscillation damping issues concurrently. Our method is distinguished by the fact that the second level of control action optimises the performance of the existing control of the rectifier stations and thus improves the performance of the substation. The controller configuration and the parameters required for this implementation are derived methodically from the MKTAFM system model. The suggested secondary control is robust and merges an input–output linearisation control method with a nonlinear control approach. This secondary controller constantly adjusts its parameters in response to changes in working circumstances or structure, thus significantly improving the MTDC controller's effectiveness.

This paper is structured as follows: the authors first gave an insight into the theoretical analysis of inter-area oscillation utilising a two-machine network. The MTDC model used was then presented alongside the control architectures, which are the main MTDC controllers as well as the secondary controller. The test system modelling and the schematic diagram were then explained. Furthermore, the methodology based on the fault analysis on the test network was given, and the results were then presented. Then, the conclusion section follows, which summarises all the work carried out in this paper.

2. Inter-Area Oscillations

Inter-area oscillations, which are oscillations occurring between different regions within a synchronously connected power system, represent the most critical threat to the reliable and stable operation of such a power system [19]. In situations where the conventional control mechanisms, due to either weak tie-lines or inadequate grid strength, are incapable of effectively dampening the system's oscillations following a disturbance, inter-area oscillation phenomena may arise [20]. The ability of an electrical power system to maintain stability in the face of these minor disturbances is referred to as small-signal stability. In contrast to small-signal perturbations, the nonlinear equations governing power systems can be linearised around a specific steady-state operating condition. Due to the inherent weak damping characteristics of inter-area oscillations, there exists a significant probability of irreversible and extensive blackouts, which can lead to severe economic losses and potential loss of human life, thereby resulting in catastrophic consequences.

The small-signal stability is evaluated using eigenvalue analysis, sometimes referred to as modal analysis. This study is primarily intended to reduce inter-area power oscillations. The theoretical background of the small-signal stability of a power system is further provided in Equations (1) to (5). The synchronous machine's inertia fluctuations are initially analysed to determine the damping coefficient needed to reduce power system oscillations as effectively as possible. The coefficient of damping signifies the adequate torque needed to control the synchronous speed. Figure 1 [20] gives a schematic of a basic two-area system consisting of two incoherent machines with inertias of the generator denoted by H_a and H_b while δ_a and δ_b indicate the rotor angle of areas A and B oscillating at two distinct frequencies and amplitudes. Equations (1) and (2) give the swing equation for individual machines, which, when subtracted, gives (3). Equation (4) is generated by dividing (3) with the inertia coefficients shortly before the second derivation. Further simplification gives (5). To avoid the complexities of swing equations for numerous synchronised machines during the stability analysis of a large system, all swing equations must be combined to minimise the total amount of iterations needed to produce the final swing equations. The total system equivalent rotating inertia is obtained by adding the rotating inertia H of each generating unit [21].

$$P_{ma} - P_{ea} = H_a \cdot \frac{2}{\omega} \cdot \frac{\partial^2 \delta_a}{\partial t^2} \tag{1}$$

$$P_{mb} - P_{eb} = H_b \cdot \frac{2}{\omega} \cdot \frac{\partial^2 \delta_b}{\partial t^2} \tag{2}$$

$$\frac{P_{ma} - P_{ea}}{H_a} - \frac{P_{mb} - P_{eb}}{H_b} = \frac{2}{\omega} \cdot \left(\frac{\partial^2 \delta_a}{\partial t^2} - \frac{\partial^2 \delta_b}{\partial t^2} \right) \tag{3}$$

$$\frac{2}{\omega} \cdot \left(\frac{H_a H_b}{H_a + H_b} \right) \cdot \left(\frac{\partial^2(\delta_a - \delta_b)}{\partial t^2} \right) = \frac{H_b P_{ma} - H_a P_{mb}}{H_a + H_b} - \frac{-H_b P_{ea} - H_b P_{eb}}{H_a + H_b} \tag{4}$$

$$\frac{2}{\omega} \cdot H_{ab} \cdot \frac{\partial^2 \delta_{ab}}{\partial t^2} = P_{abm} - P_{abe} \tag{5}$$

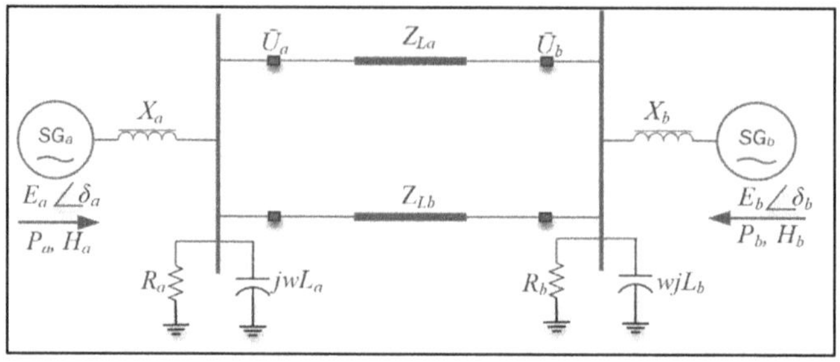

Figure 1. Schematic diagram of a two-area network with two machines [20].

The difference in rotor tilt angle between the two generators and the corresponding inertia constant is given as H_{ab}, and δ_{ab} as shown in (6) and (7), respectively. P_{mab} and P_{eab} in (8) and (9) are the respective electrical and mechanical power between the two generators. The damping constant K_D denotes the swing characteristics of the synchronous machine, and it is described using the second-order differential equation in (10). The undamped natural frequency (ω_n) and the damping coefficient (ζ) required to fully reduce the oscillation amplitude are given in (11) and (12), respectively. Since the inertia constant and the oscillation frequency are indirectly proportionate, a power network that is equipped with a high constant of inertia is less prone to faults and vice versa. This demonstrates that the output supplied power has a direct impact on the system's inertia constant. As a result, increasing the stability condition of the power system necessitates an inertia constant value that is greater than 3.5 pu.

$$H_{ab} = \frac{H_a H_b}{H_a + H_b} \tag{6}$$

$$\delta_{ab} = \delta_a - \delta_b \tag{7}$$

$$P_{abm} = \frac{H_b P_{ma} - H_a P_{mb}}{H_a + H_b} \tag{8}$$

$$P_{abe} = \frac{H_b P_{ea} - H_b P_{eb}}{H_a + H_b} \tag{9}$$

$$\frac{2}{\omega} \cdot H_{ab} \cdot \frac{\partial^2 \delta_{ab}}{\partial t^2} + K_D \frac{\partial \delta_{ab}}{\partial t} = -P \cos(\delta_o)_{ab_{max}} \tag{10}$$

$$\omega_\eta = \sqrt{\frac{\omega P \cos(\delta_o)_{ab_{max}}}{2H_{ab}}} \tag{11}$$

$$\zeta = \frac{1}{2} K_D \sqrt{\frac{\omega}{2P \cos\left(\delta_o\right)_{ab_{ab_{max}}}}} \tag{12}$$

3. System Model

3.1. Test Network Modelling

The system under investigation in this research paper is a network, depicted in Figure 2. It comprises four synchronous generators situated in two distinct areas, a 50 MW grid-tied solar PV connected at BB2, and a three-terminal line-commutated multi-terminal direct current (LMTDC). The synchronous generators employed in this study are equipped with an IEEE AC4A exciter for automatic voltage regulation (AVR) and an IEEE single-input power system stabiliser (PSS). To introduce a significant line impedance, the transmission line linking BB10 to BB11 was modified, increasing its distance from 25 km to 80 km. Consequently, a double-circuit transmission line was utilised to enhance the power transfer between area 1 and area 2. Subsequently, a real-time domain modelling and simulation approach is employed to analyse the complete generator's inter-area power and frequency oscillations. Two system faults are simulated to represent a worst-case scenario. The power output, synchronous speed, and inter-area power transmission of each generating unit are plotted during these system disturbances. Bus BB7, BB8, and BB9 are selected for power transfer quality analysis at these locations and to verify the voltage profile. This mode is classified as inter-area since it involves two coherent sets of generators oscillating at a low frequency in opposition to each other.

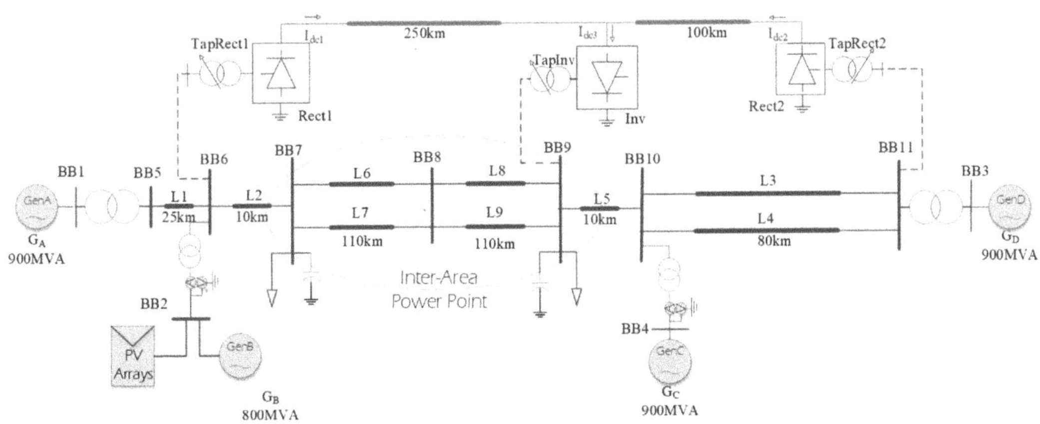

Figure 2. The Kundur two-area four-machine network.

Power Systems Computer Aided Design (PSCAD) was used to model the synchronous machines, which included the IEEE controller model for the AVR and the PSS. All generators were evaluated in terms of the true inertial constant and were linked together by a step-up transformer of 20/230 kV voltage range and an apparent power rating of 900 MVA. During the evaluation time, a constant value was pre-selected for the governor control for all four machines to obtain an in-depth impact of the DC controller, AVR, and PSS effects on minimising the disturbances caused by the fault on the complete model. The authors in [2,20,22] gave an explanation about the modelling of these machines and further provided the parameters for both the generators and the transformers and the line models.

In the dynamic simulation of this network, three operational scenarios were considered: (1) usage of only the AC line on the MKTAFM network; (2) usage of LMTDC link to interconnect bus BB6 (rectifier 1) or BB11 (rectifier 2) to the inverter station at bus BB9; and (3) implementation of a secondary controller on the rectifier 1 station. During the second and third scenarios, a single circuit was used rather than the double circuits shown in the MKTAFM network. This measure was taken to avoid excessive power transfer across the bus and to also show the scenario of replacing an existing AC line with DC lines, thereby showing the impact of it on the overall system.

The power system's dynamic characteristics are captured by the collection of nonlinear differential algebraic equations, which serve as the mathematical representation of the system. Equations (13)–(15) express these dynamics' analysis of the power system interconnections.

$$\dot{x} = f(x, y, u) \tag{13}$$

$$0 = g(x, y, u) \tag{14}$$

$$z = h(x, y, u) \tag{15}$$

The vector x in these equations represents the state variables, the vector y represents the algebraic variables, the vector u represents the input variables, and the vector z represents the output variables. These equations regulate the dynamic behaviour of the many types of models used in this study. The specific equations relating to the dynamic behaviour of these models are discussed in the following sections.

3.2. Grid-Tied Solar PV Analysis

A single solar PV source is a proportionate equivalent of several strings of PV modules connected in parallel with each of the string modules connected in series. With identical arrays of PV modules, a general analysis can be given for a single model solar cell as shown in Figure 3 using the electrical equivalent circuit [23,24]:

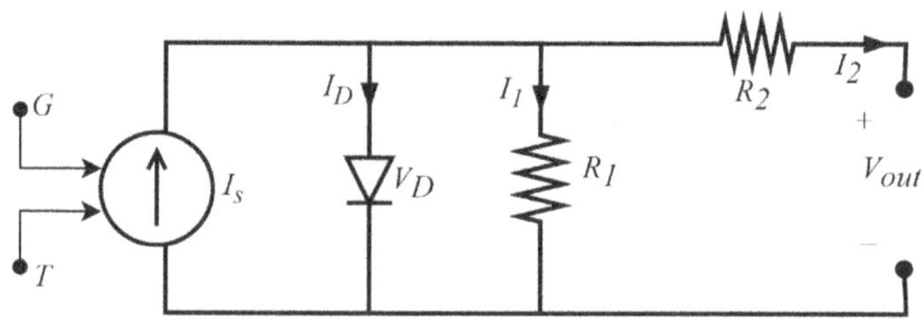

Figure 3. Solar PV cell model.

The generated photocurrent, denoted as Is, in a photovoltaic (PV) cell exhibits a linear relationship with the solar irradiance it is exposed to from sunlight. However, the presence of the anti-parallel diode results in non-linear current-voltage (I-V) characteristics of the PV cell. By applying Kirchhoff's current law to the circuit, (16) is obtained. Equation (18) is derived by substituting the relevant expressions for the diode current (I_D) and the shunt branch current (I_1). The modified non-ideal factor, expressed in (18), is directly proportional to the temperature of the cell, denoted as Tc. The photocurrent, as described in (19), is dependent on the solar radiation (G) incident on the PV cell's surface and the cell temperature (T_c). The short circuit current (I_{SCR}) is defined with respect to a reference solar radiation (G_R) and a reference cell temperature (T_R). The parameter αT represents the temperature coefficient of the photocurrent, with a value of $\alpha_T = 0.0017$ A/K for silicon solar cells. The dark saturation current (I_o), given by (20), is influenced by the cell temperature, where I_{oR} represents the dark current at the reference temperature. In these equations, q denotes the charge of an electron, k represents the Boltzmann constant, e.g., corresponds to the band-gap energy of the solar cell material, and n is the ideality factor when $kTc = q$. The instantaneous values of solar radiation (G) and the operating temperature (T) of the cell are measured in watts per square meter (W/m^2) and degrees Celsius (°C), respectively.

Considering that a PV module comprises numerous cells connected in series, and a PV array consists of both series and parallel combinations of modules, the circuit model of a single cell can be scaled up to represent any desired series/parallel configuration.

$$I_s = I_D + I_1 + I_2 \tag{16}$$

$$I_2 = I_s - I_o\left(e^{\frac{V_{out}+I_2R_2}{m}} - 1\right) - \left(\frac{V_{out} + I_2R_2}{R_1}\right) \tag{17}$$

$$m = \frac{nkT_c}{q} \tag{18}$$

$$I_s = I_{SCR} \cdot \left(\frac{G}{G_R}\right) \cdot [1 + \alpha_T(T_C - T_R)] \tag{19}$$

$$I_o = I_{oR} \cdot \left(\frac{T_C}{T_R}\right)^3 \cdot e\left[\frac{qe_g}{nk}\left(\frac{1}{T_R} - \frac{1}{T_C}\right)\right] \tag{20}$$

3.3. PV Modelling

The schematic diagram presented in Figure 4 illustrates the comprehensive configuration of a grid-connected photovoltaic (PV) model consisting of DC–DC and DC–AC converters block diagram within it (parameter details in Appendix A). The components encompassed in this scheme include PV arrays, a DC–DC boost converter, a voltage source converter for DC–AC conversion, and an AC filter. The PV subsystems are capable of functioning in two primary modes, namely, continuous conduction mode (CCM) and discontinuous conduction mode (DCM). In solar PV applications, the CCM is the preferred mode of operation [25].

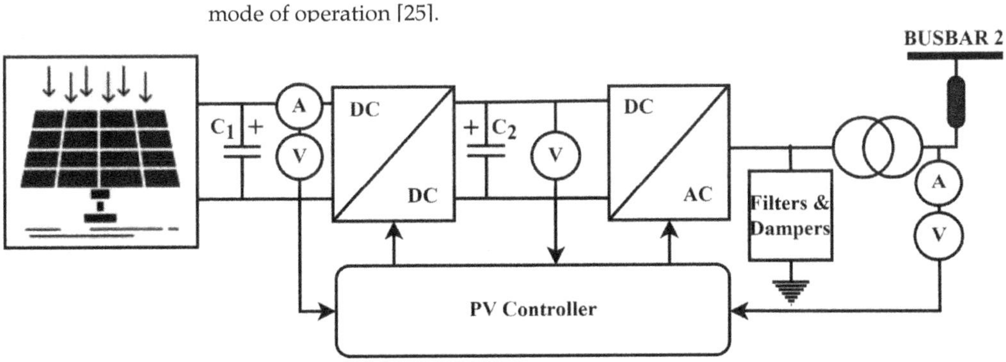

Figure 4. Grid-tied PV model.

The PV system in this study utilised a DC–DC boost converter model as shown in Figure 4. It serves as a step-up converter extensively employed for integrating low-voltage PV modules with the utility grid. Furthermore, it fulfills the role of maximum power point tracking (MPPT) under nominal utility conditions. In the boost converter, the voltage is increased by adjusting the duty cycle. The DC–DC converter plays a vital role in a solar PV application by transforming the input direct current from one voltage level to the required level. It serves as a crucial component for maximum power point (MPP) tracking. The operation of the DC–DC converter serves as the foundation for detecting the MPP, as outlined in the proposed global MPPT control algorithm. In practical scenarios, despite the implementation of MPPT, the output voltage of a PV string remains significantly low. Therefore, a front-end DC–DC converter with step-up capability is necessary for grid connection. The fundamental principle behind the MPPT algorithm relies on harnessing the voltage and current variations resulting from instantaneous power fluctuations. By analysing these variations, it becomes possible to determine the power gradient and assess whether the solar PV system is operating near the maximum power point. The maximum power delivered by the solar PV array is the product of the optimal operating voltage and current of the PV array at the maximum power output condition.

To achieve grid synchronisation, an inverter is necessary. The control of the inverter governs the power and frequency on the AC side while minimising system harmonics.

The switches within the inverter are controlled through the implementation of an inverter control algorithm [26].

3.4. MTDC Design

The standard control approach of LMTDC converters requires rectifier stations to be in DC current control (DCC) mode. Voltage-dependent current order limiter (VDCOL) and current control are further used to provide a better current order control. As shown in Figure 5, the DCC method is used at the rectifier 1 substation. The current order is generated by either selecting the reference input current deployed by the overall power controller or by selecting the current order from the VDCOL. The VDCOL gets its input from the intended filtered DC current ($I_{dcrect1}$) measured from the DC line. The lower value of any of the two is chosen as the standard current order (I_{ord}) for the generation of the firing angle for the rectifier. The same process is used at the rectifier 2 substation as illustrated in Figure 6. The difference between the rectifier 1 and the rectifier 2 controller is that the rectifier 1 substation has another current order input from the secondary controller (SC). Also, both rectifiers have different power-carrying capacities and are thus rated differently in the parameters that are used for their control signals.

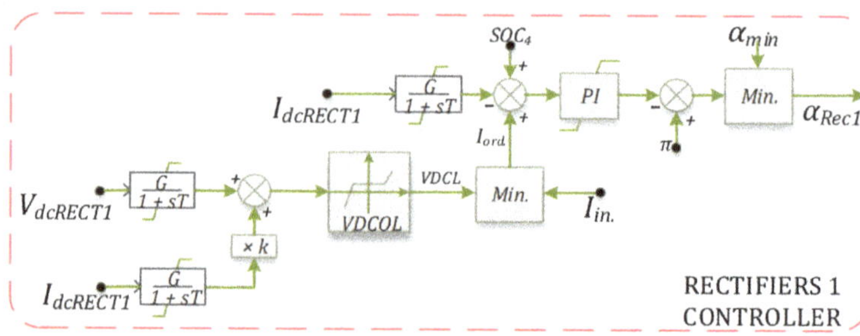

Figure 5. LMTDC rectifier 1 control diagram.

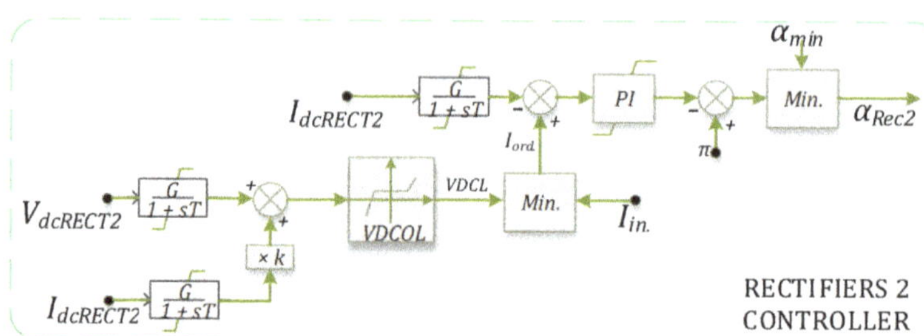

Figure 6. LMTDC rectifier 2 control diagram.

Various control approaches can be used to regulate the inverter, including extinction angle control (EAC), DCC, and DC voltage control (DVC). The reference signal is acquired in the same manner as in the rectifier control mode as shown in Figure 7. However, unlike the rectifier's DCC, the inverter's DC current is limited to the difference between the amount of the standard current order and that of the current margin (I_{marg}). In other to positively influence the power reversal process, the current margin is programmed into the inverter current control and normally set on a fixed range of 0.1–0.15 pu. A current error control (I-error) indication is commonly used between the DCC and the DVC. This generated I-error can also be inserted between the DCC and the EAC to smoothen changes between the control modes. The smallest of these generated parameters is chosen as the inverter control parameters because it is preferable to run the inverter at the smallest extinction angle possible to reduce the power losses, reduce the amount of reactive power needed to compensate for the thyristor valve, and, ultimately, reduce the harmonic distortion content.

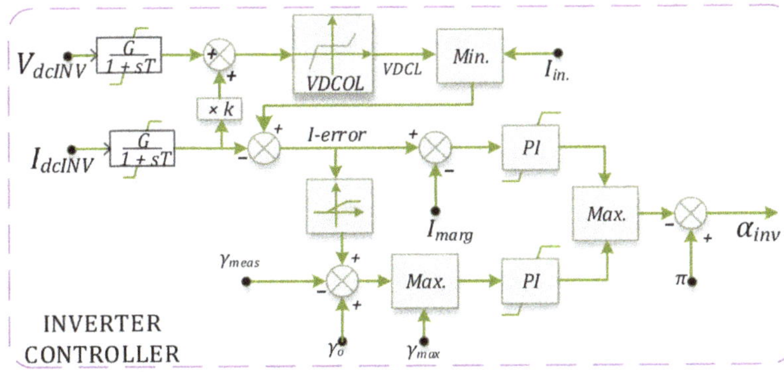

Figure 7. LMTDC inverter control diagram.

For every input of this control signal, being in per unit, it must undergo a first-order inertia transfer function to make the control switching process a smooth transition. The equation is expressed as follows:

$$\left(G_1\frac{1}{1+sT_1}\right) \tag{21}$$

The LMTDC system's fault vulnerability is entirely dependent on its general coordination; thus, each converter's controllers were preset and adjusted to fit the case study. The overall power controller regulates and oversees the LMTDC link's converter operations. It accepts a varying power order measured from the systems. This controller yields the current order after being divided by the measured DC voltage by guaranteeing that the total DC current of all converters is zero (I_{dc} = 0). The master control, seen in Figure 8, is responsible for balancing the combined converter's power and current order. This controller generates the current order for each converter system based on the measured voltage at each converter station and a preset power order. It also provides allowance and compensation in case of DC line losses.

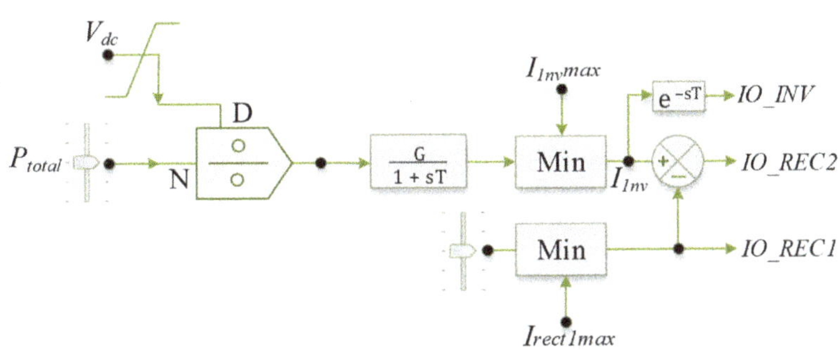

Figure 8. Overall power controller.

The LMTDC used in this study comprises a three-terminal thyristor converter. The firing angle at rectifiers and converter units is used to regulate the DC voltages. The overall power controllers generate the input value of the DC current for each of the converter's stations. The DC voltage for the rectifier and inverter stations, as well as the rectifier firing angle, are expressed below [27–29]:

$$V_{dr} = 3.12BTU\ cos\ \alpha - 0.96BX_cI_{dc} \tag{22}$$

$$V_{di} = 3.12BTU\ cos\ \beta + 0.96BX_cI_{dc} \tag{23}$$

$$\alpha_c = cos^{-1}\left(\frac{U_{dc} + I_{dcref}R_{dc} + 0.955BXI_{dcref}}{3.1213 \cdot BTU}\right) \tag{24}$$

The subscripts r and i in the (22) to (23) stand for rectifier and inverter. The rectifier and inverter DC voltage are denoted as V_{dr} and V_{di}, while the RMS L-L converter voltage is shown as U. The number of series-connected bridges of the thyristors is denoted by B. Therefore, for a 12-pulse thyristor converter, B = 2; meanwhile, a bipolar HVDC system will have a B-value of four. The firing angle (α) in (24) is the required angle at which the converter gets turned on and begins to conduct; X denotes the commutation reactance, and R_{dc} is the total resistance at the DC side of the converter. Also, T stands for the transformer ratio, and the AC voltage as each of the converter stations is depicted as U.

The classic secondary controller (SC) adopted in this study is like the power oscillation damper that has the same mode of construction as a generator's PSS (power system stabiliser), which includes a washout filter, a gain G, and 'p' phase lead-lag blocks (Figure 9). The transfer function provided in (25) defines the secondary controller equations with the parameter details in Table 1. The washout filter is typically used as a link between the measured signal and the used signal in the control loop to washout (or reject) the steady-state component of the measurements while passing the transient ones. The traditional parametric tuning method comprises two steps: (i) the calculation of T_4 and T_5 to adjust the phase of the mode's residue to 180° and (ii) the adjusting of the gain G to achieve the required damping. The output control signal from each of the transfer functions during the simulation of the whole system is plotted in Figure 10.

$$H_{soc} = \left(G_1\frac{1}{1+sT_1}\right)\cdot\left(G_2\frac{sT_2}{1+sT_3}\right)\cdot\left(G_3\frac{1+sT_4}{1+sT_5}\right)^n\cdot\left(G_4\frac{1}{1+sT_6}\right) \tag{25}$$

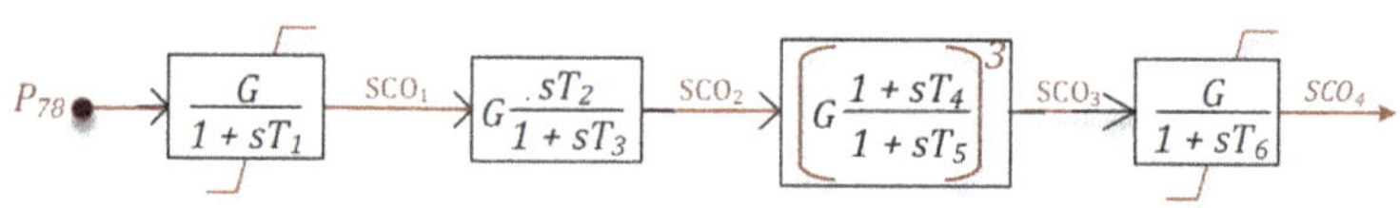

Figure 9. Secondary controller block diagram of the proposed model.

Table 1. Optimised parameter values.

Model	Parameter	Value
SC Gain SC Time (s)	G1, G2, G3, G4, T1, T2, T3, T4, T5, T6	[0.006; 0.31; 1.01; 0.24] [0.012; 1.5; 1.5; 0.514; 0.172; 0.055]

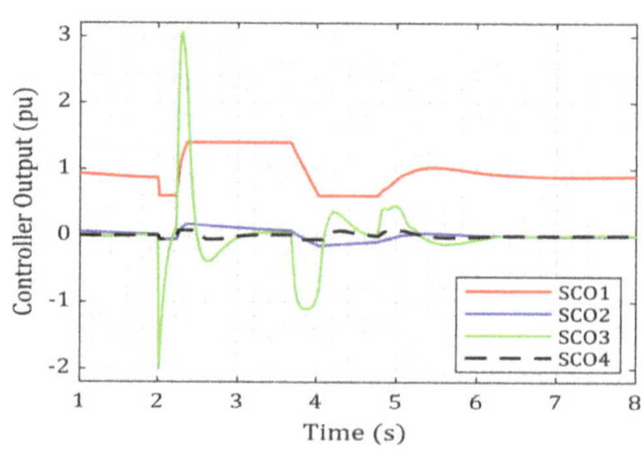

Figure 10. Measure control input during the simulation of the secondary controller.

4. Methodology and Scenarios

To study the small-signal characteristics of an MKTAFM network during the integration of solar PV systems, a dynamic RMS simulation was performed on PSCAD/EMTDC, a power systems simulation software designed for the simulation of electromagnetic transient analysis of DC networks. The base case study, which is termed the first scenario of using only the AC lines and solar PV, was first analyzed. The simulation included a fault analysis where a three-phase-to-ground (3Ph-G) fault occurred on the L7 line connecting BB7 and BB8, with a fault clearing time of 210 ms and a disturbance of 1.2 s on GenC. Both events happened concurrently at the 2.0 s mark time during the simulation, thus affecting different areas of the system to truly test the limits of the power system operation. The chosen disturbances simulated signify a worst-case scenario. The second scenario recorded a few changes, as three of the transmission lines, L4, L7, and L9, were replaced with DC lines (see Figure 2), but the same fault impedance was applied to the system when it was connected to the LMTDC link. The same situation applies to scenario 3 with LMTDC lines; however, it differs in the inclusion of the secondary controller for the quick damping of the inter-area power oscillation.

In all the scenarios covered in this study, the power generated by GenB and the solar PV link never changes, as shown in Figure 11, with GenB and the solar PV recording an active power of 622 MW and 50 MW, respectively. The goal of this study is to evaluate the controller contribution of LMTDC in reducing inter-area power oscillations even during a high penetration of renewable energy.

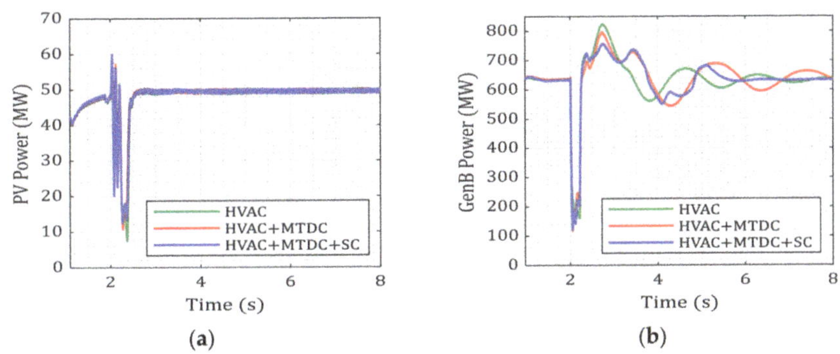

Figure 11. Active power for (**a**) solar PV, and (**b**) GenB.

5. Results

The results of the dynamic analysis using PSCAD/EMTDC simulation tools consisted of observations of the generator's active power, oscillation speed, and converter data. The magnitude and damping rate of the oscillation were studied for both situations for the inter-area power oscillation. The generator speed also provides information about the contributing element for each generator in the oscillatory phases. Since the focus is on the inter-area oscillation, the system short circuit fault time was limited to a short period of 210 ms. After these system disruptions, a positive damping coefficient was recorded during all scenarios, i.e., the systems maintained their pre-fault steady state condition, albeit with distinct amplitudes and waveform distortion.

The plot shown in Figure 12 is the active power for all the generators in the three scenarios. In the first scenario, the synchronous machine 2 (GenB) power dips lowest to a 201 MW value at the instant of the 3Ph-G fault occurrence; GenA follows thereafter. This dip in GenB is due to the near proximity to the transmission line fault following a decline in the active power produced by GenC due to a protracted generator disruption nearing 1 s simulation time on GenC. The active power recorded on this generator reached 916 MW, nearly 1.4 pu of its operating state value, due to the increase in the accelerating power after being subjected to a fault. Another observation in this plot is the inaccuracy in the power generated in each generator due to the high line impedance and losses on each of the AC lines used. Furthermore, GenD generated the largest share of the load power, followed by GenB, GenA, and then GenC. This disparity is caused by the high line impedance between

GenC and the load area. This points to the fact that even the inclusion of the solar PV link is still not sufficient to provide an adequate damping torque that can sufficiently dampen the power oscillations.

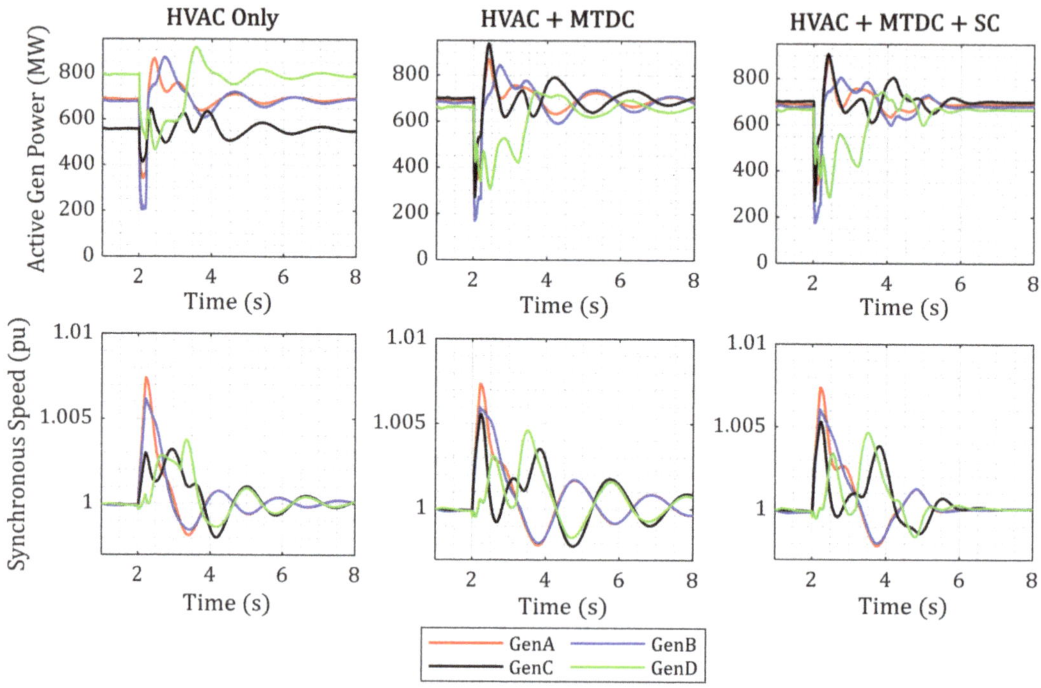

Figure 12. Active power and synchronous speed plot.

The second scenario involves the replacement of all the double transmission circuits in the system with a single transmission line and DC lines, with the DC lines forming a three-terminal DC (LMTDC) link. A performance comparison was carried out with regards to the first case study to determine which links achieved better inter-area oscillation reduction and to investigate if the LMTDC link can damp oscillation generated by solar PV during system disturbance. In Figure 12, the active power shown in this plot exhibited a close-to-average power generation. This close range of power transfer is due to the benefits offered by the LMTDC system in loss reduction and enhancement of power transfer. Therefore, after the 3Ph-G fault, due to a continuous generator fault, GenD and GenC registered the greatest power amplitude of 916 MW. These amplitudes resulted in greater inertia value in recovering the normal operating point of the complete network. GenB still recorded the lowest dip of 175 MW. However, a better power profile can be seen in this plot as all four generators recorded a close range of power. The functionality of this was made possible with the inclusion of the LMTDC link, which further helped in the active power dispatch. However, the simulation plots at this stage still contain a number of oscillations. The inclusion of the secondary controller to the rectifier 1 station totally eradicated these oscillations at 5.8 s simulation time.

Figure 12 shows the frequency swings of the generators that were observed in the system. This result comprises incoherent machines, with area A having a different inertia constant compared to area 2. This results in the generators in area A swinging with the machines in area 2. The highest oscillatory modes, up to 1.007 pu, were observed in the GenA and GenB subplots. The machine's accelerating power played a role in meeting the expected load demand in area B, which led to the maintenance of a stable operating state, albeit with a high oscillating frequency. During the first two scenarios, the generators'

power system stabiliser (PSS) and automatic voltage regulator (AVR) controllers had already provided damping torque to maintain stable operating points. However, this was insufficient in quickly generating a damping coefficient that would adequately and effectively return the entire test network to its steady-state operational state. An increase in the fault clearance time (t_c = 300 ms) would result in system collapse, where the systems can no longer maintain a stable operating point. In Figure 12, the synchronous speed for GenC and GenD both followed the same pattern of system oscillation being similar in area 2; the same applies to GenA and GenB in area 1.

In the third scenario, although initial oscillation is the same as the previous scenarios, the inclusion of the secondary controller into the KMTAFM network generated a quick damping torque for the system to quickly return to its steady state condition.

Following the fault, BB7, BB8, and BB9 in Figure 13 illustrate the bus voltages during the three scenarios. In order to fully understand the comparison between the three different scenarios based on the voltage profile, the busbar voltage was grouped together, indicating a particular bus voltage for the three different scenarios. The observation of these voltage profiles shows that BB8 is the weakest bus in the network across all three scenarios, which can be attributed to its distance from the generating plants in both areas. BB7 followed as the second weakest bus. However, the second scenario recorded a better voltage profile than when only HVAC and solar PV were used. The plot also indicates that an increase in transmission distance between BB10 and BB11 did not have a significant impact on BB9's voltage profile in contrast to the other buses examined in both scenarios. During the third scenario, the oscillations observed in the system voltage profile were completely improved, and the system's steady conditions were stabilised after approximately two cycles (5.8 s of simulation time).

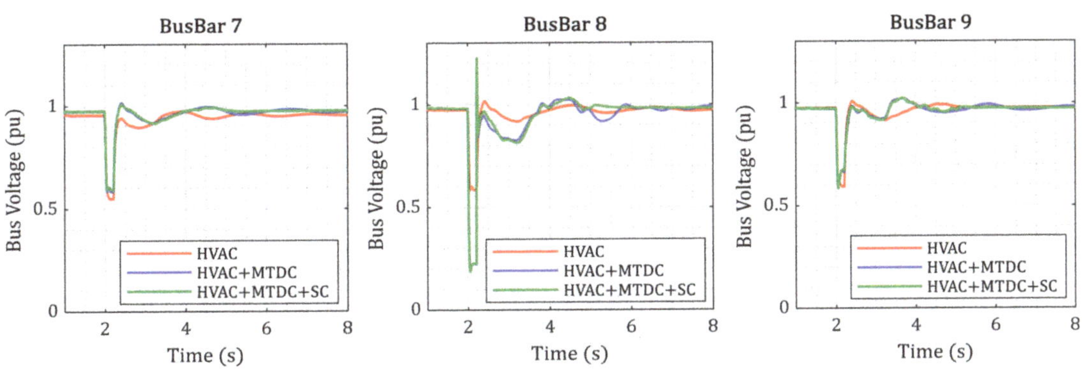

Figure 13. Bus voltage profile.

Figure 14 shows the impact of the LMTDC in reducing the oscillation after the system's disturbance. The plot was categorised into two parts, depicting the scenario with and without the secondary controller. The voltage-dependent current order limiter has a look-up graph based on the V-I characteristics of the converter that helps the converter in reducing the amount of current that flows through the link when the voltage reduces. By doing this, the converter reduces the amount of power that flows through the link during this disturbance and thus reduces the amount of reactive power consumption. The current output from the VDCOL is passed through a PI controller that further generates the firing angle for each of the converters. The scenarios without the inclusion of the secondary controller recorded a ripple that was still very much visible at the end of the simulation time. These oscillations are still further cumulated into the firing angle for each of the converters. The inclusion of the secondary controller created an immediate ripple that further provided a quick, sufficient, and robust damping torque in totally eradicating the oscillations in the current output from the VDCOL and further provided a better firing angle, and thus a better DC power output.

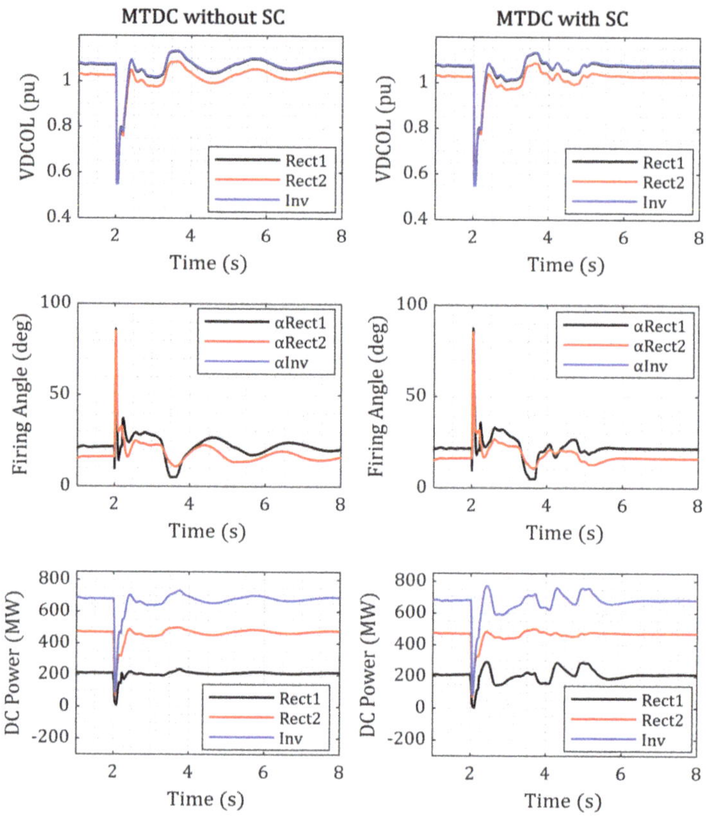

Figure 14. LMTDC converter plots.

Figure 15 demonstrates the inter-area power transfer between BB7 and BB9. The study revealed that without the presence of MTDC links, the inter-area power oscillations had a slow damping rate during the initial scenario, where only AC transmission lines were used. During the post-fault transient condition, the two areas in the study system contributed more to the oscillations, which delayed the system from reaching a steady-state condition. However, the addition of the MTDC link showed that the damping rate of the inter-area power oscillations increased significantly during the same disturbance scenario. The amplitude of the IAP transfer likewise showed a considerable positive damping rate, leading to a more stable operative condition. However, not all the inter-area oscillation was fully damped at the 8 s simulation time. Therefore, a secondary controller was further used. The result from this analysis shows a better oscillation damping, whereby all the oscillation was completely damped at exactly 5.8 s simulation time. During this scenario, the active power transfer comprised the power transmitted from the AC line L6 to line L8 and the DC power transmitted through the rectifier 1 converter link. The plot indicates that the damping ratio resulted in a complete suppression of the oscillation amplitude's value during the second period of simulation.

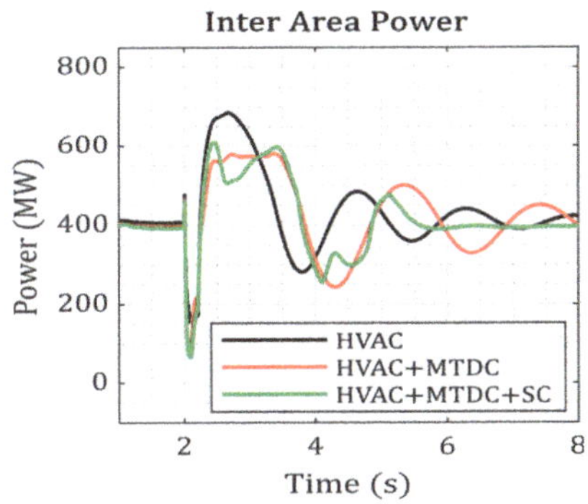

Figure 15. Inter-area power oscillation.

Performance Comparison

In this section, a comparison of various schemes, presented in Table 2, is conducted to assess the contributions of the proposed model and controllers. The highlighted reviews are based on authors who carried out inter-area power oscillation studies using the two-area four-machine (TAFM) network.

Table 2. Performance comparison of the proposed model.

Ref	Method	Performance in Comparison to This Study
[30]	Adaptive online control design for a power system stabiliser (PSS) and comparing the classic PSS controller and an online tuned PSS.	The results based on the 3Ph-G faults show little impact as both the classic and the online tuned PSS recorded some amount of oscillation even at 10 s simulation time.
[31]	They used weighted signals representing different areas to optimize a PSS. They used two optimization methods: the Harmony Search (HS) algorithm and the Teaching Learned Based Optimization (TLBO) technique.	The TLBO with the PSS provided a quick settling time of the bus voltage oscillation and recorded a settling time of 4.8 s, but fails to include the dynamics of HVDC/MTDC lines on the systems.
[32,33]	They used VSC HVDC power oscillation dampers (POD), PSS, and wide area measurement signals-based controllers designed with a novel linear quadratic gaussian (LQG) method.	The result utilising LQG POD provided a better oscillation reduction technique. However, the time taken to reach a total settling phase is around 12 s.
[34,35]	These literatures used an enhanced static synchronous compensator to damp inter-area power oscillation and also compared different FACTs devices.	The result shows an improved power oscillation; however, at a longer simulation time of ≅ 10 s. It also differs from the present study as no MTDC link was used.
[36–38]	They used a single thyristor controlled braking resistor and dual fuzzy logic for the minimization of inter-area power oscillation.	A reduced oscillation was recorded, but the 3-phase fault on the line generated a prolong oscillation.
[39,40]	They used a clustering distributed generator to provide more reserves for generators. They also used a phase wide area measurement for the control.	They recorded a quick settling time, but the DGs in the load bus provided a quick supply to the load. An MTDC link was not included.

6. Conclusions

The small-signal stability of two incoherent machines is one of the biggest problems facing power utilities. Further problems also arise during the integration of renewable energies, especially solar PV with high levels of intermittencies. Therefore, in this study, a secondary control technique that utilises a hierarchical control structure to enhance the damping of inter-area oscillation stability was introduced. The solution is multi-faceted, consisting primarily of the local controllers, voltage and rotor speed regulator, a centralised power controller for the LMTDC, and a current and extinction angle controller for the LMTDC model. These joint controllers were added to maximally improve the system performance. Firstly, a dynamic simulation model of an MKTAFM network was developed (first scenario). In the second scenario, a generator disturbance at GenD and a 210 ms 3Ph-G fault on the transmission line (L6) linking BB7 to BB8 were considered, with both faults occurring at 2 s simulation time. In the third scenario, the system response consisting of the details of the generators was monitored on a plot, and the times at which the system experienced a highly imbalanced waveform were obtained, as well as the time taken for the generator control to sufficiently dampen the system's oscillations. Afterwards, a study case involving the replacement of the double transmission lines on the network with an LMTDC system (second scenario) was implemented. A secondary control was further added to the LMTDC (third scenario). Finally, a nonlinear time-domain dynamics analysis was carried out on the LMTDC network, and the generator and the bus voltage profile were monitored.

Following the 3Ph-G fault on the test network, the poorest damped modes in each grid scenario being observable confirmed that the implementation of the secondary controller on the LMTDC provided a 100% damping of the oscillation at exactly 5.8 s simulation time with zero rates of change of the amplitudes compared to the first and second scenarios. The same is seen in the inter-area power transfer between BB7 and BB9, which initially recorded a continuous oscillation due to insufficient damping torque. The findings thus show that the proposed novel controller is durable, significantly raises the stability limit of the entire system, and improves the performance of primary controllers.

Further research will be conducted to reduce the sensitive response of the controller to large and non-uniform transmission delays and its impact if added to other LMTDC substations. Another point of focus is also to reduce the transient increase in the firing angle of the LMTDC converters during faults.

Author Contributions: Conceptualization, O.E.O. and O.M.L.; methodology, O.E.O. and O.M.L.; software, O.E.O.; validation, O.E.O. and O.M.L., formal analysis, O.E.O. and O.M.L.; investigation, O.E.O. and O.M.L.; resources, O.E.O. and O.M.L.; data curation, O.E.O. and O.M.L.; writing—original draft preparation, O.E.O. and O.M.L.; writing—review and editing, O.E.O. and O.M.L.; visualization, O.E.O. and O.M.L.; supervision, O.M.L.; project administration, O.M.L.; funding acquisition, O.M.L. All authors have read and agreed to the published version of the manuscript.

Funding: This research received no external funding.

Data Availability Statement: No data was created.

Conflicts of Interest: The authors declare no conflict of interest.

Appendix A

Table A1. Solar PV system specifications.

Parameter	Value
Maximum power	250 W
Short-circuit current per cell	2.5 A
Current at maximum power point	2.0 A
Temperature coefficient of Is	0.001
Modules connected in series per	35
Modules connected parallel pers	11,000
Cells connected in series per module	35
Reference radiation	1010
Reference cell temperature	29
Diode Ideality factor (m)	1.5
PV system rating/system base	50 MW

References

1. Sitompul, S.; Fujita, G. In Impact of State-of-Charge Control Integrated with Load-Frequency Control on Battery Energy Storage System in Islanded Microgrid System. In Proceedings of the 2021 IEEE 12th Energy Conversion Congress & Exposition-Asia (ECCE-Asia), Singapore, 24–27 May 2021; pp. 199–203.
2. Klein, M.; Rogers, G.J.; Kundur, P. A fundamental study of inter-area oscillations in power systems. *IEEE Trans. Power Syst.* **1991**, *6*, 914–921. [CrossRef]
3. Machowski, J.; Lubosny, Z.; Bialek, J.W.; Bumby, J.R. *Power System Dynamics: Stability and Control*; John Wiley & Sons: Hoboken, NJ, USA, 2020.
4. Ufa, R.; Rudnik, V.; Malkova, Y.; Bay, Y.; Kosmynina, N. Impact of renewable generation unit on stability of power systems. *Int. J. Hydrogen Energy* **2022**, *47*, 19947–19954. [CrossRef]
5. Tina, G.M.; Maione, G.; Licciardello, S. Evaluation of Technical Solutions to Improve Transient Stability in Power Systems with Wind Power Generation. *Energies* **2022**, *15*, 7055. [CrossRef]
6. He, X.; Geng, H. Transient stability of power systems integrated with inverter-based generation. *IEEE Trans. Power Syst.* **2020**, *36*, 553–556. [CrossRef]
7. Shair, J.; Li, H.; Hu, J.; Xie, X. Power system stability issues, classifications and research prospects in the context of high-penetration of renewables and power electronics. *Renew. Sustain. Energy Rev.* **2021**, *145*, 111111. [CrossRef]
8. Huang, R.; Gao, W.; Fan, R.; Huang, Q. A guided evolutionary strategy based static var compensator control approach for inter-area oscillation damping. *IEEE Trans. Ind. Inform.* **2022**, *19*, 2596–2607. [CrossRef]
9. Huang, R.; Gao, W.; Fan, R.; Huang, Q. Damping inter-area oscillation using reinforcement learning controlled TCSC. *IET Gener. Transm. Distrib.* **2022**, *16*, 2265–2275. [CrossRef]
10. Hashmy, Y.; Yu, Z.; Shi, D.; Weng, Y. Wide-area measurement system-based low frequency oscillation damping control through reinforcement learning. *IEEE Trans. Smart Grid* **2020**, *11*, 5072–5083. [CrossRef]
11. Saadatmand, M.; Gharehpetian, G.B.; Moghassemi, A.; Guerrero, J.M.; Siano, P.; Alhelou, H.H. Damping of low-frequency oscillations in power systems by large-scale PV farms: A comprehensive review of control methods. *IEEE Access* **2021**, *9*, 72183–72206. [CrossRef]
12. Solomon, E.; Khan, B.; Boulkaibet, I.; Neji, B.; Khezami, N.; Ali, A.; Mahela, O.P.; Pascual Barrera, A.E. Mitigating Low-Frequency Oscillations and Enhancing the Dynamic Stability of Power System Using Optimal Coordination of Power System Stabilizer and Unified Power Flow Controller. *Sustainability* **2023**, *15*, 6980. [CrossRef]
13. Sreedivya, K.; Jeyanthy, P.A.; Devaraj, D. Improved design of interval type-2 fuzzy based wide area power system stabilizer for inter-area oscillation damping. *Microprocess. Microsyst.* **2021**, *83*, 103957. [CrossRef]
14. Biswal, S.S.; Swain, D.R.; Rout, P.K. Inter-area and intra-area oscillation damping for UPFC in a multi-machine power system based on tuned fractional PI controllers. *Int. J. Dyn. Control* **2022**, *10*, 1594–1612. [CrossRef]
15. Huang, L.; Coulson, J.; Lygeros, J.; Dörfler, F. Decentralized data-enabled predictive control for power system oscillation damping. *IEEE Trans. Control Syst. Technol.* **2021**, *30*, 1065–1077. [CrossRef]
16. Prakash, A.; Kumar, K.; Parida, S. Energy capacitor system based wide-area damping controller for multiple inter-area modes. *IEEE Trans. Ind. Appl.* **2022**, *58*, 1543–1553. [CrossRef]
17. Alalwani, S.; Isik, S.; Bhattacharya, S. In Inter-area Oscillation Damping Controller for DFIG based Wind Power Plants. In Proceedings of the 2022 IEEE Energy Conversion Congress and Exposition (ECCE), Detroit, MI, USA, 9–13 October 2022; pp. 1–6.
18. Ahmed, F.; Foley, A.; McLoone, S.; Best, R.J.; Cameron, C.; Al Kez, D. Dynamic grid stability in low carbon power systems with minimum inertia. *Renew. Energy* **2023**, *210*, 486–506. [CrossRef]
19. Eberlein, S. *Small-Signal Stability Modelling and Optimization of Microgrids*; BoD–Books on Demand: Norderstedt, Germany, 2021; Volume 32.
20. Oni, O.E.; Swanson, A.G.; Carpanen, R.P. Small signal stability analysis of a four-machine system with placement of multi-terminal high voltage direct current link. *J. Energy S. Afr.* **2020**, *31*, 73–87. [CrossRef]
21. Zou, X. *Frequency and Damping Characteristics of Generators in Power Systems*; Virginia Tech: Blacksburg, VA, USA, 2018.
22. Oni, O.E.; Swanson, A.G.; Carpanen, R.P. In Small signal stability analysis of a four machine system with strategic placement of monopolar LCC-HVDC link. In Proceedings of the 2019 Southern African Universities Power Engineering Conference/Robotics and Mechatronics/Pattern Recognition Association of South Africa (SAUPEC/RobMech/PRASA), Bloemfontein, South Africa, 29–31 January 2019; pp. 437–443.
23. Rajapakse, A.D.; Muthumuni, D. In Simulation tools for photovoltaic system grid integration studies. In Proceedings of the 2009 IEEE Electrical Power & Energy Conference (EPEC), Montreal, QC, Canada, 22–23 October 2009; pp. 1–5.
24. Liu, S.; Dougal, R.A. Dynamic multiphysics model for solar array. *IEEE Trans. Energy Convers.* **2002**, *17*, 285–294.
25. Owais, R.; Iqbal, S.J. An Intelligent Two-Level Control of Solar Photovoltaic Power Plant for Electromechanical Oscillation Damping in Power Systems. *Arab. J. Sci. Eng.* **2023**, *48*, 6271–6292. [CrossRef]
26. Gupta, R.; Gupta, G.; Kastwar, D.; Hussain, A.; Ranjan, H. In Modeling and design of MPPT controller for a PV module using PSCAD/EMTDC. In Proceedings of the 2010 IEEE PES Innovative Smart Grid Technologies Conference Europe (ISGT Europe), Gothenburg, Sweden, 11–13 October 2010; pp. 1–6.
27. Faruque, M.; Zhang, Y.; Dinavahi, V. Detailed modeling of CIGRE HVDC benchmark system using PSCAD/EMTDC and PSB/SIMULINK. *IEEE Trans. Power Deliv.* **2005**, *21*, 378–387. [CrossRef]

28. Oni, O.E.; Swanson, A.G.; Carpanen, R.P. In Modelling and control of multiterminal LCC HVDC. In Proceedings of the 2018 IEEE PES/IAS PowerAfrica, Cape Town, South Africa, 28–29 June 2018; pp. 274–279.
29. Mankour, M.; Sami, B.S. Mitigation of commutation failure method in LCC converter based on HVDC systems by mean of modeling and simulation. *J. Ambient Intell. Humaniz. Comput.* **2020**, *14*, 5837–5852. [CrossRef]
30. Peng, J.-H.; Nair, N.-K.; Maryani, A.; Ahmad, A. Adaptive Power System Stabilizer tuning technique for damping inter-area oscillations. In Proceedings of the IEEE PES General Meeting, Minneapolis, MN, USA, 25–29 July 2010; pp. 1–6.
31. Yakout, A.H.; Attia, M.A. In Damping Inter-Area Oscillations Via Weighted Area Signals to PSSs Using TLBO & HS Algorithm. In Proceedings of the 2019 21st International Middle East Power Systems Conference (MEPCON), Cairo, Egypt, 17–19 December 2019; pp. 496–501.
32. Preece, R.; Milanović, J.V.; Almutairi, A.M.; Marjanovic, O. Damping of inter-area oscillations in mixed AC/DC networks using WAMS based supplementary controller. *IEEE Trans. Power Syst.* **2012**, *28*, 1160–1169. [CrossRef]
33. Patidar, N.; Kolhe, M.; Tripathy, N.; Sahu, B.; Sharma, A.; Nagar, L.; Azmi, A. In Optimized design of wide-area PSS for damping of inter-area oscillations. In Proceedings of the 2015 IEEE 11th International Conference on Power Electronics and Drive Systems, Sydney, Australia, 9–12 June 2015; pp. 1172–1177.
34. Bolzoni, A.; Johansson, N.; Hasler, J. In Modeling the impact of grid-forming E-STATCOMs on inter-area system oscillations. In Proceedings of the 2022 24th European Conference on Power Electronics and Applications (EPE'22 ECCE Europe), Hanover, Germany, 5–9 September 2022; pp. 1–10.
35. Fadhil, S.T.; Vural, A.M. In Comparison of dynamic performances of TCSC, Statcom, SSSC on inter-area oscillations. In Proceedings of the 2018 5th International Conference on Electrical and Electronic Engineering (ICEEE), Istanbul, Turkey, 3–5 May 2018; pp. 138–142.
36. Fayez, M.; Mandor, M.; El-Hadidy, M.; Bendary, F. Fuzzy logic based dynamic braking scheme for stabilization of inter-area. *Acta Electron. Malays. (AEM)* **2019**, *3*, 16–22. [CrossRef]
37. Fayez, M.; Mandor, M.; El-Hadidy, M.; Bendary, F. Stabilization of inter-area oscillations in two-area test system via centralized interval type-2 fuzzy-based dynamic brake control. *J. Electr. Syst. Inf. Technol.* **2021**, *8*, 1–19. [CrossRef]
38. Selim, F.; Attia, A.-F. In Power System Stabilizer with Self-Tuning Based on Hierarchical Fuzzy Logic Controller. In Proceedings of the 2022 23rd International Middle East Power Systems Conference (MEPCON), Cairo, Egypt, 13–15 December 2022; pp. 1–6.
39. Sharafi, A.; Vahidnia, A.; Jalili, M. In Application of Consensus Algorithm in Clustering of Distributed Generation Systems for Wide-Area Management and Control. In Proceedings of the 2023 IEEE International Conference on Environment and Electrical Engineering and 2023 IEEE Industrial and Commercial Power Systems Europe (EEEIC/I&CPS Europe), Madrid, Spain, 6–9 June 2023; pp. 1–6.
40. Sharafi, A.; Vahidnia, A.; Jalili, M. In Wide-Area Stabilizing Control using Distributed Generation Systems. In Proceedings of the 2023 11th International Conference on Smart Grid (icSmartGrid), Paris, France, 5–7 June 2023; pp. 1–6.

energies

MDPI

Article

Long-Term Voltage Stability Bifurcation Analysis and Control Considering OLTC Adjustment and Photovoltaic Power Station

Sheng Li *, Can Zhang and Jili Zuo

School of Electric Power Engineering, Nanjing Institute of Technology, Nanjing 211167, China; zhangcan8211@163.com (C.Z.); jilizuo0324@163.com (J.Z.)
* Correspondence: lisheng_njit@126.com

Abstract: The influence of photovoltaic (PV) output with stochasticity and uncertainty on the grid-connected system's voltage stability is worth further exploration. The long-term voltage stability of a 3-bus system with a large-scale PV power station considering the adjustment of an on-load tap changer (OLTC) was studied. In this typical system, two supercritical Hopf bifurcation (SHB) points are found using the bifurcation calculation. At the SHB point that appears first, a small sudden increase in reactive load power or a sudden increase in PV active power P_{pv} can eventually cause a voltage collapse after a long increasing oscillation. The long-term collapse phenomenon shows that SHB cannot be ignored in the PV grid-connected system. Meanwhile, the time constant of OLTC can affect the progress of long-term voltage collapse, but it has different effects under different disturbances. When P_{pv} drops suddenly at the SHB point, due to the adjustment of OLTC, the load bus voltage can recover to near the target value of OLTC after a long period of time. Similarly, the time constant of OLTC can affect the progress of long-term voltage recovery. To prevent the long-term voltage collapse when P_{pv} increases suddenly at the SHB point, a new locking-OLTC index I_{lock}, depending on the value of P_{pv} corresponding to the SHB point, and a locking OLTC method are proposed, and the voltage can be recovered to an acceptable stable value quickly. Compared with the system without OLTC, OLTC adjustment can effectively prevent long-term voltage oscillation instability and collapse, so that PV power can play a bigger role in power systems.

Keywords: photovoltaic (PV); long-term voltage stability; voltage collapse; OLTC; supercritical Hopf bifurcation (SHB)

Citation: Li, S.; Zhang, C.; Zuo, J. Long-Term Voltage Stability Bifurcation Analysis and Control Considering OLTC Adjustment and Photovoltaic Power Station. *Energies* **2023**, *16*, 6383. https://doi.org/10.3390/en16176383

Academic Editor: Ying-Yi Hong

Received: 29 June 2023
Revised: 29 August 2023
Accepted: 1 September 2023
Published: 3 September 2023

1. Introduction

From the perspective of protecting the environment and reducing smog pollution, the development of renewable energy and clean energy generation has become a mainstream trend, and vigorously developing the new energy is an important strategy for achieving the goal of "carbon peaking and carbon neutrality" [1,2]. Photovoltaic (PV) power generation is an important method of solving the problem of carbon emissions [3]. At present, PV power generation has been vigorously promoted and widely used around the world.

PV power generation has the characteristics of volatility and intermittency. When large-scale PV stations are connected to the power grid, it will inevitably have an adverse impact on the stability of the power system, increasing the complexity and uncertainty of grid operation [4,5]. Therefore, the influence of PV output on the grid-connected system's voltage stability must be studied [6–9].

Long-term voltage stability belongs to large-disturbance voltage stability. The large disturbances in the power system refer to a short circuit, a disconnection fault, or a large-scale new energy off-grid, etc. The large-disturbance voltage stability includes two aspects: the transient process and the long-term process [10]. The time frame for the transient voltage stability investigations is usually 10–15 s, while the time frame for the long-term voltage stability investigations is usually from a few minutes to tens of minutes with the role of the slow-regulating devices, such as the on-load tap changer (OLTC) and the over-excitation limiter (OEL), etc.

Currently, several studies on long-term voltage stability have been reported. The literature [11] proposed a remedial measure system to prevent medium-term and long-term voltage instabilities in power systems. When the operating point deviates from the normal operating state, the system will select a required procedure, power generation rescheduling or load shedding, based on the concept of electrical distance. According to the selected procedure, the system implements the corresponding remedial measure to bring the operating point back to a normal operating state. The literature [12] proposed a coordinated decentralized emergency control strategy for long-term voltage instability by dividing the power system into multiple local areas based on the concept of electrical distance; and designing a performance indicator based on the load bus voltage and the generator reactive power to assess the severity of disturbances and emergency risks in each local area.

The literature [13] proposed the concepts of long-term voltage stability PDR (Reactive Power Reserve) and short-term voltage stability PDR and established a dual-objective reactive power reserve optimization model to coordinate long-term and short-term voltage stability simultaneously. The literature [14] proposed a long-term voltage stability index that can identify the critical load power of voltage collapse, and a transmission-distribution distinguishing index (TDDI) is also proposed to identify whether the voltage stability limit is due to the transmission or distribution network. Both VSI-index and TDDI-index indicators are calculated based on the measured value of the phasor measurement unit, which is convenient for online applications. The literature [15] designed a long-term voltage stability monitoring index based on the measured value of voltage amplitude and the calculation value of the Thevenin equivalent angle using a voltage trajectory method. The literature [16] studied the coordinated scheduling problem of gas-electricity integrated energy systems including wind power, taking the long-term voltage stability as a constraint condition.

In recent years, machine learning and data mining technologies have developed rapidly, and machine learning algorithms have also been applied to long-term voltage stability research. The literature [17] used the random forest algorithm to predict the load margin to monitor long-term voltage stability in real-time, and a variety of different voltage stability indexes were used as the input variables of the machine learning integration model. The literature [18] proposed a data-based learning and control method based on offline knowledge accumulation and feature acquisition to solve the long-term voltage stability problem caused by emergency online events. The literature [19] proposed an auto-encoder constructed by long short-term memory networks combined with a fully connected layer, which only needs to train the data from a safe operating state to evaluate the long-term voltage stability of power systems.

OLTC is a kind of typical slow-adjustment device and it completes one-tap adjustments with a required time of 10 s to 100 s. When studying long-term voltage stability, the dynamic adjustment process of OLTC cannot be neglected. On the other hand, the dynamic load also has a greater impact on voltage stability, and the adjustment effect of OLTC on voltage stability is related to the dynamic load characteristics [20]. Generally, when the system has sufficient reactive power, OLTC adjustment can contribute to the system's voltage stability; if the system reactive power is insufficient, OLTC will have a negative voltage adjustment effect [21].

The literature [22] studied the impact of a variable-speed wind turbine on long-term voltage stability, considering the actions of OEL and OLTC. When the doubly fed induction generator adopts the grid-side converter to control the reactive power, the long-term voltage stability can be effectively improved. The literature [23] took the Nordic 32-bus test system including AVR (Automatic Voltage Regulator), OEL, and OLTC as an example to investigate the impact of PV power generation on long-term voltage stability.

With the large-scale access to new energy power stations, the impact of wind power and PV power generation on long-term voltage stability has also begun to receive attention. In a large-scale PV grid-connected system, it is worth discussing the mechanisms of sudden changes in PV power (including sudden increases and sudden drops) on long-term voltage instability, especially under extreme weather conditions. The impacts of OLTC adjustment and dynamic load characteristics on long-term voltage stability under sudden PV power changes also deserve further study.

Because bifurcation parameters correspond to uncontrollable parameters such as load power and PV power variations, bifurcation theory and its analysis method are suitable for studying voltage stability in the PV grid-connected system. The literature [24] analyzed in detail the bifurcation behaviors of a typical 3-bus system with a PV station. When the bifurcation parameter changes gradually, once the behavior of the grid-connected system changes, the system will bifurcate, which may be saddle-node bifurcation (SNB) or other forms of bifurcation, such as Hopf bifurcation (HB), including supercritical Hopf bifurcation (SHB) and subcritical Hopf bifurcation (alternatively termed unstable Hopf bifurcation (UHB)), etc. These bifurcations may have adverse impacts on the system's voltage stability. For example, at the UHB point, a small increase in load power or a sudden change in PV active power can cause a long-term voltage collapse or a long-term voltage oscillation [24,25]. In the literature [25], an index-based predictive control approach for UHB was proposed. However, the impact of OLTC adjustment is not considered in the literature [24,25]. This may cause a one-sided bifurcation analysis result, as well as a poor understanding of the mechanism of long-term voltage instability.

The biggest factor affecting PV active power is the intensity of solar light. When encountering extreme weather conditions, such as solar eclipses, snowstorms, and sandstorms, etc., it is highly likely that a sudden and significant drop or even complete loss of PV power will occur, but whether it leads to long-term voltage instability should be further investigated.

This paper mainly investigates, in detail, the long-term voltage instability mechanism of the PV grid-connected system, considering OLTC adjustments when the PV output has a sudden change by using bifurcation theory, explores OLTC's key influencing factors on long-term voltage instability, and considers the influence of load dynamic characteristics at the same time. In addition, this paper attempts to design a novel long-term voltage stability index that can predict the long-term voltage instability phenomena, so as to provide an adaptive control scheme.

2. Long-Term Voltage Stability Bifurcation Analysis Based on Matcont

2.1. Brief Introduction of the Hopf Bifurcation Theory

The PV grid-connected system can be described by the single-parameter family form of ordinary differential equations (ODEs).

$$\dot{x} = f(x, \mu) \tag{1}$$

where x represents the state variables; μ represents the bifurcation parameter such as the load power or the PV power, etc., and f is a continuously differentiable function.

The eigenvalues of the state matrix A of Equation (1) at the system's equilibrium point (x_0, μ) can describe the grid-connected system's dynamic stability. With the gradual change of μ, if a pair of conjugate eigenvalues $\alpha(\mu) \pm \mathrm{j}\beta(\mu)$ of A pass through the imaginary axis in the complex plane, and satisfy the next equation at $\mu = \mu_0$ [26]:

$$\begin{cases} \alpha(\mu_0) = 0 \\ \frac{d\alpha(\mu_0)}{d\mu} \neq 0 \end{cases} \tag{2}$$

Then a HB happens at the equilibrium point (x_0, μ_0) in the system, and the point corresponding to the pure imaginary eigenvalues $\pm \mathrm{j}\beta(\mu_0)$ is namely the HB point [26].

According to the stability of the limit cycle arising from the HB point, HB can be divided into SHB (the corresponding first Lyapunov coefficient (FLC) is less than 0) and UHB (the corresponding FLC is greater than 0) [27].

2.2. Model of the 3-Bus System with OLTC and PV

As shown in Figure 1a, the traditional 3-bus system (not including the OLTC and PV stations), has been widely used in traditional voltage stability bifurcation analysis [28,29]. Now a PV station and an OLTC station are introduced into the system. n is the transformation ratio of OLTC, in general, 0.8 pu $\leq n \leq$ 1.2 pu. The PV station is connected to the load bus to investigate the impact of PV power on the load bus's voltage stability. For facilitating the bifurcation calculation, the PV transmission line is omitted. P_{pv} + jQ_{pv} is the PV power. $E_{\mathrm{m}}\angle\delta_{\mathrm{m}}$ is the equivalent generator potential, $E_0\angle 0$ is the infinite bus voltage, and $V\angle\delta$ is the load bus voltage. Y_{m} and Y_0 are the grid equivalent admittance modulus, θ_{m} and θ_0 are the corresponding admittance angles, and Y_{m} includes the equivalent admittance of OLTC. The load P_{D} + jQ_{D} adopts the Walve dynamic load model [28,29]. The load P_1 + jQ_1 is a constant power load. C is the parallel capacitor bank. In the traditional 3-bus system without OLTC and PV, C is used to raise the load bus voltage. Due to the intermittency of P_{pv}, this setting is retained in this case.

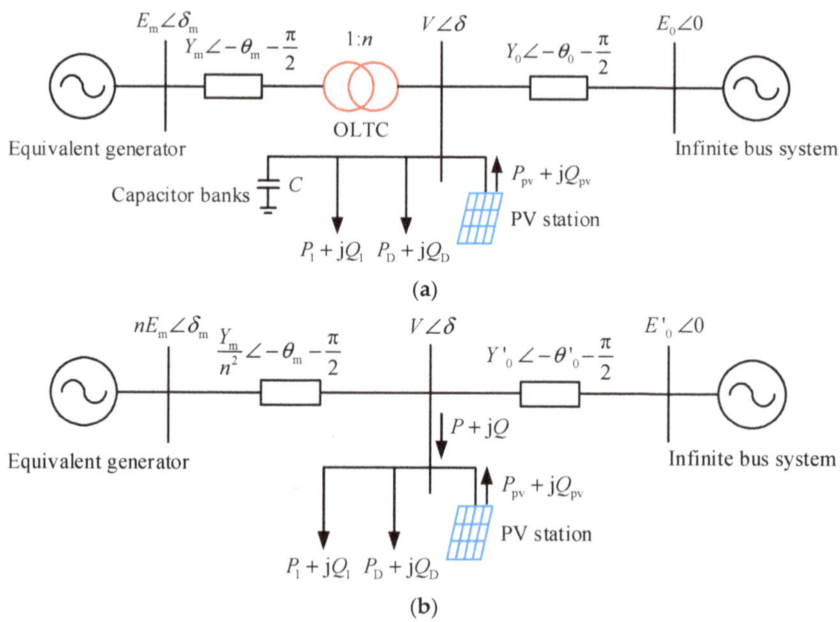

Figure 1. The 3-bus system with OLTC and PV. (**a**) The initial model. (**b**) The simplified model.

To simplify Figure 1a, the generator branch is converted to the load bus side, and the capacitor banks and the infinite system branch are subjected to a Thevenin equivalent transformation, as shown in Figure 1b.

The inverter of the PV station adopts a 1-order dynamic model [30], and its dynamic characteristic equation is as follows [31]:

$$\begin{cases} \dot{i}_d = \frac{1}{T_p}\left[\frac{1}{V}(Q_{\mathrm{pv}}\cos\delta - P_{\mathrm{pv}}\sin\delta) - i_d\right] \\ \dot{i}_q = \frac{1}{T_q}\left[\frac{1}{V}(P_{\mathrm{pv}}\cos\delta + Q_{\mathrm{pv}}\sin\delta) - i_q\right] \end{cases} \tag{3}$$

where i_d, i_q are the d-axis and q-axis components of the PV inverter's output current, and T_p, T_q are the responsive time constants of the PV inverter.

The PV station is set as a PQ-type station operating in a constant power factor mode [9], so we can assume:

$$Q_{\text{pv}} = kP_{\text{pv}} \tag{4}$$

where k is the tangent of the PV station's power factor angle. For example, if k is 0.2, the PV power factor is 0.98.

The discrete model of OLTC is as follows [32]:

$$n_{k+1} - n_k = df(V_{\text{ref}} - V) \tag{5}$$

where d is the tap step length, and V_{ref} is the load bus reference voltage (namely the adjustment target value of OLTC).

The function $f(V_{\text{ref}} - V)$ can be expressed as follows:

$$f(V_{\text{ref}} - V) = \begin{cases} +1 & V_{\text{ref}} - V \geq \Delta V \\ 0 & |V_{\text{ref}} - V| \leq \Delta V \\ -1 & V_{\text{ref}} - V \leq -\Delta V \end{cases} \tag{6}$$

where ΔV is the adjustment dead zone.

In order to facilitate the bifurcation calculation with the numerical bifurcation analysis software, in this paper, OLTC adopts a continuous adjustment dynamic model, ignoring the adjustment dead zone and the tap time delay [33]:

$$\dot{n} = \frac{1}{T_n}(V_{\text{ref}} - V) \tag{7}$$

where T_n is the time constant of OLTC (s). To approach the discrete model, T_n can be expressed as $T_n = \tau/d$, τ is the tap time delay.

To simplify the analysis process, the equivalent generator uses the 2-order model:

$$\begin{cases} \dot{\delta}_{\text{m}} = \omega \\ \dot{\omega} = \frac{1}{M}(P_{\text{m}} - P_{\text{G}} - D\omega) \end{cases} \tag{8}$$

where δ_{m} and ω are the generator power angle and the rotor angular velocity, respectively, M is the inertia, D is the damping coefficient, and P_{m} and P_{G} are the mechanical power and output electric power of the generator, respectively.

Derived from Figure 1b, the output active power of the generator can be given below:

$$P_G = -E_{\text{m}}^2 Y_{\text{m}} \sin\theta_{\text{m}} - \frac{E_{\text{m}} V Y_{\text{m}}}{n} \sin(\delta - \delta_{\text{m}} - \theta_{\text{m}}) \tag{9}$$

The power supplied by the grid to all loads is:

$$\begin{cases} P = -E_0' V Y_0' \sin(\delta + \theta_0') - \frac{E_{\text{m}} V Y_{\text{m}}}{n} \sin(\delta - \delta_{\text{m}} + \theta_{\text{m}}) + (Y_0' \sin\theta_0' + \frac{Y_{\text{m}}}{n^2} \sin\theta_{\text{m}}) V^2 \\ Q = E_0' V Y_0' \cos(\delta + \theta_0') + \frac{E_{\text{m}} V Y_{\text{m}}}{n} \cos(\delta - \delta_{\text{m}} + \theta_{\text{m}}) - (Y_0' \cos\theta_0' + \frac{Y_{\text{m}}}{n^2} \cos\theta_{\text{m}}) V^2 \end{cases} \tag{10}$$

The dynamic load model adopts the Walve load model, then the ODEs of the system shown in Figure 1 can be given below:

$$\begin{cases} \dot{\delta}_{\text{m}} = \omega \\ \dot{\omega} = \frac{1}{M}(P_{\text{m}} - P_{\text{G}} - D\omega) \\ \dot{\delta} = \frac{1}{k_{q\omega}}(-k_{qV}V - k_{qV2}V^2 + Q - Q_0 - Q_1 + kP_{\text{pv}}) \\ \dot{V} = \frac{1}{Tk_{q\omega}k_{pV}}\left[\begin{array}{l} k_{p\omega}k_{qV2}V^2 + (k_{p\omega}k_{qV} - k_{q\omega}k_{pV})V \\ +k_{p\omega}(Q_0 + Q_1 - Q - kP_{\text{pv}}) - k_{q\omega}(P_0 + P_1 - P - P_{\text{pv}}) \end{array}\right] \\ \dot{n} = \frac{1}{T_n}(V_{\text{ref}} - V) \\ \dot{i}_d = \frac{1}{T_p}\left[\frac{P_{\text{pv}}}{V}(k\cos\delta - \sin\delta) - i_d\right] \\ \dot{i}_q = \frac{1}{T_q}\left[\frac{P_{\text{pv}}}{V}(\cos\delta + k\sin\delta) - i_q\right] \end{cases} \tag{11}$$

where T is the dynamic load time constant (s), $k_{p\omega}$, $k_{q\omega}$, k_{pV}, $k_{qV,}$ and k_{qV2} are the load coefficients of the Walve load, and P_0 and Q_0 are the constant powers of the Walve load.

The values of the power supply and network parameters are shown in Table 1, and the load parameters are shown in Table 2 (part of the data from the literature [24,29]). In this paper, all electrical quantities are in per-unit value and the phase angle's unit is rad, the time's unit is s. We set τ = 10 s and d = 0.625%, then $T_n = \tau/d$ = 1600 s, as listed in Table 1.

Table 1. The power supply and network parameters.

E_{m}/pu	P_{m}/pu	M/pu	D/pu	E_0/pu	Y_{m}/pu	θ_{m}/rad	Y_0/pu	θ_0/rad	C/pu	T_n/s	V_{ref}/pu	S_{base}
1	1	0.3	0.05	1	5	−0.0872	20	−0.0872	12	1600	1	100 MVA

Table 2. The load and PV parameters.

T/s	$k_{p\omega}$	$k_{q\omega}$	k_{pV}	k_{qV}	k_{qV2}	P_0/pu	Q_0/pu	P_1/pu	T_p/s	T_q/s	k
8.5	0.4	−0.03	0.3	−2.8	2.1	0.6	1.3	1.2	10	10	0.2

2.3. The Equilibrium Point Curves

Keep P_{pv} = 1 pu, and take Q_1 as the bifurcation parameter, the numerical bifurcation calculation software Matcont (Version 6.2, A. Dhooge, W. Govaerts, etc., Universiteit Gent, Belgium; Utrecht University, The Netherlands) was used to calculate the equilibrium point curves and bifurcation points of the ODEs shown in Equation (11). We can obtain the equilibrium point curves of the system, as shown in Figure 2. In the Q_1–n curve shown in Figure 2a, the lower half of the curve is a stable equilibrium point curve. Before Q_1 reaches the SNB point named "LP" (corresponding to Q_1 = 12.6347 pu), two Hopf bifurcation points appear: H_1 (corresponding to Q_1 = 10.3418 pu) and H_2 (corresponding to Q_1 = 11.9544 pu), FLC's values are both less than 0 (FLC_{H1} = −0.00876 and FLC_{H2} = −2.21), so the two Hopf bifurcation points are both SHB points.

Now keep Q_1 = 10.3418 pu and take P_{pv} as the bifurcation parameter. The P_{pv}–n curve is shown in Figure 2b, and the lower half of the curve is the stable equilibrium point curve. H_1 (corresponding to P_{pv} = 1 pu) and H'_2 (corresponding to P_{pv} = 16.7645 pu) are both SHB points. The installed capacity of a large-scale PV station can reach several pu (namely several hundred MW) currently, so it is possible that the system operates at the SHB point H_1, and the impact of this SHB point on the system's long-term voltage stability is noteworthy.

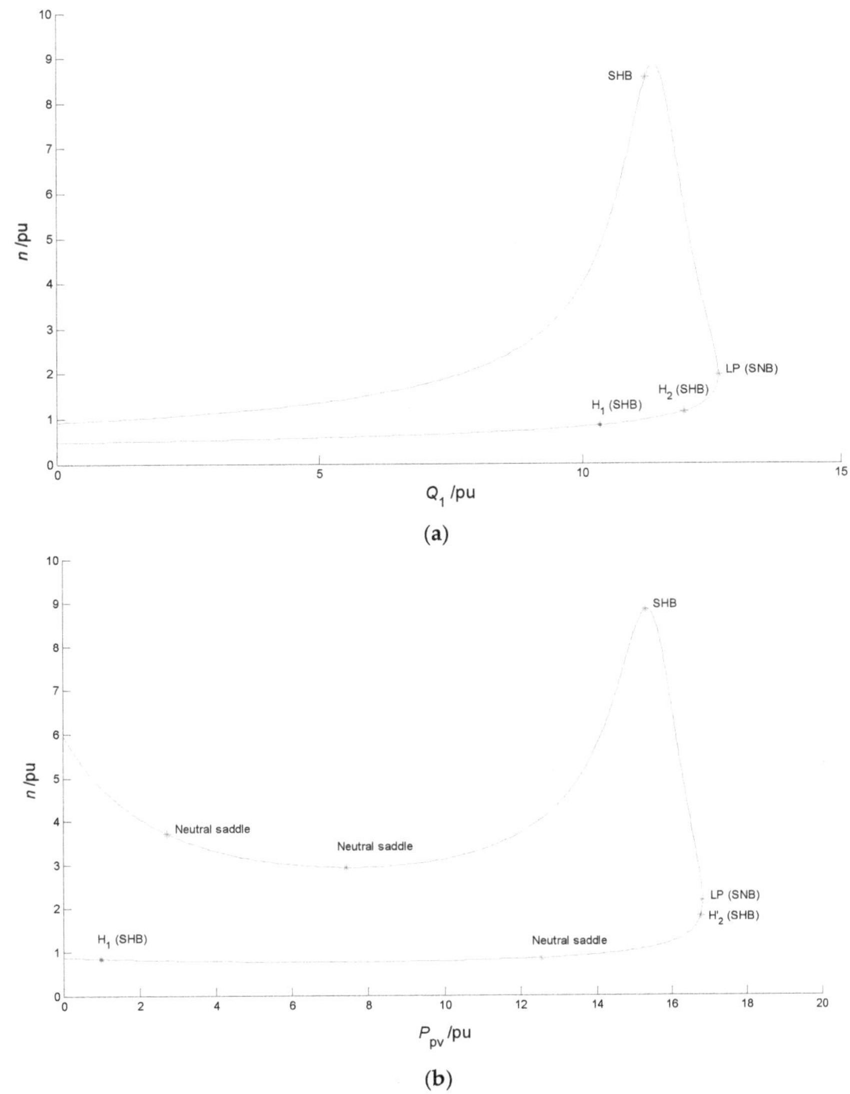

Figure 2. The equilibrium point curves. (**a**) Q_1–n curve (P_{pv} = 1 pu). (**b**) P_{pv}–n curve (Q_1 = 10.3418 pu).

2.4. Impact of Load Power Disturbance on Long-Term Voltage Stability

2.4.1. Sudden Increase in the Reactive Load Q_1—Long-Term Voltage Collapse

Take the SHB point H_1 as the initial operating point, and the parameters of this bifurcation point are: (δ_m ω δ V n i_d i_q Q_1) = (0.3057 0 0.1227 1 0.8478 0.0761 1.0170 10.3418). Assume that Q_1 has a forward small disturbance (a small increase) at time t = 50 s suddenly, and it increases from 10.3418 pu to 10.3918 pu. That is, the increment ΔQ_1 is 0.05 pu. The time-domain change curves are shown in Figure 3. The load bus voltage V eventually collapses at about t = 975 s after a long increasing oscillation, as shown in Figure 3a.

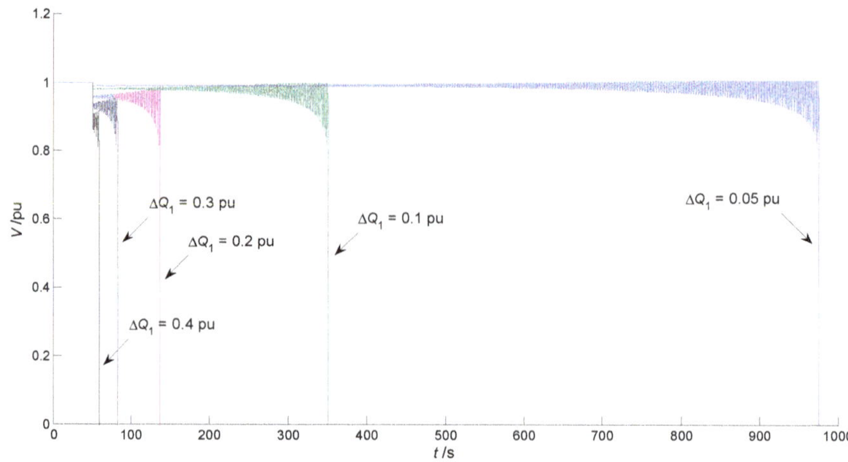

(**a**)

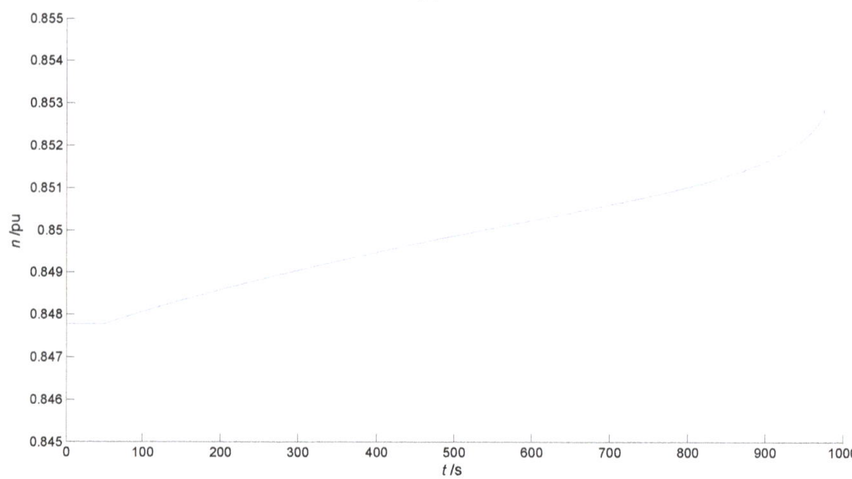

(**b**)

Figure 3. The time-domain change curves when Q_1 increases suddenly from the SHB point H_1 (T_n = 1600 s). (**a**) t–V curves at different increments of Q_1. (**b**) t–n curve (ΔQ_1 = 0.05 pu).

Figure 3b is the time-domain change curve of the OLTC transformation ratio n. After the disturbance occurs, the transformation ratio n keeps increasing to increase V, but the voltage collapse is unavoidable, and n no longer increases.

Now assume that starting from the SHB point H_1, the load power increment ΔQ_1 is 0.1 pu~0.4 pu, respectively. It can be seen from Figure 3a that the larger ΔQ_1 is, the faster V collapses, and the time frame becomes smaller as the long-term voltage collapse gradually evolves into the short-term voltage collapse. Therefore, it is recommended that the maximum value of Q_1 should not exceed the value of Q_1 corresponding to the SHB point H_1.

In general, it is believed that if there is a UHB point in power systems, it will pose a threat to the voltage stability and lead to oscillation-type voltage instability and collapse [24,25]. For this example, it can be seen from the above analyses that at the SHB point, a forward small or large disturbance of the load power can also cause long-term voltage oscillation instability and collapse. This phenomenon regarding SHB deserves more attention.

2.4.2. Impact of OLTC's Time Constant—Key Influencing Factor Investigation

The impact of OLTC's time constant T_n is investigated on long-term voltage collapse when the system operates at the SHB point H_1. Changing the tap step length d or the tap time delay τ can change the time constant of OLTC. Now set τ = 10 s and d = 1.25%, then $T_n = \tau/d$ = 800 s. Set ΔQ_1 = 0.05 pu, the time to reach the voltage collapse increases, as shown in Figure 4. Moreover, set τ = 20 s and d = 0.625%, $T_n = \tau/d$ = 3200 s, it can be seen from Figure 4 that the time taken to reach voltage collapse gets shorter. Therefore, increasing the time constant T_n can shorten the process of the long-term voltage collapse, that is, a larger T_n is not conducive to voltage stability.

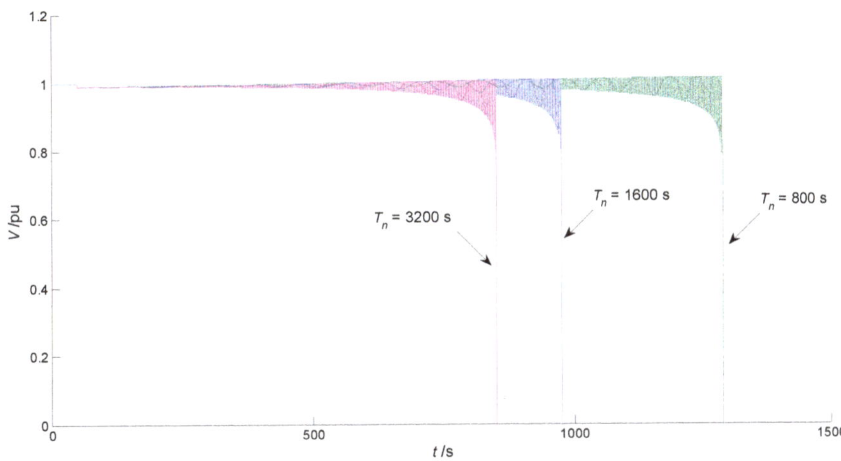

Figure 4. t–V curves at different time constants of OLTC when Q_1 increases suddenly from the SHB point H_1 (ΔQ_1 = 0.05 pu).

2.5. Impact of PV Power Disturbance on Long-Term Voltage Stability

2.5.1. Sudden Drop of PV Power—Long-Term Voltage Recovery

Assume that the system shown in Figure 1 is operating at the SHB point H_1, the initial P_{pv} is 1 pu, and Q_1 is maintained at 10.3418 pu. Considering extremely bad weather conditions, such as a total solar eclipse, at time t = 50 s, P_{pv} suddenly drops from 1 pu to 0 pu, 0.5 pu, and 0.8 pu, respectively. The drop depth dP_{pv} is 1 pu, 0.5 pu, and 0.2 pu, respectively. It can be seen from Figure 5a that the load bus voltage V drops after the disturbance occurs, but an OLTC adjustment can finally recover V to near the target value V_{ref} (1 pu) after a long period of time. As shown in Figure 5b, the drop depth dP_{pv} is 1 pu and it is big enough. When the disturbance occurs, the transformation ratio n continues to increase to raise the load bus voltage and eventually stabilizes at a value of 0.884 pu after a very long time.

It can be determined by calculation that when P_{pv} drops suddenly, OLTC can all adjust V to near the target value, and V can be recovered faster under the smaller drop depth dP_{pv}.

2.5.2. Sudden Increase in PV Power—Long-Term Voltage Collapse

Now assume that the system shown in Figure 1 is operating at the SHB point H_1, P_{pv} increases suddenly from 1 pu to 1.1 pu, 1.3 pu, and 1.5 pu, respectively. The increment ΔP_{pv} is 0.1 pu, 0.3 pu, and 0.5 pu, respectively. It can be seen from Figure 6 that the voltage V collapses after a long period of time. The larger ΔP_{pv} is, the faster V collapses, but the time frame is still in the category of long-term voltage stability.

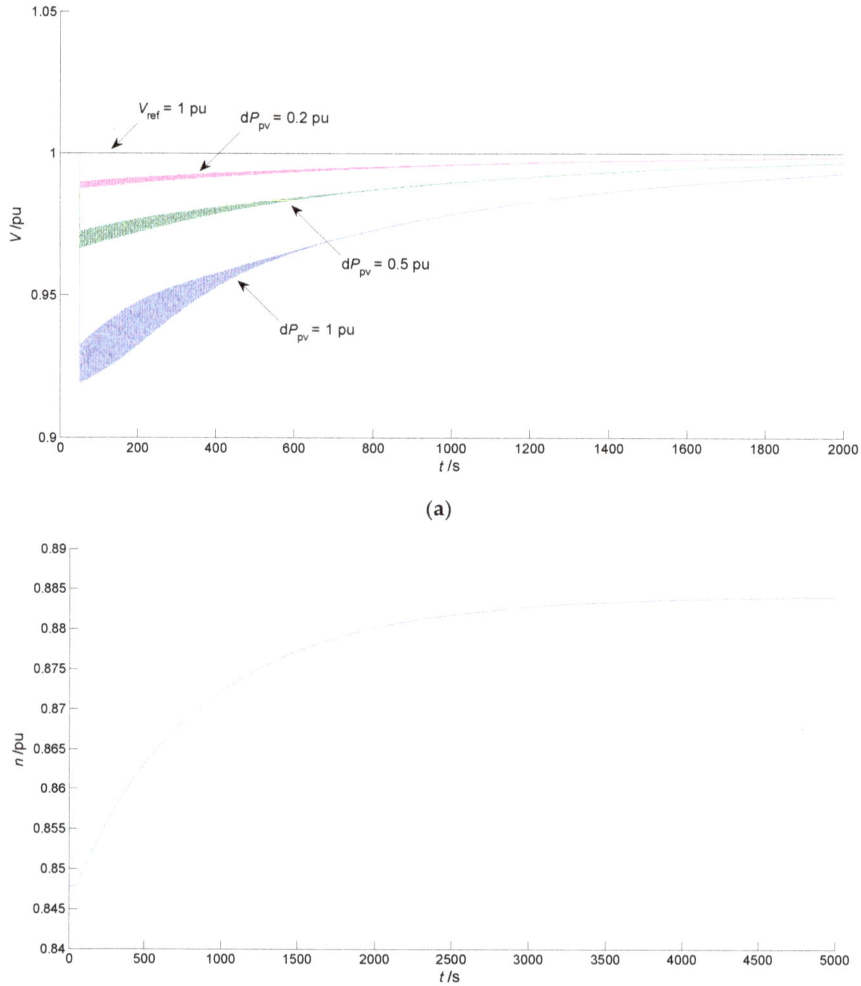

(**a**)

(**b**)

Figure 5. The time-domain change curves when P_{pv} drops suddenly from the SHB point H_1 (T_n = 1600 s). (**a**) t–V curves at different drop depths of P_{pv}. (**b**) t–n curve (dP_{pv} = 1 pu).

According to the above analyses, in the absence of any control measures taken, P_{pv} is preferably not greater than the PV active power corresponding to the SHB point H_1 ($P_{pv.H1}$). In practice, P_{pv} is generally less than the PV-installed capacity P_{pvN}, hence a solution is to ensure that $P_{pvN} \leq P_{pv.H1}$. However, this solution limits the use of PV power generation.

2.5.3. Impact of OLTC's Time Constant—Key Influencing Factor Investigation

Different time constants T_n can make different impacts on long-term voltage stability when P_{pv} has a sudden change. In Figure 7, when P_{pv} drops from 1 pu to 0 pu, a smaller T_n is conducive to the recovery of the voltage V. However, in Figure 8, when P_{pv} increases from 1 pu to 1.3 pu, a smaller T_n is not conducive to voltage stability.

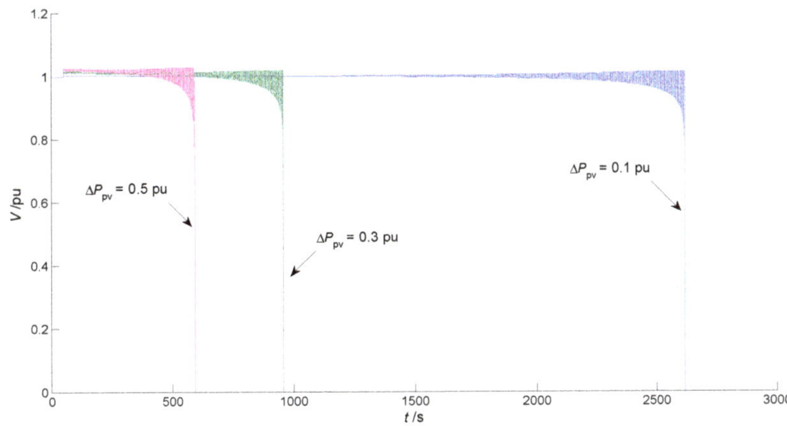

Figure 6. t–V curves when P_{pv} increases suddenly from the SHB point H_1 (T_n = 1600 s).

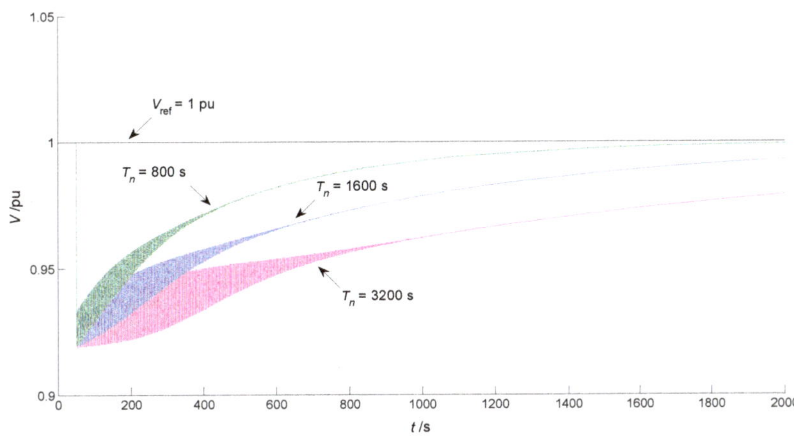

Figure 7. t–V curves at different time constants of OLTC when P_{pv} drops suddenly from the SHB point H_1 ($\mathrm{d}P_{\mathrm{pv}}$ = 1 pu).

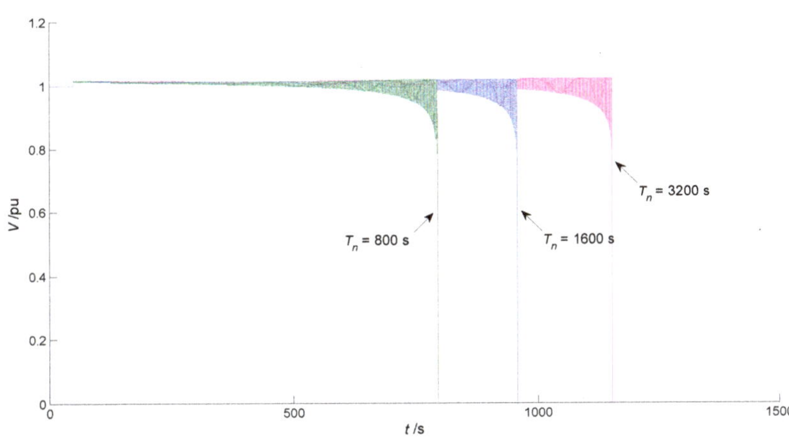

Figure 8. t–V curves at different time constants of OLTC when P_{pv} increases suddenly from the SHB point H_1 (ΔP_{pv} = 0.3 pu).

3. Comparison with the PV Grid-Connected 3-Bus System without OLTC

As shown in Figure 1, the PV grid-connected 3-bus system without OLTC is equivalent to the system with OLTC being locked in n = 1 pu, and the system's ODEs take the first four equations of Equation (11) [24]. Keeping P_{pv} = 1 pu, the bifurcation calculation results show that on the stable half of the Q_1–V equilibrium point curve, there is a UHB point H_1 (corresponding to Q_1 = 11.1309 pu), a SHB point H_2 (corresponding to Q_1 = 11.4999 pu), and a SNB point LP (corresponding to Q_1 = 11.5069 pu) [24].

Taking the UHB point H_1 as the initial operating point, since it is a subcritical Hopf bifurcation point, a small sudden increase in the reactive load Q_1 can cause long-term voltage collapse. As shown in Figure 9, assuming that P_{pv} is maintained at 1 pu, Q_1 increases suddenly from 11.1309 pu to 11.1809 pu at time t = 50 s (namely ΔQ_1 = 0.05 pu), and the load bus voltage V collapses at about t = 708 s.

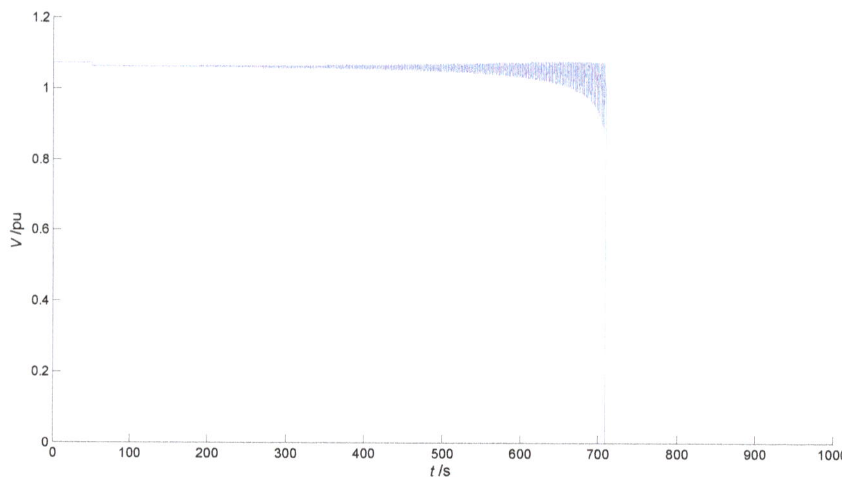

Figure 9. t–V curve when Q_1 increases suddenly from the UHB point H_1 (ΔQ_1 = 0.05 pu).

Compared with Figure 3a, due to the lack of OLTC adjustment, the time to reach voltage collapse has been reduced, but it still belongs in the category of long-term voltage stability.

Assume that Q_1 remains unchanged (keeping 11.1309 pu), and the impact of the sudden change of P_{pv} on long-term voltage stability is investigated. As shown in Figure 10, P_{pv} drops suddenly from 1 pu to 0 pu (namely dP_{pv} = 1 pu) at time t = 50 s. After V drops, the increasing oscillation is carried out firstly and the continuous oscillation (namely the constant-amplitude oscillation) begins at about t = 350 s after the disturbance occurs, and the oscillation amplitude of V is (0.9942 − 0.9344) × 100% = 5.98% > 5%. For the continuous oscillation, if the voltage's oscillation amplitude is greater than or equal to 5%, it can be considered that the load bus is experiencing long-term voltage oscillation instability.

The smaller the drop depth dP_{pv} is, the longer the time for the voltage V to begin the continuous oscillation is, but the larger the oscillation amplitude of V will be [24]. For example, when P_{pv} suddenly drops from 1 pu to 0.8 pu (namely dP_{pv} = 0.2 pu), the time for the load voltage V to begin the continuous oscillation is about t = 4800 s after the disturbance occurs, and the oscillation amplitude of V reaches (1.072 − 0.8562) × 100% = 21.58% >> 5%.

On the other hand, as shown in Figure 10, when P_{pv} suddenly increases from 1 pu to 1.5 pu (namely ΔP_{pv} = 0.5 pu), after the voltage V increases, it quickly enters a continuous oscillation, and the oscillation amplitude is (1.105 − 1.099) × 100% = 0.6%, far less than 5%, which does not pose a threat to voltage stability.

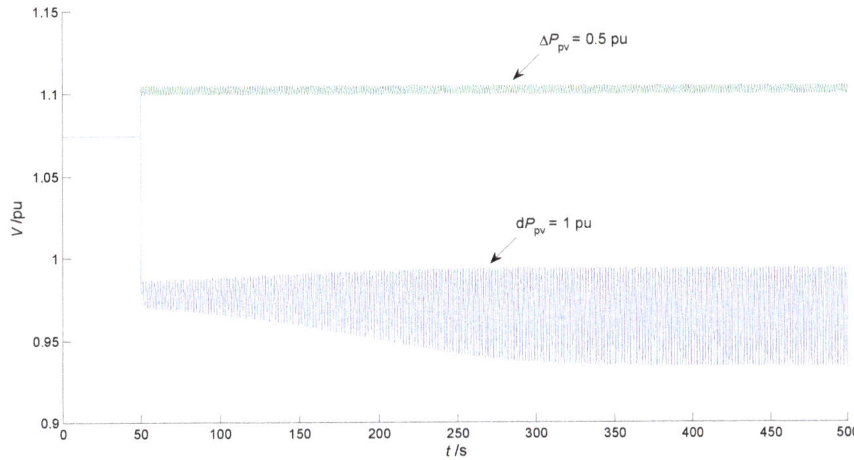

Figure 10. t–V curves when P_{pv} changes suddenly from the UHB point H_1 (dP_{pv} = 1 pu and ΔP_{pv} = 0.5 pu).

The operation comparison results of the system with OLTC and without OLTC are listed in Table 3. According to the comparative analyses, in order to avoid the long-term voltage oscillation instability when P_{pv} has a sudden drop, it is advisable to adopt the scheme of installing OLTC. For the sudden increase in P_{pv}, measures such as limiting the installed capacity of the PV station or locking OLTC can be used to solve the potential threat.

Table 3. Comparison of the PV grid-connected 3-bus system with OLTC and without OLTC, taking the HB point H_1 as the initial operating point.

The Content of Comparison	With OLTC	Without OLTC
Bifurcation type of point H_1	SHB	UHB
Reactive load Q_1 of H_1 (pu)	10.3418 pu	11.1309 pu
Voltage amplitude of H_1 (pu)	1 pu (V_{ref})	1.0744 pu
Sudden small increase of Q_1	Long-term voltage collapse	Long-term voltage collapse
Sudden drop of P_{pv}	Long-term voltage recovery	Long-term voltage oscillation instability (Oscillation amplitude $\geq$ 5%)
Sudden increase of P_{pv}	Long-term voltage collapse	Long-term voltage oscillation without instability (Oscillation amplitude << 5%)

4. Design of the Index and Control Methods for Preventing Long-Term Voltage Collapse

4.1. Prevention of Voltage Collapse Caused by a Sudden Increase in PV Power

For the 3-bus system with OLTC and PV shown in Figure 1, the installed capacity P_{pvN} can be limited according to the value of P_{pv} at the SHB point H_1 ($P_{pv.H1}$) so that P_{pv} does not exceed $P_{pv.H1}$. However, in practice, P_{pvN} may be much larger than $P_{pv.H1}$, so limiting the installed capacity is not an economic method.

In Figure 10 and Table 3, at the UHB point, when P_{pv} increases suddenly from 1 pu to 1.5 pu (ΔP_{pv} = 0.5 pu), long-term voltage oscillation with minimal amplitude will occur in the system without OLTC. Therefore, the locking OLTC method (namely maintaining a constant transformation ratio) can be used when P_{pv} increases suddenly in the system with OLTC, in order to prevent long-term voltage collapse.

For a better implementation of this locking OLTC method, a locking-OLTC index is designed:

$$I_{\text{lock}} = 1 - \frac{P_{\text{pv}}}{P_{\text{pv.H1}}} \leq 5 \sim 10\% \tag{12}$$

Under the premise of a constant Q_1, where the value of the I_{lock} index is less than or equal to 0, OLTC should be locked in time. To keep a certain margin, the threshold can even be raised to 5~10%.

For example, when the system with OLTC operates at the SHB point H_1, assuming that Q_1 is maintained at $Q_{1.H1}$ and P_{pv} fluctuates, once the I_{lock} index value is less than or equal to 5~10%, the locking OLTC method can be used to avoid long-term voltage collapse.

Assume that the system with OLTC operates at the SHB point H_1, and Q_1 is maintained at 10.3418 pu. When P_{pv} increases suddenly from 1 pu to 1.1 pu, 1.3 pu, and 1.5 pu, respectively, I_{lock} is −0.1, −0.3, and −0.5, respectively. Now OLTC is locked in n = 0.9 pu, it can be seen from Figure 11 that V can quickly recover to an acceptable stable value.

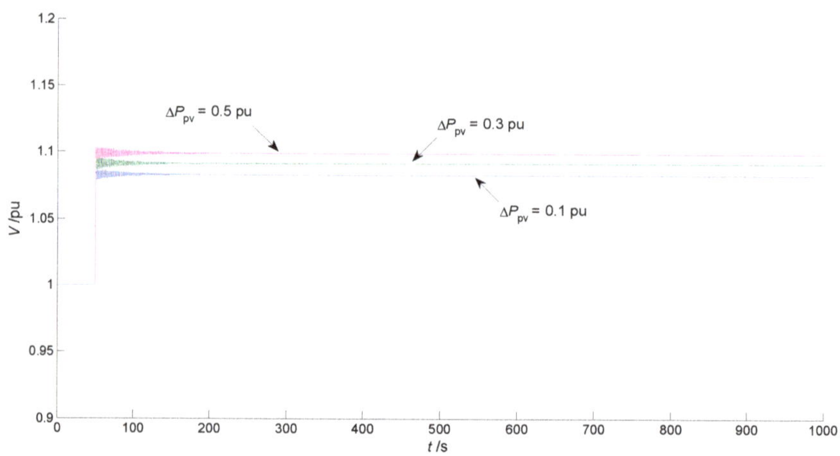

Figure 11. t–V curves when P_{pv} changes suddenly from the SHB point H_1 (OLTC is locked in n = 0.9 pu).

The flow chart of the locking OLTC method is shown in Figure 12. Since the voltage collapse caused by the sudden increase in P_{pv} is a long process (up to hundreds of seconds or more), the method of locking OLTC is completely feasible.

Considering the fluctuation of P_{pv}, the I_{lock} index should be calculated and judged twice or even more consecutively, with a delay of 0.5~1 min between the two consecutive calculations and judgments. Once the two (or more) consecutive judgment results are the same, the command of locking OLTC should be issued, and n will be locked at a constant value.

4.2. Prevention of Voltage Collapse Caused by a Sudden Increase in Load Power

The traditional load-margin index [7] can be modified as follows:

$$I_{\text{LMHB}} = 1 - \frac{Q_1}{Q_{1.\text{H1}}} \tag{13}$$

where $Q_{1.\text{H1}}$ is the reactive power of the SHB point H_1.

Once the I_{LMHB} index value is less than 0, the load-shedding method can be taken to improve the load margin. A detailed load-shedding scheme is referred to in the literature [7]. Since the impact of the sudden increase in P_{pv} at the SHB point can be solved by using the locking OLTC method, the value of I_{LMHB} can be 0.

Load shedding is certainly a conservative method, and it may also cause a small-amplitude oscillation in the load bus voltage.

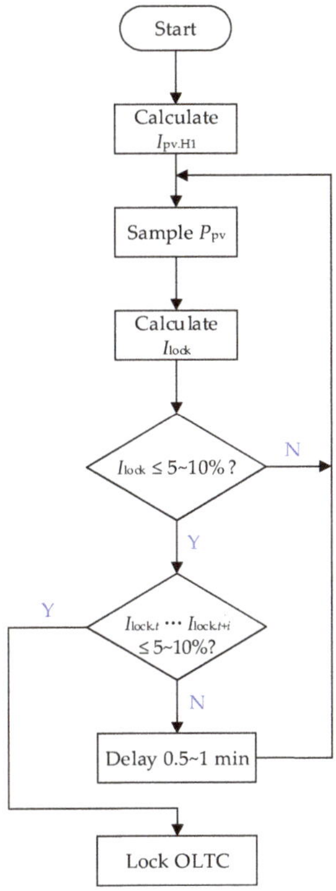

Figure 12. The flow chart of the locking OLTC method (t and i are the sampling time and times).

5. Conclusions

The typical 3-bus system containing an OLTC and a PV station is investigated for long-term voltage stability using bifurcation theory, and a new index and a locking OLTC method are used to prevent long-term voltage collapse. Using the bifurcation calculation and the time-domain calculation, and by comparison with the system without OLTC, the innovation conclusions are stated as follows:

(1) There are two SHB points and an SNB point on the stable half of the equilibrium point curve in the 3-bus system with OLTC and PV, and at the SHB point, a small sudden increase in reactive load power Q_1, or a sudden increase in PV active power P_{pv} can all cause long-term voltage collapse. This phenomenon suggests that SHB is also potentially threatening to the voltage stability.

(2) Aiming to change of P_{pv} at the SHB point suddenly, when P_{pv} drops suddenly, an OLTC adjustment can recover the load bus voltage to near the target value after a long period of time; when P_{pv} increases suddenly, locking OLTC and taking a suitable ration n can recover the voltage to a stable value. The impact of PV power disturbance at the SHB point on long-term voltage stability can be eliminated through the OLTC adjustment, and the system's long-term voltage stability will focus on the impact of load power disturbance.

(3) The time constant T_n of OLTC is an important influencing factor in long-term voltage stability, as it can affect the process of long-term voltage collapse and long-term voltage recovery. However, under different disturbances, the magnitude of T_n can cause different effects.

(4) A new locking OLTC index I_{lock} based on PV active power was proposed. It can be taken as the benchmark of locking OLTC when P_{pv} increases suddenly at the SHB point.

Moreover, a modified load-margin index I_{LMHB} is proposed to prevent long-term voltage collapse caused by the small sudden increase in load power at the SHB point.

Obviously, the discrete model of OLTC and the continuous changes of P_{pv} are not considered in the bifurcation analysis of the present case, and the calculation workload of bifurcation is very heavy for the multi-bus grid-connected system. The above topics require further study.

Author Contributions: Conceptualization, S.L.; methodology, S.L. and C.Z.; software, S.L., C.Z. and J.Z.; validation, S.L. and C.Z.; formal analysis, S.L. and C.Z.; investigation, S.L., C.Z. and J.Z.; resources, S.L. and J.Z.; data curation, S.L. and C.Z.; writing—original draft preparation, S.L. and C.Z.; Writing—review & editing, S.L. All authors have read and agreed to the published version of the manuscript.

Funding: This work was supported by the University Student Innovation & Entrepreneurship Training Program Project of Jiangsu Province (202211276009Z), and the Scientific Research Foundation of Nanjing Institute of Technology (ZKJ202102).

Data Availability Statement: The data presented in this study are available on request from the corresponding author.

Conflicts of Interest: The authors declare no conflict of interest.

References

1. Zhou, X.; Zhao, Q.; Zhang, Y.; Sun, L. Integrated energy production unit: An innovative concept and design for energy transition toward low-carbon development. *CSEE J. Power Energy Syst.* **2021**, *7*, 1133–1139.
2. Guo, J.; Ma, S.; Wang, T.; Jing, Y.; Hou, W.; Xu, H. Challenges of developing a power system with a high renewable energy proportion under China's carbon targets. *iEnergy* **2022**, *1*, 12–18.
3. Landera, Y.G.; Zevallos, O.C.; Neves, F.A.S.; Neto, R.C.; Prada, R.B. Control of PV systems for multimachine power system stability improvement. *IEEE Access* **2022**, *10*, 45061–45072.
4. Sajadi, A.; Rañola, J.A.; Kenyon, R.W.; Hodge, B.M.; Mather, B. Dynamics and stability of power systems with high shares of grid-following inverter-based resources: A tutorial. *IEEE Access* **2023**, *11*, 29591–29613.
5. Campanhol, L.B.G.; Silva, S.A.O.D.; Oliveira, A.A.D.; Bacon, V.D. Power flow and stability analyses of a multifunctional distributed generation system integrating a photovoltaic system with unified power quality conditioner. *IEEE Trans. Power Electr.* **2018**, *34*, 6241–6256.
6. Kawabe, K.; Ota, Y.; Yokoyama, A.; Tanaka, K. Novel dynamic voltage support capability of photovoltaic systems for improvement of short-term voltage stability in power systems. *IEEE Trans. Power Syst.* **2017**, *32*, 1796–1804.
7. Li, S.; Wei, Z.; Ma, Y. Fuzzy load-shedding strategy considering photovoltaic output fluctuation characteristics and static voltage stability. *Energies* **2018**, *11*, 779.
8. Rahman, S.; Saha, S.; Islam, S.N.; Arif, M.T.; Mosadeghy, M.; Haque, M.E.; Oo, A.M.T. Analysis of power grid voltage stability with high penetration of solar PV systems. *IEEE Trans. Ind. Appl.* **2021**, *57*, 2245–2257.
9. Li, S.; Lu, Y.; Ge, Y. Static voltage stability zoning analysis based on a sensitivity index reflecting the influence degree of photovoltaic power output on voltage stability. *Energies* **2023**, *16*, 2808.
10. General Administration of Quality Supervision, Inspection and Quarantine of the People's Republic of China; Standardization Administration of the People's Republic of China. *Code on Security and Stability for Power System (GB 38755–2019)*; Standards Press of China: Beijing, China, 2019; p. 13.
11. Khoshkhoo, H.; Yari, S.; Pouryekta, A.; Ramachandaramurthy, V.K.; Guerrero, J.M. A remedial action scheme to prevent mid/long-term voltage instabilities. *IEEE Syst. J.* **2021**, *15*, 923–934.
12. Islam, S.R.; Sutanto, D.; Muttaqi, K.M. Coordinated decentralized emergency voltage and reactive power control to prevent long-term voltage instability in a power system. *IEEE Trans. Power Syst.* **2015**, *30*, 2591–2603.
13. Sun, Q.; Cheng, H.; Song, Y. Bi-Objective reactive power reserve optimization to coordinate long- and short-term voltage stability. *IEEE Access* **2018**, *6*, 13057–13065.
14. Matavalam, A.R.R.; Singhal, A.; Ajjarapu, V. Monitoring long term voltage instability due to distribution & transmission interaction using unbalanced μPMU & PMU measurements. *IEEE Trans. Smart Grid.* **2020**, *11*, 873–883.
15. Matavalam, A.R.R.; Ajjarapu, V. Long term voltage stability Thevenin index using voltage locus method. In Proceedings of the 2014 IEEE PES General Meeting, National Harbor, MD, USA, 27–31 July 2014; pp. 1–5.
16. Wang, C.; Ju, P.; Wu, F.; Lei, S.; Pan, X. Long-term voltage stability-constrained coordinated scheduling for gas and power grids with uncertain wind power. *IEEE Trans. Sustain. Energy* **2022**, *13*, 363–377.
17. Dharmapala, K.D.; Rajapakse, A.; Narendra, K.; Zhang, Y. Machine learning based real-time monitoring of long-term voltage stability using voltage stability indices. *IEEE Access* **2020**, *8*, 222544–222555.

18. Cai, H.; Ma, H.; Hill, D.J. A data-based learning and control method for long-term voltage stability. *IEEE Trans. Power Syst.* **2020**, *35*, 3203–3212.
19. Yang, H.; Qiu, R.C.; Tong, H. Reconstruction residuals based long-term voltage stability assessment using autoencoders. *J. Mod. Power Syst. Clean Energy* **2020**, *8*, 1092–1093.
20. Peng, Z.; Hu, G.; Han, Z. Dynamic analysis of the power system voltage stability affected by OLTC. *Proc. CSEE* **1999**, *19*, 61–65, 78.
21. Sun, Y.; Wang, Z. Modeling of OLTC and its impact on the voltage stability. *Autom. Electr. Power Syst.* **1998**, *22*, 10–13.
22. Londero, R.R.; Affonso, C.d.M.; Vieira, J.P.A. Long-term voltage stability analysis of variable speed wind generators. *IEEE Trans. Power Syst.* **2015**, *30*, 439–447.
23. Munkhchuluun, E.; Meegahapola, L. Impact of the solar photovoltaic (PV) generation on long-term voltage stability of a power network. In Proceedings of the 2017 IEEE Innovative Smart Grid Technologies-Asia (ISGT-Asia), Auckland, New Zealand, 4–7 December 2017; pp. 1–6.
24. Li, S.; Wei, Z.; Sun, G.; Gao, P.; Xiao, J. Voltage stability bifurcation of large-scale grid-connected PV system. *Electr. Power Autom. Equip.* **2016**, *36*, 17–23.
25. Li, S.; Wei, Z.; Ma, Y.; Cheng, J. Prediction and control of Hopf bifurcation in a large-scale PV grid-connected system based on an optimised support vector machine. *J. Eng.* **2017**, *14*, 2666–2671. [CrossRef]
26. Canizares, C.A.; Mithulananthan, N.; Milano, F.; Reeve, J. Linear performance indices to predict oscillatory stability problems in power systems. *IEEE Trans. Power Syst.* **2004**, *19*, 1104–1114. [CrossRef]
27. Cutsem, T.V.; Vournas, C. Bifurcations. In *Voltage Stability of Electric Power Systems*; Springer: New York, NY, USA, 1998; pp. 153–160.
28. Wang, Y.; Chen, H.; Zhou, R. A nonlinear controller design for SVC to improve power system voltage stability. *Int. J. Electr. Power Energy Syst.* **2000**, *22*, 463–470.
29. Jing, Z.; Xu, D.; Chang, Y.; Chen, L. Bifurcations, chaos, and system collapse in a three node power system. *Int. J. Electr. Power Energy Syst.* **2003**, *25*, 443–461.
30. Tamimi, B.; Cañizares, C.; Bhattacharya, K. System stability impact of large-scale and distributed solar photovoltaic generation: The case of Ontario, Canada. *IEEE Trans. Sustain. Energy* **2013**, *4*, 680–688. [CrossRef]
31. Li, S.; Wei, Z.; Sun, G.; Wang, Z. Stability research of transient voltage for multi-machine power systems integrated large-scale PV power plant. *Acta Energiae Solaris Sin.* **2018**, *39*, 3356–3362.
32. Bao, L.; Duan, X.; He, Y. Analysis of the effect of OLTC on the voltage stability through time domain simulation. *J. Huazhong Univ. Sci. Tech.* **2000**, *28*, 61–63.
33. Duan, X.; He, Y.; Chen, D. Dynamic analysis of the relation between on-load tap changer and voltage stability. *Autom. Electr. Power Syst.* **1995**, *19*, 14–19.

energies

MDPI

Review

A Comprehensive Review on the Modelling and Significance of Stability Indices in Power System Instability Problems

Chandu Valuva [1,†], Subramani Chinnamuthu [1,*], Tahir Khurshaid [2,†] and Ki-Chai Kim [2,*]

1 Department of Electrical and Electronics Engineering, SRM Institute of Science and Technology, Kattankulathur 603203, India; chanduv43@gmail.com
2 Department of Electrical Engineering, Yeungnam University, Gyeongsan 38541, Republic of Korea; tahir@ynu.ac.kr
* Correspondence: csmsrm@gmail.com (S.C.); kckim@ynu.ac.kr (K.-C.K.)
† These authors equally contributed to this work.

Abstract: Many technological advancements in the modern era have made actual use of electrical power and the constrained operating of power systems within stability limits. Some expeditious load variations and rising power demands initiate complications in voltage stability and can put stress on performance, leading to voltage instability. Voltage Stability Indices can be used to perform voltage stability assessment. This review evaluates various VSIs based on mathematical derivations, assumptions, critical values, and methodology. VSIs determine the maximum loadability, voltage collapse proximity, stability margin, weak areas, and contingency ranking. Stability indices can also specify the optimal placing and sizing of Distributed Generators. Thus, VSIs play a vital role in power system voltage stability. This review is a comprehensive survey of various indices and analyses their accuracy in determining the instability of power systems. Voltage stability is a crucial concern in operating a reliable power system, and the systematic evaluation of voltage stability is essential in a power system. This review considered and analyzed 34 indices from 138 articles from the literature for their significant performance in various power system stability problems. Of 33 indices, were 22 derived from transmission line parameters, referred to as line indices, and 12 from bus and line parameters, referred to as bus indices.

Keywords: power system stability; voltage stability analysis; stability indices; power losses; optimal placement; weakest bus

Citation: Valuva, C.; Chinnamuthu, S.; Khurshaid, T.; Kim, K.-C. A Comprehensive Review on the Modelling and Significance of Stability Indices in Power System Instability Problems. *Energies* **2023**, *16*, 6718. https://doi.org/10.3390/en16186718

Academic Editor: Ying-Yi Hong

Received: 28 July 2023
Revised: 6 September 2023
Accepted: 14 September 2023
Published: 20 September 2023

1. Introduction

Voltage instability is a crucial phenomenon that effects power systems because it is the main reason for blackouts and voltage collapse. Some unexpected consequences like an unexpected increase in load, line outage, and generator tripping may provoke an excessive load demand in the power system, leading to voltage instability. Consequently, a power system may be disrupted without appropriate action, resulting in cascading failures and large-scale blackouts. The notable blackouts around the world have primarily been caused by voltage instability. From [1], From 1965 to 2005, it can be seen that 22 significant backouts occurred; out of these 22 blackouts, 11 had voltage instability as a leading cause. Voltage instability is also known as voltage collapse; moreover, when instability is developing, the continuity of such events affects the system, resulting in low voltage and even more power outages [2].

According to an IEEE Power System Engineering Committee, "Voltage stability is the ability of a system to maintain voltage so that, when increasing load admittance, load power increases, and so that both power & voltage are manageable". The consolidation definition of the IEEE and CIGRE is: "Voltage stability refers to the ability of a power system to maintain steady voltages at all buses in the system after being subjected to a disturbance from a given initial operating condition".

Power system voltage instability may be due to the inadequacy of the source to deliver sufficient reactive power or to faulty power transmission lines not transmitting the desired reactive power. Voltage instability may affect the stability of a wide area or local area, resulting in the voltage collapse of a whole power system—mainly reliable active power delivered from the generators and reactive compensators like shunt capacitors. Voltage instability in a network is most probably the cause of unexpected load demand, transformer tripping, being over the limit of the generators' reactive power, or defects in the on-load tap changing transformers. Consequently, the regular tracking and forecasting of instability are essential in the power system. In this respect, definite methods for suppressing voltage instability are:

1. Strengthen the weak bus while planning the power system, like accurately selecting distributed generators and balancing voltage.
2. Temporarily defending the supply by load shedding or improving the power factor by shunt capacitor switching.
3. Enhancing the Voltage Stability Margin (VSM) with FACTS devices.
4. Blocking on-load tap changers (OLTCs) to operate the transformer.
5. Build the generation station near to the load centre and reduce the transmission line length.
6. For executive controllers, make use of system voltage fluctuation controllers, Shunt Compensation, step-up transformer controllers, and Automatic Voltage Regulators (AVRs).
7. For Real-time Performance: Scheduling the generation according to load demand, evaluating voltage stability, and protecting the load shedding.
8. For safeguarding systems: Instant Load Tap Changer (LTC) control, Contingency of load demand, and High Voltage Direct Current (HVDC).

Indeed, over the past few decades, the reliability and stability of power systems have gone through considerable improvements. Stability analysis has evolved from simple static and transient stability analyses to more sophisticated and comprehensive methods.

Voltage Stability Indices (VSIs) generally measure voltage stability. The stability indices' performance is essential in analyzing and illustrating the consequences of power system operation. Furthermore, the indices assist in predicting eventual improvement and determining long-term advancement. Especially in the last three decades, numerous techniques and methodologies have been developed for easy stability analysis. The performances of various indices for determining voltage stability have been explained in this review. For assessing a system's stability, several indices consider the system impedance. However, few indices are unconstrained by impedance and performed using the current and voltage of the system. In real-time operation, it is impossible to determine the accurate impedance of a system because of troposphere conditions and inadequate knowledge of the system. Therefore, the accuracy of the impedance-operated indices is repeatedly accompanied by errors.

A taxonomy of comprehensive characteristics, variations in accuracy, and some parameters being inaccurate is presented below, as well as the terminology concerning VSIs, including mathematical derivation, instability conditions, and assumptions of specific parameters. This taxonomy can serve to make researchers aware of the advancements in the VSIs, and the respective operations including voltage stability enhancement, strengthening the weak bus, locating the reactive compensation devices, reducing the cost-effectiveness and location of FACTS devices, and counteracting the prevention of voltage collapse.

Determining the VSI values for a system is carried out in three modes of operation. One is the online mode: In this mode of operation, the engineer observes the voltage stability and operates accordingly to maintain the system's stability. The statistics used for this in this paper were composed of mathematical models and power system measurements. The second mode of the stability indices is the offline mode: In this mode of operation, the designer gathers the simulated or historical data. The preserved data identify the voltage instability intensity and voltage collapse and do not contain real-time data on voltage instability. The real-time method is the third mode of stability indices, and is obtained using a Phasor Measurement Unit, Power Flow Analysis, and Artificial Intelligence.

This article aims to thoroughly review and scrutinize numerous Voltage Stability Indices. This review indicates the limitations, strengths, underlying principles, and applicability of these indices by sharply examining them. This article provides clarity to researchers, engineers, and power system operators in choosing suitable stability indices depending on the particular characteristics of a system and functional needs. Researchers will gain synthesized knowledge of Voltage Stability Indices, as this article presents the diverse range of stability indices with mathematical formulations, highlighting their peculiar characteristics and underlying principles. This review supports decision-making for system operators with the knowledge needed to estimate instabilities and ensure voltage stability, sustainable energy supply, and grid reliability.

2. Voltage Stability Indices

Voltage Stability Indices are values of an arithmetic sequence adopted for evaluating voltage stability. The indices specify the accessible and equitable regulation of the system's stability so that voltage levels are within sustainable limits. Indices predominantly assist power engineers in determining voltage stability, predicting voltage collapse, and evaluating the significance of system characteristics. This review elaborates on the numerous methodologies for assessing voltage stability. Furthermore, categorizing the different VSIs in [3–5] as line VSIs, Bus VSIs, and overall VSIs. Jacobian matrix-based indices and PMU-based indices fall under overall VSIs.

The overall VSIs' precision is superior compared to the other stability indices. The overall VSIs are challenging to determine and take more computation time. Line VSIs obtain easy and simple indices that are reasonable in determining all optimization issues through the proficiency of these indices and identifying the critical line and weak bus. These VSIs are derived from a mathematical equation by considering the standard two-bus concept. With these mathematical derivations, the system examines the stability of various stability indices. Similar aspects of voltage collapse points determine the VSIs' formation. Figure 1 represents an illustration of the interconnected network with a single line in a standard two-bus model.

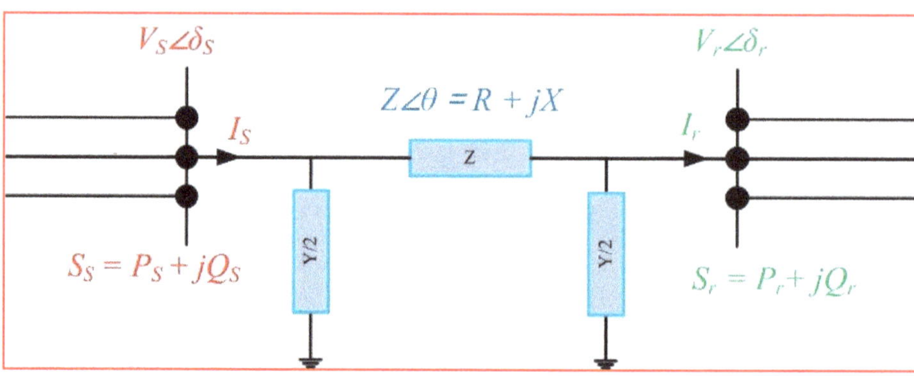

Figure 1. Standard two-bus model.

Jacobian matrix-based VSIs specify the scope of system stability, determine the voltage collapse point, and evaluate the VSM. However, they consume more time for computation, and in general, mathematical topology variations compel the transformation of the Jacobian matrix, causing the matrix to need to be again; therefore, they are inconvenient in real-time applications. Additionally, Jacobian matrix indices tend to increase the operational time of DG placing and sizing issues. However, the VSIs established by system variables claim lesser computational time and are acceptable for real-time operations. The detriment of these stability indices is that they are less accurate in evaluating the VSM. Consequently, they are not as accurate in determining the weak line and critical bus. In most applications, like placement of the DG and sizing issues, VSIs prevail for identifying the weak bus and critical line or stimulating the deterrents for voltage instability.

$$S_r = P_r + jQ_r = V_r I^* \tag{1}$$

$$I = \left(\frac{S_r}{V_r}\right)^* = \frac{(P_r - jQ_r)}{V_r\angle - \delta_r} \tag{2}$$

V_s, V_r: magnitude of bus voltage at sending and receiving bus; S_s, P_s, Q_s: apparent, real, and reactive power at sending bus; S_r, P_r, Q_r: apparent, real, and reactive power at receiving bus; δ_s, δ_r: voltage angle at sending and receiving bus; R, X, Z, θ = line resistance, line reactance, line impedance, and line impedance angle.

Considering the Kirchhoff Voltage Law (KVL), the current equation for two buses is,

$$I = \frac{V_s\angle\delta_s - V_r\angle\delta_r}{R + jX} \tag{3}$$

Substituting Equation (3) in Equation (2), we obtain:

$$\frac{V_s\angle\delta_s - V_r\angle\delta_r}{R + jX} = \frac{(P_r - jQ_r)}{V_r\angle - \delta_r} \tag{4}$$

$$V_s V_r\angle(\delta_s - \delta_r) - V_r^2 = P_r R - jQ_r R + jP_r X + Q_r X \tag{5}$$

Let $\delta_s - \delta_r = \delta$ and simplify Equation (5) as:

$$V_s V_r\angle\delta - V_r^2 = P_r R - jQ_r R + jP_r X + Q_r X \tag{6}$$

Transforming $V_s V_r \angle\delta$ into a rectangular form, we obtain:

$$V_s V_r \cos\delta + jV_s V_r \sin\delta - V_r^2 = P_r R - jQ_r R + jP_r X + Q_r X \tag{7}$$

Segmenting Equation (7) into real and imaginary parts, we obtain:

$$V_s V_r \cos\delta - V_r^2 = P_r R + Q_r X \tag{8}$$

$$V_s V_r \sin\delta = -Q_r R + P_r X \tag{9}$$

Readjusting Equations (8) and (9), we obtain:

$$P_r = \frac{-Q_r X + V_s V_r \cos\delta - V_r^2}{R} \tag{10}$$

$$Q_r = \frac{P_r X - V_s V_r \sin\delta}{R} \tag{11}$$

Substituting Equation (11) in Equation (10) and vice versa, we obtain:

$$P_r = \frac{-\left(\frac{P_r X - V_s V_r \sin\delta}{R}\right) X + V_s V_r \cos\delta - V_r^2}{R} \tag{12}$$

$$Q_r = \frac{\left(\frac{-Q_r X + V_s V_r \cos\delta - V_r^2}{R}\right) X - V_s V_r \sin\delta}{R} \tag{13}$$

Rearrange Equations (12) and (13) allows us to obtain:

$$V_r^2 - V_sV_r\left(\cos\delta + \frac{X}{R}\sin\delta\right) + P_r\left(R + \frac{X^2}{R}\right) = 0 \tag{14}$$

$$V_r^2 + V_sV_r\left(\frac{R}{X}\sin\delta - \cos\delta\right) + Q_r\left(\frac{R^2}{X} + X\right) = 0 \tag{15}$$

For a few VSIs, the line resistance (R) and reactance (X) increase with the impedance (Z) and impedance angle (θ); therefore, reducing the current equation, Equation (3) is redefined as:

$$I = \frac{V_s\angle\delta_s - V_r\angle\delta_r}{Z\angle\theta} \tag{16}$$

Substituting Equation (16) into Equation (1), the receiving end's apparent power concerning the real and reactive power is:

$$P_r + jQ_r = V_r\left(\frac{V_s\angle\delta_s - V_r\angle\delta_r}{Z\angle\theta}\right)^* \tag{17}$$

Assuming $\delta_s - \delta_r = \delta$ then Equation (17) is simplified as:

$$P_r + jQ_r = \frac{V_sV_r}{Z}\angle(\theta - \delta) - \frac{V_r^2}{Z}\angle\theta \tag{18}$$

Equating the real and imaginary parts of Equation (18), we obtain:

$$P_r = \frac{V_sV_r}{Z}\cos(\theta - \delta) - \frac{V_r^2}{Z}\cos\theta \tag{19}$$

$$Q_r = \frac{V_sV_r}{Z}\sin(\theta - \delta) - \frac{V_r^2}{Z}\sin\theta \tag{20}$$

Simplifying Equations (19) and (20), we obtain:

$$V_r^2 cos\theta - V_sV_r\cos(\theta - \delta) + P_rZ = 0 \tag{21}$$

$$V_r^2 sin\theta - V_sV_r\sin(\theta - \delta) + Q_rZ = 0 \tag{22}$$

Most of the stability indices were derived mathematically; the quadratic equation's voltage discriminant is greater than or equal to zero. The characteristics vary for the index, so different assumptions are considered for deriving the stability indices.

2.1. Line Voltage Stability Indices

2.1.1. Voltage Stability Load Index (VLSI)

The Voltage Stability Load Index (VLSI) derives from the mathematical derivation [6] of a standard two-bus system and Thevenin equivalent circuit. Figure 1 presents the voltage equation considered in this index, obtained from the network. Moreover, we determine the power flow by performing a power flow analysis directed toward speedy converging. Considering the square of the sending-end current, receiving-end active and reactive power with losses, we obtain:

$$|I_s|^2 = \frac{P_s^2 + Q_s^2}{V_s^2} \tag{23}$$

$$P_r = P_s - P_{loss} \tag{24}$$

$$Q_r = Q_s - Q_{loss} \tag{25}$$

$$P_{loss} = \left(\frac{P_r^2 + Q_r^2}{V_r^2}\right) R_s \tag{26}$$

$$Q_{loss} = \left(\frac{P_r^2 + Q_r^2}{V_r^2}\right) X_s \tag{27}$$

We then substitute Equations (24)–(27) in Equation (23) to simplify it, resulting in:

$$|I|^2 = \frac{\left[P_r + \left(\frac{P_r^2+Q_r^2}{V_r^2}\right) R_s\right]^2 + \left[Q_r + \left(\frac{P_r^2+Q_r^2}{V_r^2}\right) X_s\right]^2}{V_s^2} \tag{28}$$

The analyzed equation is further reduced by equating it to zero and obtaining the sending end voltage equation:

$$V_s^2 = V_r^2 + 2(P_r R_s + Q_r X_s) + \left(\frac{P_r^2 + Q_r^2}{V_r^2}\right)\left(R_s^2 + X_s^2\right) \tag{29}$$

$$V_r^4 + V_s^2\left[2(P_r R_s + Q_r X_s) - V_s^2\right] + \left(P_r^2 + Q_r^2\right)\left(R_s^2 + X_s^2\right) = 0 \tag{30}$$

If the obtained receiving end voltage quadratic equation contains real roots, then the discriminant is greater than or equal to zero; hence:

$$8P_r Q_r R_s X_s - 4V_r^2(P_r R_s + Q_r X_s) + V_s^4 - 4\left(P_r^2 X_s^2 + Q_r^2 R_s^2\right) \geq 0 \tag{31}$$

$$\frac{4\left[V_s^2(P_r R_s + Q_r X_s) + (P_r X_s - Q_r R_s)^2\right]}{V_s^4} \leq 1 \tag{32}$$

Accordingly, L represents the voltage stability load index,

$$L = \frac{4\left[V_s^2(P_r R_s + Q_r X_s) + (P_r X_s - Q_r R_s)^2\right]}{V_s^4} \tag{33}$$

since $V_s V_r cos(\delta_s - \delta_r) - V_r^2 = P_r R_s + Q_r X_s$ & $V_s V_r sin(\delta_s - \delta_r) = P_r X_s - Q_r R_s$

$$\text{Therefore } L = \frac{4\left[V_s V_r \cos(\delta_s - \delta_r) - V_r^2 \cos(\delta_s - \delta_r)^2\right]}{V_s^2} \tag{34}$$

$$L = \frac{4\left[V_s V_r \cos(\delta) - V_r^2 \cos(\delta)^2\right]}{V_s^2} \tag{35}$$

Hence, the value L should be less than 1.0 for the system's stability. Whenever the index value is more significant than 1.0, corresponding to Equation (32), the voltage emerges as imaginary and collapses. To defend the load instability, the load index value must be $L_i \leq 0$, and if L_i is nearer to 1.0, then bus 'i' approaches the stability limit.

2.1.2. Line Stability Index (L_{mn})

The Line Stability Index is determined in [7] and derived from the voltage quadratic equations discriminant. Considering this index and Equation (20), we can derive V_r:

$$V_r = \frac{V_s \sin(\theta - \delta) \pm \left\{[V_s \sin(\theta - \delta)]^2 - 4ZQ_r sin\theta\right\}^{0.5}}{2sin\theta} \tag{36}$$

Assuming $Zsin\theta = x$,

$$V_r = \frac{V_s \sin(\theta - \delta) \pm \left\{ [V_s \sin(\theta - \delta)]^2 - 4xQ_r \right\}^{0.5}}{2sin\theta} \tag{37}$$

The above voltage equation should have real roots to obtain absolute values of V_r in terms of Q_r. Therefore, the following condition is to manipulate the stability of the system.

$$\begin{gathered} \left\{ [V_s sin(\theta - \delta)]^2 - 4xQ_r \right\} \geq 0 \\ \frac{4xQ_r}{[V_s sin(\theta - \delta]^2} = L_{mn} \leq 1 \end{gathered} \tag{38}$$

L_{mn} represents the stability index of the line. Voltage collapse is predicted rigorously depending on the stability index's value. If the obtained value of L_{mn} is lesser than the unity, then the system is in a stable condition; otherwise, if the value is more significant than the unity, then the system exhausts the stability; moreover, the voltage collapses.

2.1.3. Line Stability Factor (LQP)

The Line Stability Factor (LQP) developed in [8] makes use of a similar theory of the discriminant of the voltage quadratic equation, primarily from receiving end real power:

$$P_r = P_s - R\frac{(P_s^2 - Q_s^2)}{V_s^2} \tag{39}$$

Reorganizing the power equation as $R\frac{P_s^2}{V_s^2} - P_s + P_r + \left(P_r + R\frac{Q_s^2}{V_s^2}\right) = 0$ allows us to obtain the roots of P_s:

$$1 - 4\left(\frac{R}{V_s^2}\right)\left(P_r + R\frac{Q_s^2}{V_s^2}\right) \geq 0 \tag{40}$$

From Equation (40), it is clear that the system loses its stability when it violates this condition. Further, deriving the reactive power is similar to the derivation of active power:

$$1 - 4\left(\frac{X}{V_s^2}\right)\left(Q_r + X\frac{P_s^2}{V_s^2}\right) \geq 0 \tag{41}$$

$$\text{LQP} = 4\left(\frac{X}{V_s^2}\right)\left(Q_r + X\frac{P_s^2}{V_s^2}\right) \tag{42}$$

Since LQP < 1, the system is stable. Here, for deriving the index, it is assumed that the lines are lossless ($R/X << 1$) and neglect the shunt admittance.

2.1.4. Voltage Collapse Proximity Index (VCPI)

The VCPI [9] evaluates the line voltage stability depending on the theory of maximum power. The analysis assumes a frequent occurrence and varying of the absolute impedance value by keeping ϕ constant. With this inference, the performance may not be accurate, but it is easy to solve. During the operation, the power factor remained constant.

Along with the load demand increase, the current increases with a decrease in Z_r, resulting in a receiving-end voltage drop:

$$I = \frac{V_s}{\sqrt{\left[(Z_s cos\theta + Z_r cos\varnothing)^2 + (Z_s sin\theta + Z_r sin\varnothing)^2\right]}} \tag{43}$$

$$V_r = Z_r I = \frac{V_s}{\sqrt{\left[1 + (Z_r/Z_s)^2 + 2(Z_r/Z_s)cos(\theta - \varnothing)\right]}} \tag{44}$$

Hence, the receiving-end power is $P_r = V_r I cos\phi$ and $Q_r = V_r I sin\phi$
Simplifying Equation (44) with that for power, we obtain:

$$P_r = \frac{(V_s)^2/Z_s}{1+(Z_r/Z_s)^2+2(Z_r/Z_s)cos(\theta-\varnothing)}\frac{Z_r}{Z_s}cos\varnothing \quad (45)$$

$$Q_r = \frac{(V_s)^2/Z_s}{1+(Z_r/Z_s)^2+2(Z_r/Z_s)cos(\theta-\varnothing)}\frac{Z_r}{Z_s}sin\varnothing \quad (46)$$

Correspondingly, the power loss is $P_l = I^2 Z_s cos\theta$ and $Q_l = I^2 Z_s sin\theta$.
Substituting the current equation, we then obtain:

$$P_l = \frac{(V_s)^2/Z_s}{1+(Z_r/Z_s)^2+2(Z_r/Z_s)cos(\theta-\varnothing)}cos\theta \quad (47)$$

$$Q_l = \frac{(V_s)^2/Z_s}{1+(Z_r/Z_s)^2+2(Z_r/Z_s)cos(\theta-\varnothing)}sin\theta \quad (48)$$

Considering the boundary condition $\partial P_r/\partial Z_r = 0$ allows us to calculate the maximum active power delivered at the receiving end. Hence, this condition results in the impedance $Z_r/Z_s = 1$. The maximum power transfer obtained by substituting this ratio in Equation (45) is:

$$P_{r(max)} = \frac{V_s^2}{Z_s}\frac{cos\varnothing}{4cos^2\frac{(\theta-\varnothing)}{2}} \quad (49)$$

Similarly, adopting this approach, the following power equation can be derived:

Maximum transferable reactive power $Q_{r(max)} = \frac{V_s^2}{Z_s}\frac{sin\varnothing}{4cos^2\frac{(\theta-\varnothing)}{2}}$ (50)

In line with the maximum active power loss $P_{l(max)} = \frac{V_s^2}{Z_s}\frac{cos\theta}{4cos^2\frac{(\theta-\varnothing)}{2}}$ (51)

In line with the maximum reactive loss $Q_{l(max)} = \frac{V_s^2}{Z_s}\frac{sin\theta}{4cos^2\frac{(\theta-\varnothing)}{2}}$ (52)

Considering these maximum limits, Voltage Collapse Proximity Indicators can be designed as:

$$VCPI\ (1) = \frac{P_r}{P_{r(max)}} = \frac{Real\ power\ tranferred\ to\ the\ receiving\ end}{Maximum\ real\ power\ that\ can\ be\ transferred} \quad (53)$$

$$VCPI\ (2) = \frac{Q_r}{Q_{r(max)}} = \frac{Reactive\ power\ tranferred\ to\ the\ receiving\ end}{Maximum\ reactive\ power\ that\ can\ be\ transferred} \quad (54)$$

$$VCPI\ (3) = \frac{P_l}{P_{l(max)}} = \frac{Real\ powerin\ the\ line}{Maximum\ possible\ real\ power\ that\ loss} \quad (55)$$

$$VCPI\ (4) = \frac{Q_r}{Q_{r(max)}} = \frac{Reactive\ power\ loss\ in\ the\ line}{Maximum\ possible\ reactive\ power\ loss} \quad (56)$$

The system may collapse if the values of the VCPIs are more significant than the unity. Therefore, *VCPI* (1) = *VCPI* (2) and *VCPI* (3) = *VCPI* (4), since only active or reactive terms were considered instead of all proximity indices.

2.1.5. Voltage Stability Index (L_p)

The line stability index proposed in [10] is a similar mathematical formulation to earlier line VSIs. Basically, from the active and reactive power equations, i.e., from Equations (19) and (20), δ can be derived as:

$$\delta = \cos^{-1}\frac{V_s^2 - (P_sR + Q_sX)}{V_sV_r} \tag{57}$$

Reorganizing Equation (19) and deriving the voltage equation V_r, we obtain:

$$V_r = \frac{V_s\cos(\theta-\delta) \pm \left\{[V_s\cos(\theta-\delta)]^2 - 4RP_r\right\}^{0.5}}{2cos\theta} \tag{58}$$

Hence, the V_r equation should possess real roots for obtaining the absolute values of V_r expressed in terms of P_r. Accordingly, it is then simplified to be enough to withstand the following conditions:

$$\left\{[V_s\cos(\theta-\delta)]^2 - 4RP_r\right\} \geq 0 \tag{59}$$

$$\frac{4RP_r}{[V_s\cos(\theta-\delta)]^2} = L_p \leq 1 \tag{60}$$

L_P is a line stability index that determines the condition of the transmission line and indicates instability limits. If the calculated value of L_p is more significant to the unity, the system approaches instability. The voltage collapses when the system exceeds the critical limits; moreover, the V_r becomes imaginary.

2.1.6. Fast Voltage Stability Index (FVSI)

The Fast Voltage Stability Index is derived in [11] primarily by considering the current through the line and then calculating the absolute roots of a receiving-end voltage (Equation (15)):

$$V_r = \frac{\left(\frac{R}{X}sin\delta + cos\delta\right)V_s \pm \sqrt{\left[\left(\frac{R}{X}sin\delta + cos\delta\right)V_s\right]^2 - 4\left(X + \frac{R^2}{X}\right)Q_r}}{2} \tag{61}$$

For determining the absolute roots of V_r, the V_r discriminant is greater than or equal to zero.

$$\frac{4Z^2Q_rX}{V_s^2(Rsin\delta + Xcos\delta)^2} \leq 1 \tag{62}$$

The angular difference δ is minute; thus, $\delta \approx 0$, $Rsin\delta \approx 0$, and $Xcos\delta \approx X$. As a result, Equation (62) can be simplified as:

$$FVSI = \frac{4Z^2Q_r}{V_s^2X} \tag{63}$$

The line is in-transit to the instability limit if the $FVSI$ value is adjacent to 1. With further increases in a stability index value, that line may encounter an unexpected voltage drop accompanying system collapse.

2.1.7. Voltage Stability- Load Bus Index (VSLBI)

VSLBI evaluates the voltage stability by adopting the PMU [12]. Considering the maximum power condition, this stability index is subject to a voltage-drop ΔVr over the transmission impedance Zr that is equivalent to load bus voltage Vr:

$$\Delta V_r = V_r \tag{64}$$

Accordingly, to approach the voltage collapse exposure during the constant power loads, the Voltage Stability Load Bus Index (VLSBI) is represented as:

$$VSLBI = \frac{V_r}{\Delta V} \tag{65}$$

If the obtained value of the VSLBI is more significant than the unity, then the system is considered stable; if the VSLBI is less than 1, then the system is unstable, and the system may collapse.

2.1.8. Voltage Stability Margin Index (VSMI)

The VSMI, proposed in [13], depends on the correlation between the power transfer from the line and the angular difference between the sending- and receiving-end buses. To derive the fundamental equations of real and reactive power, the receiving end voltage is:

$$V_r = V_s \frac{cos(\varnothing_r + \delta)}{cos(\varnothing_r)} \tag{66}$$

$$P_r = \frac{1}{2}\frac{V_s^2}{X}\left(\frac{sin(\varnothing_r + 2\delta)}{cos(\varnothing_r)} - tan(\varnothing_r)\right) \tag{67}$$

Since $tan(\varnothing_r) = \frac{Q_r}{P_r}$, considering Equations (66) and (77), the power P_r maximum value can be calculated for whatever the value of Q_r/P_r is. The respective angular differences of the sending- and receiving-end buses (δ), and the receiving-end voltage (V_r), are expressed as:

$$P_{rmax} = \frac{1}{2}\frac{V_s^2}{X}\left(\frac{1}{cos(\varnothing_r)} - tan(\varnothing_r)\right) \tag{68}$$

$$V_{rmax} = \frac{cos\left(\frac{\pi}{2} + \varnothing_r\right)}{cos(\varnothing_r)} V_s \tag{69}$$

$\delta_{rmax} = \frac{\frac{\pi}{2} - \varnothing_r}{2}$ for any ratio of Q_r/P_r

With the correlation between the voltage stability and angular difference, the voltage stability margin that is determined depends on what degree of δ is adjacent to δ_{rmax}. Therefore, whatever the operating constraints, the receiving end index can be represented as:

$$VSMI = \frac{\delta_{rmax} - \delta}{\delta_{rmax}} \tag{70}$$

The calculated value of the *VSMI* should be higher than zero to sustain the system's stability and protection.

2.1.9. Voltage Collapse Proximity Index (*VCPI_1*)

The significant motive of *VCPI_1* is that, during the voltage collapse, the voltage drop at the Thevenin impedance equals the load voltage. The proximity index derived in [14] is denoted as *VCPI_1* to prevent confusion from the *VCPI* notation. An easy arithmetic prediction indicates that, during the critical situation, the generator phasor voltage ($\overline{V_s}$) is double the load phasor voltage, $\overline{V_r}$. Hence, for the estimation of the possibility of voltage instability, *VCPI_1* is represented as:

$$VCPI\_1 = V_r cos\delta - 0.5V_s \tag{71}$$

If the *VCPI* is greater than or equal to zero, then the system is stable, otherwise it is unstable. In this index, the lines are complicated. However, considering that the sending-end bus connects to the ideal voltage source, the receiving end with the equivalent Thevenin impedance is neglected.

2.1.10. Critical Voltage (Vcr)

The Critical Voltage is an intelligible index that is derived from a single load, an infinite bus system utilizing the Eigen-value theorem, and load flow equations [15]. In this paper, it is used for adapting the power factor $cos\phi$ to be constant. From the active and reactive power flow equations, the bus voltage angle δ is eliminated, and the resulting equation is:

$$V_r^4 + \left[2Z(V_r cos\theta + Q_r sin\theta) - V_s^2\right]V_r^2 + \left(P_r^2 + Q_r^2\right)Z^2 = 0 \quad (72)$$

Corresponding to favorable circumstances, Equation (72) contains a specific solution; moreover, the correlation between the $\overline{V_s}$, $\overline{V_r}$, $\overline{I}$ vectors is developed as:

$$V_{cr} = V^+ = V^- = \frac{V_s}{2cos\theta} \quad (73)$$

The load bus critical voltage is denoted by V_{cr} (P-V Curve nose point). V^+ and V^- are the P-V curves' upper and lower parts, respectively. Nearer to this, V^+ and V^- are superposed values of V_{cr}. The indices concern the maximum power limit as a stability limiting point. Evaluating the simple system by the constant MVA load and cosϕ power factor, Equation (72) is:

$$S^2 + 2S\frac{V^2}{Z}cos(\theta - \varnothing) + \frac{V^4 - V_s^2V_r^2}{Z^2} = 0 \quad (74)$$

$$\text{Apparent Power } S = \frac{P}{cos\varnothing} \quad (75)$$

To obtain the maximum S, dS/dV is equated to zero. Accordingly:

$$V_{cr} = \frac{V_s}{\sqrt{2(1 + cos(\theta - \varnothing))}} \quad (76)$$

After the phasor relationship, further reducing Equation (76) allows us to obtain:

$$V_{cr} = \frac{E}{2cos\theta} \quad (77)$$

2.1.11. Power Transfer Stability Index (PTSI)

The Power Stability Index proposed in [16] is derived with the help of a standard two-bus Thevenin-equivalent system; here, the slack bus and the load bus were connected on a single branch, as represented in Figure 2. The current delivery to load and load power is:

$$\overline{I} = \frac{\overline{V}_{Thev}}{\overline{Z}_{Thev} + \overline{Z}_L} \quad (78)$$

$$\overline{S}_L = \overline{Z}_L\overline{I}\overline{I^*} = \overline{Z}_L|\overline{I}|^2 \quad (79)$$

Substituting Equation (78) in (79), we obtain:

$$\overline{S}_L = \overline{Z}_L\left|\frac{\overline{V}_{Thev}}{\overline{Z}_{Thev} + \overline{Z}_L}\right|^2 \quad (80)$$

Assuming that $\overline{Z}_L = Z_L\angle\varphi$ and $\overline{Z}_{Thev} = Z_{Thev}\angle\theta$, and alternating them into Equation (80), we obtain:

$$\overline{S}_L = Z_L\angle\varphi\left|\frac{V_{Thev}}{Z_{Thev}\angle\theta + Z_L\angle\varphi}\right|^2 \quad (81)$$

$$S_L = \frac{V_{Thev}^2 Z_L}{|Z_{Thev}\angle\theta + Z_L\angle\varphi|^2} \quad (82)$$

$$S_L = \frac{V_{Thev}^2 Z_L}{Z_{Thev}^2 + Z_L^2 + 2Z_{Thev}Z_L\cos(\theta-\varphi)} \quad (83)$$

To obtain the maximum value of the apparent power of the load, the S_L is differentiated and equated to zero, $\partial S_L/\partial Z_L = 0$.

$$S_{Lmax} = \frac{V_{Thev}^2}{2Z_{Thev}(1+2\cos(\theta-\varphi))} \quad (84)$$

The power margin at the load bus distance when it is approaching voltage collapse is $S_{Lmax} - S_L$. If $Z_L = Z_{Thev}$,, the power margin is equal to zero; moreover, it indicates no power transfer and may cause voltage collapse. Conversely, a voltage may collapse if the ratio of power is:

$$\frac{S_L}{S_{Lmax}} = 1 \quad (85)$$

Substituting Equations (83) and (84) in Equation (85), the power transfer stability index is:

$$PTSI = \frac{2S_L Z\ (1+cos(\theta-\varphi))}{V_s^2} \quad (86)$$

The value of the PTSI is determined with Equation (86), considering the load's power, impedance, Thevenin Voltage, Thevenin Impedances, and the Angular phase of the load. The obtained PTSI value lies between 0 and 1. Voltage collapse occurs when the index value reaches 1.

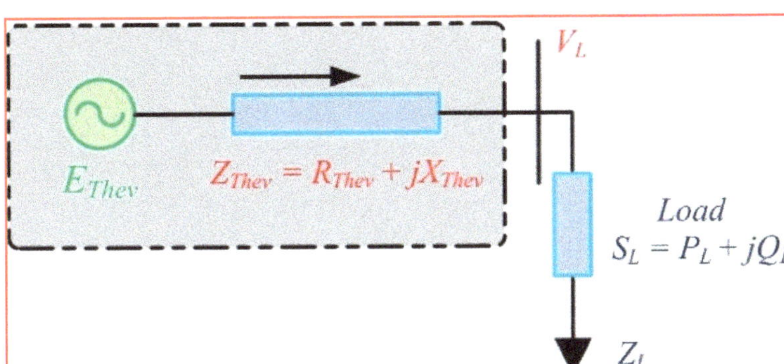

Figure 2. Simple two-bus Thevenin Equivalent Circuit.

2.1.12. Voltage Stability Index (VSI_1)

The stability index predicts the voltage stability at the steady state operation; this index is proposed in [17]; however, for easy identification, the notation of the stability index is represented as VSI_1. Initially, this index calculates the maximum active, reactive, and apparent power transfer. To obtain the receiving-end voltage (V_r), considering the Equation (19) for active power (P_r) and Equation (20) for reactive power (Q_r), we obtain:

$$V_r = \sqrt{\frac{V_s^2}{2} - (QX+PR) \pm \sqrt{A}} \quad (87)$$

Here, $= \frac{V_s^2}{4} - (QX+PR)V_s^2 - (PX-QR)^2$, A ≥ 0.

Transmission lines containing a higher *X/R* ratio and ignoring the resistance value, resulting in P_{max}, Q_{max}, and S_{max}, can be derived as:

$$P_{max} = \frac{QR}{X} - \frac{V_S^2 R}{2X^2} + \frac{|Z_L| V_s \sqrt{V_S^2 - 4QX}}{2X^2} \tag{88}$$

$$P_{max} = \sqrt{\frac{V_S^4}{4X^2} - Q\frac{V_S^2}{X}} \tag{89}$$

$$Q_{max} = \frac{PX}{R} - \frac{V_S^2 X}{2R^2} + \frac{|Z_L| V_s \sqrt{V_S^2 - 4PR}}{2R^2} \tag{90}$$

$$Q_{max} = \frac{V_S^2}{4X} - \frac{P^2 X}{V_S^2} \tag{91}$$

$$S_{max} = \frac{V_s^2[|Z_L| - (\sin(\theta)X + \cos(\theta)R)]}{2(\cos(\theta)X - sin(\theta)R)^2} \tag{92}$$

$$S_{max} = \frac{(1 - \sin(\theta))\, V_s^2}{2cos(\theta)^2 X} \tag{93}$$

The derived maximum powers identified that the maximum power transfer increases with an increase in the voltage magnitude (V_r) or a decrease in the impedance (Z_L). Equivalently, P_{max} decreases with an increase in Q, and Q_{max} decreases with an increase in P.

$$P_{margin} = P_{max} - P, Q_{margin} = Q_{max} - Q, S_{margin} = S_{max} - S \tag{94}$$

From the maximum powers, the corresponding load margins can be obtained with Equation (94). Hence, the derived VSI from the obtained load margin is represented in Equation (95). If the obtained VSI is small, the load bus is adjacent to the stability margin and the lesser load margin that is available. The VSI equals zero if the load bus reaches its stability margin point.

$$VSI = min\left(\frac{P_{margin}}{P_{max}}, \frac{Q_{margin}}{Q_{max}}, \frac{S_{margin}}{S_{max}}\right) \tag{95}$$

2.1.13. Novel Line Stability Index (NLSI)

The NLSI is determined with a similar power transmission approach [18], considering the quadratic Equations (8) and (9):

$$V_r = \frac{V_s cos\delta \pm \sqrt{V_s^2 cos^2\delta - 4(P_r R + Q_r X)}}{2} \tag{96}$$

The discriminant should be greater than or equal to 0 to obtain the absolute value of V_r.

$$\frac{PR + QX}{0.25V_s^2 cos^2\delta} \leq 1 \tag{97}$$

As the angular difference (δ) between receiving and sending ends is usually very small, $cos\delta \approx 1$. Subsequently, the Novel Line Stability Index can be represented as:

$$NLSI = \frac{R_{sr} P_r + X_{sr} Q_r}{0.25V_s^2} \tag{98}$$

If the value of the NLSI of any line is adjacent to 1, that line is moving closer to the stability limit; consequently, the system may collapse. Hence, the NLSI allows us to find the stability limit, i.e., the index value is lesser than the unity.

2.1.14. Stability Index (SI)

For load flow analysis, the quadratic equation is preferable for calculating the sending-end voltage of a line and can be simplified in general form as:

$$V_r^4 + 2V_r^2(PR + QX) - V_s^2V_r^2 + \left(P^2 + Q^2\right)|Z|^2 = 0 \tag{99}$$

Further, the receiving-end line active and reactive power is derived as:

$$P = \left[-cos(\theta)V_r^2 \pm \sqrt{cos^2(\theta)V_r^4 - V_r^4 - \lceil Z \rceil^2 Q^2 - 2V_r^2QX + V_s^2V_r^2}\right]/|Z| \tag{100}$$

$$Q = \left[-sin(\theta)V_r^2 \pm \sqrt{sin^2(\theta)V_r^4 - V_r^4 - \lceil Z \rceil^2 P^2 - 2V_r^2PR + V_s^2V_r^2}\right]/|Z| \tag{101}$$

Therefore, to obtain the absolute value of the active and reactive power, the corresponding equation should satisfy the following:

$$cos^2(\theta)V_r^4 - V_r^4 - \lceil Z \rceil^2 Q^2 - 2V_r^2QX + V_s^2V_r^2 \geq 0 \tag{102}$$

$$sin^2(\theta)V_r^4 - V_r^4 - \lceil Z \rceil^2 P^2 - 2V_r^2PR + V_s^2V_r^2 \geq 0 \tag{103}$$

Adding both equations, we obtain:

$$2V_s^2V_r^2 - V_r^4 - 2V_r^2(PR + QX) - |Z|^2\left(P^2 + Q^2\right) \geq 0 \tag{104}$$

The above equation illustrates that equation values decrease with increased power and impedance. Moreover, Equation (104) considers a stability index [19].

$$SI(r) = 2V_s^2V_r^2 - V_r^4 - 2V_r^2(PR + QX) - |Z|^2\left(P^2 + Q^2\right) \tag{105}$$

Therefore, the classic stability principle determines the stability index for individual lines of radial distribution networks. The minimum stability index value is the most vulnerable to collapse. If the index is equal to zero, then the system collapses. Here, also, neglecting the shunt admittance to determine the stability index will lead to collapse.

2.1.15. Voltage Stability Margin (VSM)

VSM technique [20] is when the admissible rise in load power occurs during the current operating mode or the critical mode. Considering Figure 1, the voltage (V_s) and load apparent power (S_L) are derived as follows:

$$V_r = \frac{V_sZ_r}{\left[Z_r^2 + Z_0^2 + Z_LZ_0cos(\theta_0 - \theta_r)\right]^{1/2}} \tag{106}$$

$$S_r = \frac{V_r^2Z_r}{\left[Z_r^2 + Z_0^2 + Z_LZ_0cos(\theta_0 - \theta_r)\right]} \tag{107}$$

Concerning the critical point, the load power attains the maximum value and can be written as:

$$S_{cr} = \frac{V_s^2}{2Z_0[1 + cos(\theta_0 - \theta_r)]} \tag{108}$$

Z_r is the available load impedance, and the variation between Z_r and Z_0 is assumed as an invulnerability limit. Accordingly, the VSM in terms of impedance is as follows:

$$VSM_Z = \frac{Z_r - Z_0}{Z_0} \tag{109}$$

The allowable power increases until the stability limit is contingent to a higher degree on the angle of the load power factor with a similar VSM_s. Therefore, it can be enhanced to signify the margin concerning the load apparent power, VSM_s.

$$VSM_s = \frac{S_{cr} - S_L}{S_{cr}} \tag{110}$$

Hence, the margin of the apparent load power is derived from the VSM. Two assumptions for the stability margin are (i) that the line shunt admittance neglected and (ii) a constant power factor. The system is unstable when the apparent load power is more significant than the critical one. Even if the value of the VSM is less than zero, the system is unstable.

2.1.16. Voltage Reactive Power Index (VQI)

The line index proposed in [21] has a similar methodology to that of L_p. To insert the relation between the voltage (V_s) and reactive power (Q_r) comprised from Equation (19) into Equation (18), we obtain:

$$|V_r V_s Y_{sr}|.cos(\theta - \delta) - |V_r|^2 |Y_{sr}|.cos(\theta) - jQ_r = |V_r V_s Y_{sr}| \angle(\theta - \delta) - |V_r|^2 |Y_{sr}| \angle\theta \tag{111}$$

$$|V_r|^2 - |V_r V_s| \frac{sin(\theta - \delta)}{sin(\theta)} + \frac{Q_r}{|Y_{sr}| sin(\theta)} = 0 \tag{112}$$

As the value of δ is negligible and minimized to zero, then the entire term of $(sin(\theta - \delta)/sin(\theta))$ is omitted:

$$|V_r|^2 - |V_r V_s| + \frac{Q_r}{|Y_{sr}| sin(\theta)} = 0 \tag{113}$$

Replacing the term Y_{rs} $sin(\theta)$ with $B_{rs,}$, the corresponding equation is:

$$|V_r|^2 - |V_r V_s| + \frac{Q_r}{|B_{sr}|} = 0 \tag{114}$$

Extracting the real roots of the quadratic equation in terms of V_r, we obtain:

$$V_r = \frac{|V_r| \mp \sqrt{|V_r|^2 - \frac{4Q_r}{|B_{sr}|}}}{2} \tag{115}$$

When the discriminant of V_r equates to zero, the respective equation contains one distinct real root, or two equal roots. Hence, the respective real roots are:

$$|V_r|^2 - \frac{4Q_r}{|B_{sr}|} \le 0 \Longrightarrow \frac{4Q_r}{|B_{sr}||V_r|^2} \le 1 \tag{116}$$

The value of V_r ranges from 0 to 1, illustrating the limitation of real roots as the voltage stability limits. Moreover, real roots should be less than 1 and greater than 0, or the stability is affected. Thus, the derived equation evaluates the system stability and identifies the voltage collapse point as the voltage reactive power index.

$$VQI = \frac{4Q_r}{|B_{sr}||V_r|^2} \le 1$$

If the obtained value of VQI is nearer to 1, then the system is within stability limits. Otherwise, a system beyond the stability limits leads to instability. VQI also identifies the instability point of occurrence and voltage collapse.

2.1.17. Line Collapse Proximity Index (LCPI)

An improved LCPI was developed in [22] to overcome the limitations of specific indices [7,8,11] which ignored the line charging reactance while deriving their values. Since the line charging reactance is a critical factor supporting voltage stability, this proximity index was derived considering the definite transmission line model as well as the consequences of active and reactive power flows through lines. The definite transmission line model primarily represents an equivalent circuit of a two-port network considering the ABCD matrix. Thus, the proposed proximity index is derived from the ABCD matrix.

$$\begin{bmatrix} V_s \\ I_s \end{bmatrix} = \begin{bmatrix} A & B \\ C & D \end{bmatrix} \begin{bmatrix} V_r \\ I_r \end{bmatrix} \tag{117}$$

Hence, the transmission line parameters for a two-port circuit are A, B, C, and D.

$$\begin{gathered} A = \left(1 + Z \times \tfrac{Y}{2}\right),\ B = Z,\ C = Y \times \left(1 + Z \times \tfrac{Y}{4}\right),\ D = A \\ \text{Receiving-end current}:\ I_r = (P_r - jQ_r)/V_r^* = (P_r - jQ_r)/V_r\angle - \delta_r \end{gathered} \tag{118}$$

Considering the receiving-end current, the sending-end voltage V_s is derived as:

$$V_s\angle\delta_s = A\angle\alpha^* V_r\angle\delta_r + B\angle\beta^* I_r\angle 0^0 \tag{119}$$

A and B are magnitudes, and their corresponding phase angles are α and β, respectively. Combining the current with the voltage equation, we obtain:

$$V_s\angle\delta_s = A\angle\alpha^* V_r\angle\delta_r + B\angle\beta^* (P_r - jQ_r)/V_r\angle - \delta_r \tag{120}$$

$$V_s V_r\angle\delta = A\angle\alpha^* V_r^2 + B\angle\beta^* (P_r - jQ_r) \tag{121}$$

Here, $\delta = \delta_s - \delta_r$. Ignoring the imaginary part and considering the real part of the above quadratic equation, we obtain:

$$V_r^2(Acos\alpha) - V_r(V_s cos\delta) + (P_r Bcos\beta + Q_r Bsin\beta) = 0 \tag{122}$$

Obtaining the roots of the quadratic equation allows us to obtain:

$$V_r = \frac{-V_s cos\delta \pm \sqrt{(V_s cos\delta)^2 - 4Acos\alpha(P_r Bcos\beta + Q_r Bsin\beta)}}{2Acos\alpha} \tag{123}$$

It should contain the actual non-zero values of the Voltage equation and the actual non-zero roots obtained from the voltage equation discriminant.

$$(V_s cos\delta)^2 - 4Acos\alpha(P_r Bcos\beta + Q_r Bsin\beta) > 0 \tag{124}$$

Therefore, to maintain the system as stable and prevent voltage collapse, the following conditions should be satisfied:

$$\frac{4Acos\alpha(P_r Bcos\beta + Q_r Bsin\beta)}{(V_s cos\delta)^2} < 1 \tag{125}$$

$$\therefore LCPI = \frac{4Acos\alpha(P_r Bcos\beta + Q_r Bsin\beta)}{(V_s cos\delta)^2} \tag{126}$$

Therefore, the index value should be less than 1 to maintain the system stability, i.e., LCPI < 1. At no load, the LCPI value is zero, and the LCPI value is adjacent to the unity during the system instability.

2.1.18. New Voltage Stability Index (NVSI)

The NLSI derived in [23] is a similar concept of the power transmission quadratic equation represented in Equation (8). Determining the NLSI is similar to the process for determining L_p. Neglecting δ, Equations (19) and (20) represented as:

$$\left(V_r^2\right)^2 + \left(2Q_rX - V_s^2\right)V_r^2 + X^2\left(P_r^2 + Q_r^2\right) = 0 \tag{127}$$

The Voltage Equation V_r is a second-order equation. The limitation to possessing at least one solution is: $\left(2Q_rX - V_s^2\right)^2 - 4X^2\left(P_r^2 + Q_r^2\right) \geq 0$

$$\begin{gathered}\frac{2X\sqrt{\left(P_r^2+Q_r^2\right)}}{4Q_rX-V_s^2} \leq 1 \\ \therefore NVSI = \frac{2X\sqrt{\left(P_r^2+Q_r^2\right)}}{4Q_rX-V_s^2}\end{gathered} \tag{128}$$

Hence, to sustain stability, the value of the NLSI must be less than 1.00 in order toobtain the NLSI by neglecting the line resistance and line shunt admittances.

2.1.19. Integrated Transmission Line Transfer Index (ITLTI)

An integrated transmission line transfer index depends on the radial topology. The ITLTI's purpose is identifying the power transfer; moreover, it identifies the weak line depending on the transmission line parameters, consequent to the unity, lagging, and leading power factor states. The transmission line in Figure 1 depicts a constant sending-end voltage $V_s\angle\delta$ transferring the power S_s. Hence, the functional ABCD parameters represent the receiving-end power S_r, Voltage $V_r\angle\delta$ with the respective power factor angle θ_r. Further, the receiving-end complex power is derived as:

$$S_r = V_rI_r = -\frac{AV_r^2}{B}\angle(\beta-\alpha) + \frac{V_sV_r}{B}\angle(\beta-\delta) \tag{129}$$

Equivalently, the sending-end complex power is represented as:

$$S_s = V_sI_s = -\frac{AV_s^2}{B}\angle(\beta-\alpha) + \frac{V_sV_r}{B}\angle(\beta+\delta) \tag{130}$$

Considering the radial transmission line [24], we obtain two-power circles with the same radius and two different centers. Receiving-end complex power circle polar coordinates are represented in Figure 3 with a variable power angle δ.

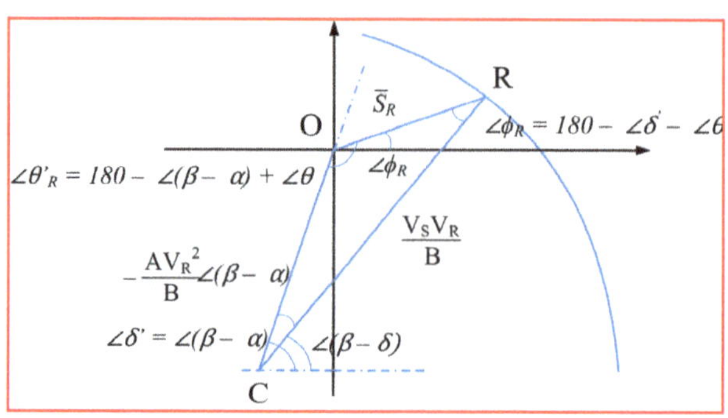

Figure 3. The receiving-end circle.

Considering the receiving-end circle and ΔOCR, we obtain:

$$\begin{gathered} \angle C = \angle\delta' = \angle(\beta-\alpha) - (\beta-\delta) = \angle(\delta-\alpha) \\ \angle O = \angle\theta_r' = 180 - \angle(\beta-\alpha) + \angle\theta_r \\ \angle R = \angle\varnothing_r = 180 - \angle\delta' - \angle\theta_r' \\ \frac{S_r}{sin\delta'} = \frac{\frac{V_s V_r}{B}}{sin\theta_r'} = \frac{\frac{AV_r^2}{B}}{sin\varnothing_R} \end{gathered} \tag{131}$$

Therefore, the relation between the receiving-end and sending-end voltage can be derived by $V_r = \frac{V_s}{A}\frac{sin\varnothing_r}{sin\theta_r'}$. Accordingly, an increased power angle at distinct power factors decreases the receiving-end voltage. The receiving-end power magnitude is:

$$S_r = \frac{V_s^2}{AB}\frac{sin(\theta_r' + \delta')sin\delta'}{(sin\theta_r')^2} \tag{132}$$

To deliver the maximum power, S_r is partially derived using δ' and is equated to zero, which determines the critical value of the power angle.

$$\delta_{r\_CR} = 90 - \frac{\theta_r'}{2} + \alpha \tag{133}$$

Critical voltage value:

$$V_{r(CR)} = \frac{V_s}{2A}\frac{1}{sin\frac{\theta_r'}{2}} \tag{134}$$

Maximum Power Transfer:

$$S_{r(max)} = \frac{V_s^2}{4AB}\frac{1}{\left(sin\frac{\theta_r'}{2}\right)^2} \tag{135}$$

The maximum limit of the receiving-end power for different power angles is as follows:

$$S_r = S_{r(max)}\frac{sin(\theta_r' + \delta')sin\delta'}{\left(\cos\frac{\theta_r'}{2}\right)^2} \tag{136}$$

Moreover, the integrated transmission line transfer index (ITLTI) is expressed as:

$$S_{r\_index} = \frac{sin(\theta_r' + \delta')\ sin\delta'}{\left(\cos\frac{\theta_r'}{2}\right)^2} \tag{137}$$

Hence, the ITLTI determines the system's stability for all healthy operations. Determining the S_{r_index} with varying power angles, the maximum value should be 1; if it exceeds 1, the system is unstable. Therefore, the index value is maintained at less than 1. The loadability concerning the critical power angle increased with an increase in the receiving-end power factor, irrespective of leading or lagging operations. Separating the active and reactive power of S_r, the active power P_r is expressed as follows:

$$P_r = -\frac{AV_r^2}{B}cos(\beta-\alpha) + \frac{V_s V_r}{B}cos(\beta-\delta) \tag{138}$$

2.1.20. Critical Boundary Index (CBI)

The Critical Boundary Index derived in [25] considers the active and reactive power variations and utilizes arithmetical methods to determine the critical boundaries. This index is preferable because of its high accuracy in predicting system performance and analyzing and adding the actual (Equation (8)) and imaginary (Equation (9)) parts of the power flow equation (Equation (7)).

$$\left(P_rR_{sr} + Q_rX_{sr} + V_r^2\right)^2 + (P_rX_{sr} - Q_rR_{sr})^2 = V_s^2V_r^2 \tag{139}$$

$$\left(V_r^2\right)^2 + 2\left(R_{sr}P_r + X_{sr}Q_r - \frac{V_s^2}{2}\right)V_r^2 + \left(R_{sr}^2 + X_{sr}^2\right)\left(P_r^2 + Q_r^2\right) = 0 \tag{140}$$

$$V_r^2 = -\left(R_{sr}P_r + X_{sr}Q_r - \frac{V_s^2}{2}\right) \pm \sqrt{A} \tag{141}$$

Imagining that the intensity of the receiving-end and sending-end voltages is more significant than zero, and rearranging Equation (141), we obtain the following:

$$A = \sqrt{\left(R_{sr}P_r + X_{sr}Q_r - \frac{V_s^2}{2}\right)^2 - (R_{sr}^2 + X_{sr}^2)(P_r^2 + Q_r^2)} \tag{142}$$

Hence, the above equation examines that the power transmission had limits for a specific time duration. Those limits are acknowledged as power and voltage stability limits. These limits are logical only when the magnitude of A is 0. Hence, determining the Q_r when $A = 0$ is written as follows:

$$Q_r = \frac{\pm\sqrt{X_{sr}^2V_s^4 + R_{sr}^2V_s^4 - 4X_{sr}^2R_{sr}V_s^2P_r - 4X_{sr}^3V_s^2P_r}}{2R_{sr}^2} + \frac{2X_{sr}R_{sr}P_r - X_{sr}V_s^2}{2R_{sr}^2} \tag{143}$$

We plot the power stability limit P-Q characteristics with Q_r. The P-V characteristic is the nose curve. The power system stability in Figure 4 includes the unstable and stable boundaries, starting from a current stable point K (P_o, Q_o) and reaching a critical boundary point C (X, Y). Considering the Lagrange Multipliers, Equation (141) simplifies as:

$$C(X,\ Y) = \left(R_{sr}X + X_{sr}Y - \frac{V_s^2}{2}\right)^2 - \left(R_{sr}^2 + X_{sr}^2\right)\left(X^2 + Y^2\right) = 0 \tag{144}$$

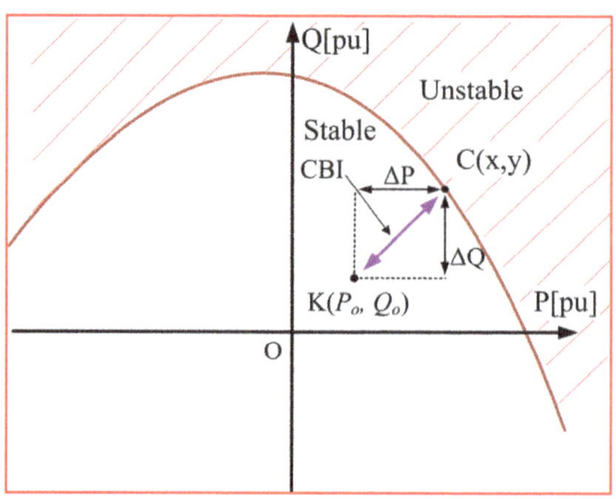

Figure 4. P-Q Characteristics.

The function f (X, Y) is the length between the current stable point K (P_o, Q_o) and the adjacent point of voltage collapse C (X, Y). The displacement of f (X, Y) is shown as:

$$f^2 = (X - P_0)^2 + (Y - Q_0)^2 \tag{145}$$

Here, the ΔP_{ik}, ΔQ_{ik}, stability evaluation index, and displacement in the stability curve are represented in Figure 4. Further, as shown in Figure 1, the operating point between the sending-end and receiving-end transmission line can evaluated by:

$$\begin{gathered}\Delta P_{sr} = X - P_0 \;\&\; \Delta Q_{sr} = Y - Q_0 \\ CBI_{sr} = \sqrt{\Delta P_{sr}^2 + \Delta Q_{sr}^2}\end{gathered} \tag{146}$$

Due to its high-accuracy prediction, this index is preferable. If the index is close to zero, it is the worst-stability system.

2.1.21. Line Voltage Stability Index (LVSI)

A novel LVSI developed to overcome existing stability indices' drawbacks, LVSI [26], is derived from the quadratic voltage equation. This index is also derived considering the transmission line ABCD parameters; moreover, the inclusion of line charging capacitance and resistance is ignored by existing stability indices. Consequently, this index evaluates the voltage sensitivity accurately, subject to all circumstances, and predicts the voltage collapse point. The receiving-end active power in terms of transmission line parameters is obtained using:

$$P_r = \frac{V_s V_r cos(\beta\delta)}{B} - \frac{AV_r^2 \cos(\beta - \alpha)}{B} \tag{147}$$

Rearranging the power equation according to a quadratic equation of voltage is obtained as:

$$V_r^2 - \frac{V_s V_r cos(\beta - \delta)}{Acos(\beta - \alpha)} + \frac{P_r B}{Acos(\beta - \alpha)} \tag{148}$$

The derivation of the voltage concerning the active power to obtain the sensitivity is shown as:

$$\frac{dV_r}{dP_r} = \frac{-B}{2V_r A\cos(\beta - \alpha) - V_s \cos(\beta - \delta)} \tag{149}$$

Hence, the sensitivity should be negative to maintain the system's stability.

$$\frac{-B}{2V_r A\cos(\beta - \alpha) - V_s \cos(\beta - \delta)} < 0 \tag{150}$$

$$V_s \cos(\beta - \delta) - 2V_r A\cos(\beta - \alpha) < 0 \tag{151}$$

Further simplifying Equation (151), the stability index can be shown as:

$$LVSI = \frac{2V_r A\cos(\beta - \alpha)}{V_s \cos(\beta - \delta)} > 1 \tag{152}$$

It is highly stable if the index value is 2 at no load. If the index value is 1 at the maximum load, it is a collapse point. Therefore, the system to maintain the stability index value must be greater than the unity.

2.1.22. New Line Voltage Stability Index (BVSI)

The index BVSI derived in [27] depends on the power transmission approach. Considering Equation (14), the quadratic equation can be rearranged as follows:

$$RV_r^2 - V_s V_r (Rcos\delta + Xsin\delta) + P_r\left(R^2 + X^2\right) = 0 \tag{153}$$

Deriving the quadratic equation roots, we obtain:

$$V_r = \frac{V_s(Rcos\delta + Xsin\delta) \pm \sqrt{(V_s(Rcos\delta + Xsin\delta))^2 - 4R(P_r(R^2 + X^2))}}{2R} \tag{154}$$

For determining the absolute and non-zero values of V_r, the discriminant is made greater than or equal to 0:

$$(V_s(Rcos\delta + Xsin\delta))^2 - 4R\left(P_r\left(R^2 + X^2\right)\right) \geq 0 \quad (155)$$

$$BVSI = \frac{4RP_rZ^2}{(V_s(Rcos\delta + Xsin\delta))} \leq 1 \quad (156)$$

Therefore, the index value is less than the unity for the system to be at the stability limit, such that the index value lies between 0 and 1. If the BVSI value exceeds 1, the corresponding line encounters a voltage drop, and the system may collapse. The BVSI line ignores the shunt admittance and the effects of reactive power on stability.

2.2. Bus Voltage Stability Indices

The Bus Voltage Stability Indices evaluate the transmission bus's voltage stability and do not derive the weak characteristics of the potential voltage issues across the system. Therefore, Bus VSIs are unable to be utilized for determining weak performances.

2.2.1. L-Index

The L-Index [28] is derived from the power flow and standard two-bus system and obtains a detailed analysis. The L index determines peculiar characteristics of power system susceptibility, measuring significant active power, restructuring weak areas or buses, predicting voltage collapse, and identifying instability. Considering the matrix F_{ij} of an ith column and jth row, and elements obtained from the Y matrix, α_L, and α_G are the load bus and generator bus. The linear transmission system represents a Hybrid matrix (H).

$$\begin{vmatrix} V^L \\ V^G \end{vmatrix} = H.\begin{vmatrix} I^L \\ V^G \end{vmatrix} = \begin{vmatrix} Z^{LL} & F^{LG} \\ K^{GL} & Y^{GG} \end{vmatrix} \cdot \begin{vmatrix} I^L \\ V^G \end{vmatrix} \quad (157)$$

Here, V^L, I^L: Voltage vector, Current vector at load nodes; V^G, I^G: Voltage vector, Current vector at generator nodes; Z^{LL}, F^{LG}, K^{GL}, Y^{GG}: H submatrices. By partial inversion, the H is obtained from the Y-matrix. For any load node j, $j \in \alpha_L$, and the corresponding voltage equation V_j is:

$$V_j = \sum\nolimits_{i \in \alpha_L} Z_{ji}.I_i + \sum\nolimits_{i \in \alpha_G} E_{ji}V_i \quad (158)$$

$$V_j^2 + \underline{V}_{oj}\underline{V}_j^* = \frac{\underline{S}_j^{+*}}{\underline{Y}_{jj}^{+}} \quad (159)$$

Substituting the equivalent voltage $\underline{V}_{oj}$, admittance $\underline{Y}_{jj}{}^{+}$, and power transferred $\underline{S}_i{}^{+}$:

$$\underline{V}_{oj} = -\sum\nolimits_{i \in \alpha_G} \underline{F}_{ji}\underline{V}_i \quad (160)$$

$$\underline{Y}_{jj}^{+} = \frac{1}{\underline{Z}_{jj}} \quad (161)$$

Hence, $\underline{S}_j{}^{+}$ contains two parts:

$$\underline{S}_j^{+} = \underline{S}_j + \underline{S}_j^{corr} \quad (162)$$

$$\underline{S}_j^{corr} = \left(\sum\nolimits_{\substack{i \in \alpha_L \\ i \neq j}} \frac{\underline{Z}_{ji}^*}{\underline{Z}_{jj}^*} \cdot \frac{\underline{S}_i}{\underline{V}_i} \right) .\underline{V}_j \quad (163)$$

The exceptional characteristic of Equation (159) is that its formation is similar to the admittance equation. Dissimilarities exist between the equivalent voltage $\underline{V}_{oj}$ and power transferred $\underline{S}_j$. The voltage $\underline{V}_{oj}$ varies slightly, not remaining as a constant value, since the generator voltage remains mostly constant at varying loads. Adopting these limitations can evaluate and regulate local indicators, and L_j can allow us to design for every node j corresponding to the line.

$$L_j = \left|1 + \frac{\underline{V}_{oj}}{\underline{V}_j}\right| = \left|\frac{\underline{S}_j^+}{\underline{Y}_{jj}^{+*} V_j^2}\right| = \ldots.. \tag{164}$$

Under stable operations, the $L_j \leq 1$ condition should never disobey any node j. Therefore, the global indicator L represents the stability of the entire subsystem, expressed as:

$$\begin{gathered} L = \underset{j \in \alpha_L}{\text{MAX}}\{L_j\} \\ L = \underset{j \in \alpha_L}{\text{MAX}}\left|1 - \frac{\sum\limits_{i \in \alpha_G} \underline{F}_{ji} \underline{V}_i}{\underline{V}_j}\right| \end{gathered} \tag{165}$$

Here, a series of load nodes is α_L, and a series of generator nodes is α_G. If the L index value is 0, the system is stable; if the value is 1, the system is approaching instability.

2.2.2. Voltage Instability Proximity Index (VIPI)

The proximity index derived in [29] for estimating instability depends upon coordinating various load flow solutions and instability. The VIPI is a scalar index that analyses the stability margin, considering the angle separating the critical and specified-value vectors. The power flow equation in terms of rectangular coordinates is as follows:

$$Y_s = Y(x) \tag{166}$$

Here, Y_s is the specified value; $x = (e_1, f_1, e_2, f_2, \ldots, e_n, f_n)^T$; e_i = Bus voltage real value; f_i = Bus voltage imaginary value. Assume two types of voltage vectors (Figure 5): x is the operating solution, and x^* is the fictitious solution; comply with the uniform specified-value vector. Employing critical vector (a) and deflection vector (b), x and x^* can be represented as $x = a + b$, $x^* = a - b$. Further simplifying them, $a = \frac{x+x^*}{2}$, $b = \frac{x-x^*}{2}$.

As this proximity index is the angle separating critical vector $Y(a)$, and specified value vector Y_s, the VIPI is as follows:

$$VIPI = \theta = \cos^{-1}\frac{Y_s^T Y(a)}{\|Y_s\|.\|Y(a)\|} \tag{167}$$

Here, θ is the angle separating two vectors; critical vector Y(a) is in the space of node specification.

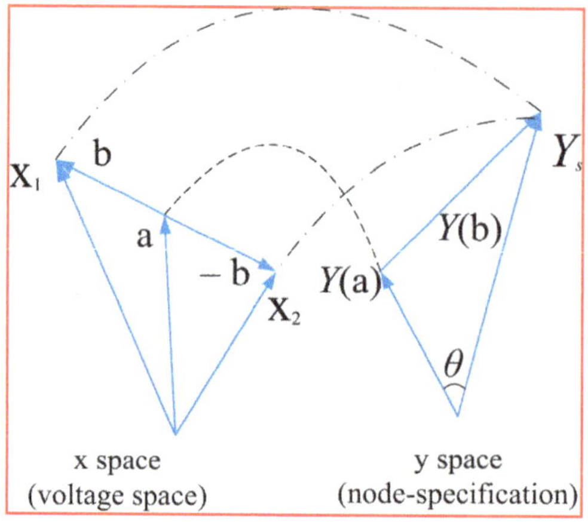

Figure 5. Concept of VIPI in the Node-Specification Space.

2.2.3. Voltage Collapse Proximity Index ($VCPI_{BUS}$)

$VCPI_{BUS}$, proposed in [30], is derived out of the standard power flow equation, and considering a network's voltage phasor measurement and admittance matrix determines the VCPI at every bus. The complex power at bus k of an N bus network is represented as:

$$S_k^* = |V_k|^2 - (|V_k|cos\delta_k - j|V_k|sin\delta_k)\left[\sum_{\substack{m=1\\ m\neq k}}^{N}(||V_m'|cos\delta_m' + j|V_m'|sin\delta_m'|)\right]Y_{kk} \quad (168)$$

The term V_m' in the above equation is expressed as:

$$V_m' = \frac{Y_{km}}{\sum_{\substack{j=1\\ j\neq k}}^{N} Y_{kj}} V_m \quad (169)$$

Hence, the right-hand term in Equation (168) is complex in the form of $a - jb$. Let the two equations contain two unknown terms (V_k, δ).

$$f_1(|V_k|, \delta) = |V_k|^2 - \sum_{\substack{m=1\\ m\neq k}}^{N} |V_m'||V_k|cos\delta \quad (170)$$

$$f_1(|V_k|, \delta) = \sum_{\substack{m=1\\ m\neq k}}^{N} |V_m'||V_k|sin\delta \quad (171)$$

A partial derivative matrix is attained by simplifying two equations for evaluating the unknowns with the Newton-Raphson Methodology. During the voltage collapse, the matrix determinant is equal to 0, which results as:

$$\frac{|V_k|cos\delta}{\sum_{\substack{m=1\\ m\neq k}}^{N}|V_m'|} = \frac{1}{2} \quad (172)$$

Further simplifying Equation (172), the voltage collapse prediction index (VCPI) determined at bus k is:

$$VCPI_{k^{th}bus} = \left|1 - \frac{\sum_{\substack{m=1\\ m\neq k}}^{N}|V_m'|}{V_k}\right| \quad (173)$$

Therefore, the above equation is the voltage collapse condition for bus k. The VCPI value fluctuates between 0 and 1, where 0 represents a stable system, and 1 represents voltage collapse.

2.2.4. S Difference Criterion (SDC)

The S Difference Criterion (SDC) [31] needs a protective relay to perform arithmetic derivation and utilizes the successive evaluation of apparent power (S) at the relay point. During the voltage instability environs, the inflation of sending-end power flow delivers the transmission losses. Consequently, the rise in the sending-end apparent power does not afford a rise in receiving-end power. It specifies that $\Delta S = 0$ at the relay point during the voltage instability. Hence, the transfer of apparent power to the receiving end is:

$$\underline{S}_j^{(k)} = \underline{V}_j^{(k)} \underline{I}_{ji}^{(k)*} \quad (174)$$

Considering the change in a time period Δt, the time interval changes to $t_k = t_k + \Delta t$. A respective rise in apparent power is as follows:

$$\underline{S}_j^{(k+1)} = \underline{S}_j^{(k)} + \Delta\underline{S}_j^{(k+1)} = \left(V_j^{(k)} + \Delta\underline{V}_j^{(k+1)}\right).\left(\underline{I}_{ji}^{(k)} + \Delta\underline{I}_{ji}^{(k+1)}\right)^* = \underline{S}_j^{(k)} + \Delta\underline{V}_j^{(k+1)}\underline{I}_{ji}^{(k)*} + \underline{V}_j^{(k)}\Delta\underline{I}_{ji}^{(k+1)*} + \underbrace{\Delta\underline{V}_j^{(k+1)}\Delta\underline{I}_{ji}^{(k+1)*}}_{\approx 0} \quad (175)$$

Here, the term $\Delta\underline{V}_j^{(k+1)}\Delta\underline{I}_{ji}^{(k+1)*}$ is very small and can be ignored.

$$\Delta\underline{S}_j^{(k+1)} = \Delta\underline{V}_j^{(k+1)}\underline{I}_{ji}^{(k)*} + \underline{V}_j^{(k)}\Delta\underline{I}_{ji}^{(k+1)*} = 0 \quad (176)$$

$$1 + \frac{\Delta\underline{V}_j^{(k+1)}\underline{I}_{ji}^{(k)*}}{\underline{V}_j^{(k)}\Delta\underline{I}_{ji}^{(k+1)*}} = 1 + ae^{j\varphi} = 1 + a(cos\varphi + jsin\varphi) = 0 \quad (177)$$

Thus, the SDC is a positive real value; then, considering the magnitude of Equation (177), we obtain:

$$SDC = \left|1 + ae^{j\varphi}\right| = \left|1 + \frac{\Delta\underline{V}_j^{(k+1)}\underline{I}_{ji}^{(k)*}}{\underline{V}_j^{(k)}\Delta\underline{I}_{ji}^{(k+1)*}}\right| \quad (178)$$

Intending to achieve voltage collapse, the value of $\Delta\underline{S} = 0$, and correspondingly, the SDC equates to 0. The relay operates if the SDC value is below a predetermined threshold.

2.2.5. Impedance Stability Index (ISI)

The ISI bus stability index depends on Tellegen's rule, an index derived in [32] made simple in evaluating Thevenin's variables. Let a bus r with a complex load (S_r) and impedance (Z_r) be linked to a power system; the remaining system is expressed as Thevenin's equivalent with the variables E and Z_{th}. Let the current equation be:

$$\underline{S}_r/\underline{V}_r = I_r{}^* = ((\underline{E} - \underline{V}_r)/\underline{Z}_{th})^* \quad (179)$$

$$\underline{V}_r(\underline{E} - \underline{V}_r)^* - \underline{S}_r\underline{Z}_{th}{}^* = 0 \quad (180)$$

Two voltage solutions empower the phasor of Equation (180) with a power S_r; if the two solutions are equal, it allows maximum power transfer.

$$\underline{V}_r = (E - \underline{V}_r)^* \quad (181)$$

Equation (181) is fundamental to a local voltage stability approach that conveys predetermined consequences through circuit theory.

$$|\underline{Z}_r| = |\underline{Z}_{th}| \tag{182}$$

The system delivers the maximum power when the load-impedance magnitude equals Thevenin's impedance magnitude. The adjacent Thevenin network $\hat{N}$ possesses an identical analysis to the base network N. Hence, the derivation is similar to putting a hat (ˆ) on every term, which signifies the adjoint network parameters. The divergence order of Tellegen's theorem is equivalent to the following:

$$\left(\hat{\underline{I}}_r^* \Delta \underline{V}_r - \hat{\underline{V}}_r \Delta \underline{I}_r\right) = 0 \tag{183}$$

Here, $\Delta \underline{V}_r$ and $\Delta \underline{I}_r^*$ Express the advancement of the complex voltage and current at bus 'r' regarding the V_r and $I_r{}^*$ base values. The adjacent network load impedance $\hat{\underline{Z}}_r$ is described as a ratio of phasor voltage $\hat{\underline{V}}_r$ to phasor current $\hat{\underline{I}}_r$.

$$\hat{Z}_r = |\hat{\underline{Z}}_r| = |\hat{\underline{V}}_r / \hat{\underline{I}}_r| \tag{184}$$

The load impedance specifies that it is the base case; moreover, it is determined at an instant earlier than that of the network that is regulated to power disturbances.

$$\hat{Z}_r = |\underline{V}_r / \underline{I}_r| \tag{185}$$

For determining the Thevenin's impedance, the current $\hat{\underline{I}}_r^*$ equation is substituted in Equation (183) from Equation (179).

$$\left((\hat{\underline{E}} - \hat{\underline{V}}_r) / \hat{\underline{Z}}_{th}\right)^* \Delta \underline{V}_r - \hat{\underline{V}}_r \Delta \hat{\underline{I}}_r^* = 0 \tag{186}$$

$$\hat{\underline{Z}}_{th}^* = \left((\hat{\underline{E}} - \hat{\underline{V}}_r)^* / \left(\hat{\underline{V}}_r \Delta \hat{\underline{I}}_r^*\right)\right) \Delta \underline{V}_r \tag{187}$$

Similarly, from Equation (181), the condition for the maximum power transfer of adjacent network can be expressed as:

$$\hat{\underline{V}}_r = (\hat{\underline{E}} - \hat{\underline{V}}_r)^* \tag{188}$$

Analyzing Equations (187) and (188), during the system collapse point, the equivalent Thevenin's impedance is expressed as:

$$\hat{\underline{Z}}_{th}^* = \Delta \underline{V}_r / \Delta I_r^* \tag{189}$$

$$\hat{\underline{Z}}_{th} = \Delta \hat{\underline{V}}_r^* / \Delta \underline{I}_r \tag{190}$$

However, the above two equations are conjugate with each other; for discarding the conjugate values, Thevenin's impedance absolute value is as follows:

$$\hat{Z}_{th} = |\hat{\underline{Z}}_{th}| = |\Delta \underline{V}_r / \Delta \underline{I}_r| \tag{191}$$

The contingency during the regular loading is:

$$\hat{Z}_r \gg \hat{Z}_{th} \tag{192}$$

During the commencement of voltage instability, the dissimilarities among these impedances converge to 0. Moreover, at that instant of collapse, the impedances are identical, i.e., $\hat{Z}_{th} = \hat{Z}_r$.

$$|\Delta \underline{V}_r / \Delta \underline{I}_r| = |\underline{V}_r / \underline{I}_r| \tag{193}$$

This index precisely determines Thevenin's impedance by evaluating the advancement in voltage and current following the base-case exposure to disturbances. Furthermore, the load impedance is evaluated through the ratio of the voltage to the current. Using Equations (192) and (193), a modest regulated impedance stability index (ISI) for identifying the voltage stability margin is:

$$ISI = \left(\hat{Z}_r - \hat{Z}_{th}\right) / \hat{Z}_r \tag{194}$$

Further simplifying by substituting the impedances in terms of voltage and current, the stability index derives as:

$$ISI = 1 - |\underline{I}_r \Delta \underline{V}_r| / |\underline{V}_r \Delta \underline{I}_r| \tag{195}$$

2.2.6. Voltage Stability Index (VSI_{BUS})

The VSI bus index relies on system elements to identify the distance between the operating and collapse points. This index [33] derives from power flow equations that are similar to the assumptions of the SDC. Let 'i' be one local bus of the system. The complex load power of bus i is:

$$S_i = S_i \angle \theta_i = (P_i + jQ_i) = V_i I_i^* \tag{196}$$

The complex bus voltage is $V_i = V_i \angle \delta_{i,}$ and the complex load current is $\boldsymbol{I}_i = I_i \angle (\delta_i - \theta_i)$. Employing Taylor's theorem, the interrelation among the advancement difference between V_i and I_i as a consequence of advancement evolution in S_i is expressed as:

$$\Delta S_i = \frac{\partial S_i}{\partial I_i} \Delta I_i + \frac{\partial S_i}{\partial V_i} \Delta V_i + higher\ order\ terms \tag{197}$$

Meanwhile, ignoring the higher-order terms,

$$\Delta S_i = V_i \Delta I_i + I_i \Delta V_i \tag{198}$$

Usually, if the load increments, there is a corresponding increase in load current and decrease in load voltage. This means that, if the value of ΔS_i is positive, Current ΔI_i is positive and the ΔV_i Voltage is negative. Although the bus reaches a critical point or voltage collapse, adding additional load on a bus might not be able to raise the apparent power S_i of the load due to rapid decrement in voltage in contrast to the current increment. Going through this case converges ΔS_i to zero. Accordingly, the power ΔS_i through the voltage-stability limit is expressed as:

$$0 \leq 1 + \left(\frac{I_i}{V_i}\right)\left(\frac{\Delta V_i}{\Delta I_i}\right) \tag{199}$$

Considering that the power of Equation (199) is made to α (>1.0), an actual value, it determines almost the same level of linear characteristics. Therefore, excluding the general loss, the voltage stability index at the 'i' bus is as follows:

$$VSI_i = \left[1 + \left(\frac{I_i}{V_i}\right)\left(\frac{\Delta V_i}{\Delta I_i}\right)\right]^{\alpha} \tag{200}$$

Hence, at no load, the VSI varies from 1, and at voltage collapse, the value varies to 0.

2.2.7. Z_L/Z_S Ratio

The Z_L/Z_s ratio index proposed in [34] depends on a similar technique to the ISI. The correspondence index is the load bus impedance (Z_L) ratio to Thevenin's impedance (Z_s). The voltage stability margin was generally evaluated from different scenarios. The evaluation of the stability margin depends on the determination of the voltage stability. For an equivalent circuit, the derivation of the apparent load power concerning load admittance is as follows:

$$\frac{dS}{dY} = \frac{V^2\left[1-(Z_SY)^2\right]}{1+(Z_SY)^2+2Z_SYcos\beta} \tag{201}$$

Whereas $Y = 1/Z_L$, dividing both sides of the equation with V^2:

$$\frac{dS^*}{dY^*} = \frac{1-(Z_SY)^2}{1+(Z_SY)^2+2Z_SYcos\beta} \tag{202}$$

While $dS^*/dY^* = (Y/S)\,(dS/dY)$, thenthe Equation (202) together evaluates (Z_L/Z_s):

$$\frac{Z_L}{Z_S} = \frac{M+1}{-Mcos\beta+\left[(Mcos\beta)^2-M^2+1\right]^{0.5}} \tag{203}$$

Here, $M = (dS^*/dY^*)$, and the relevance between Z_L/Z_S and M relies upon the angle β. The estimation of β is a phase angular difference between the source and the load, i.e., $\beta = \varphi_S - \varphi_L$. The definite value of φ_S operates on the existent form of the system. The parameter M is determined by evaluating the derivative of the apparent load power to admittance:

$$M = \left(\frac{dS^*}{dY^*}\right) = \frac{(S_2 - S_1)(Y_2 + Y_1)}{(S_2 + S_1)(Y_2 - Y_1)} \tag{204}$$

Here, S and Y refer to the load power and admittance, respectively. Moreover, suffixes 1 and 2 refer to the starting of a change (time, t_1) and completion of the change (time, t_2), respectively. The variation between the two-time measurements needs to be around 500 ms. If the index value remains above 1, it is a reliable condition.

2.2.8. Equivalent Node Voltage Collapse Index (ENVCI)

The index ENVCI depends on the ESM (Equivalent System Model) and local voltage phasors [35]. The ENVCI has numerous advantages like accurate evaluation and modeling; moreover, this index is convenient for performing real-time and online operations. The outgoing power at node N of a single-line equivalent model (Figure 6) should perform the consequent equation of plain power flow.

$$P_n + jQ_n = \vec{V}_n.\left(\frac{\vec{E}_k - \vec{V}_n}{Z_{kn}}\right)^* \tag{205}$$

Let the phasor voltage equations for two nodes of a single-line equivalent network in terms of rectangular coordinates be $\vec{E}_k = e_k + jf_k$ and $\vec{V}_n = e_n + jf_n$, then the corresponding line impedance is $Z_{kn} = R_{kn} + jX_{kn}$. Separating the real terms and imaginary terms of the power flow equation, we obtain:

$$\begin{aligned} P_nR_{kn} + Q_nX_{kn} &= e_n(e_k - e_n) + f_n(f_k - f_n) \\ P_nX_{kn} - Q_nR_{kn} &= f_ke_n - e_kf_n \end{aligned} \tag{206}$$

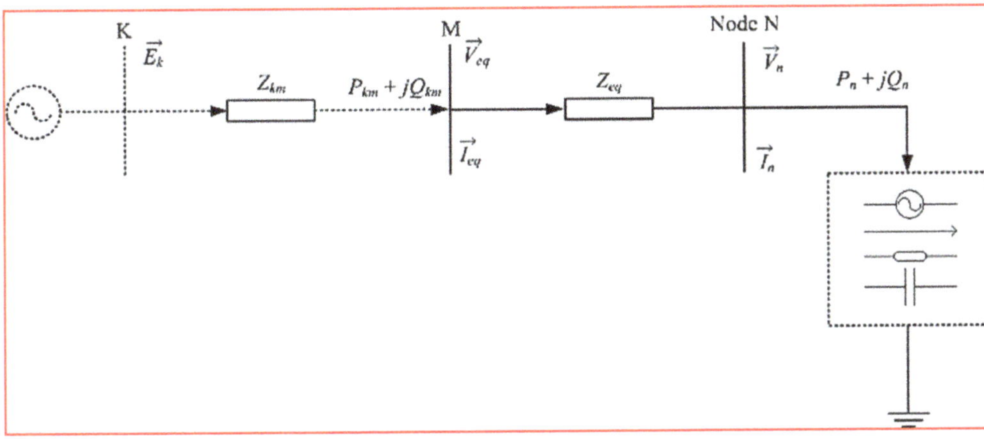

Figure 6. Equivalent System Model.

Hence, the power flow equation, Equation (206), is crucial for determining the phasor voltage at the receiving node, describing the system's performance. In contrast, the phasor voltage of the sending node is known. Whenever the respective ESMs of all equivalent lines possess an arithmetical solution of the receiving node, signifying that they contain every node voltage of the system, at that instant, the system should possess a complete power flow solution; moreover, this should influence the voltage stability. The system's stability relies on evaluating Equation (206) at every node. The evaluation of Equation (206) can be examined by the Jacobina matrix singularity:

$$J = \begin{bmatrix} e_k - 2e_n & f_k - 2f_n \\ f_k & -e_k \end{bmatrix} \tag{207}$$

$$det(J) = 2(e_k e_n + f_k f_n) - \left(e_k^2 + f_k^2\right) = 0 \tag{208}$$

Therefore, Equation (208) specifies a novel stability index that is known as the Equivalent Node Voltage Collapse Index:

$$ENVCI = 2(e_k e_n + f_k f_n) - \left(e_k^2 + f_k^2\right) \tag{209}$$

The ENVCI expression in terms of polar coordinates is:

$$ENVCI = 2E_k V_n cos\theta_{kn} - E_n^2 \tag{210}$$

Determining the ENVCI requires only the phasor voltage of two nodes. Every node contains ESM. The system converges to a collapse point if the ENVCI of a minimum of one node is approaching 0; the respective node is the weak node that induces system instability.

2.2.9. Power Stability Index (PSI)

The power stability index [36] was developed to identify the optimal position of the Distributed Generator (DG), including critical buses that are nearer voltage collapse. This index derives from a standard two-bus system, which is stable within the unity margin. Let the power flow in a standard two-bus system without DG be:

$$S_L = P_L + jQ_L = V_r I_r^* \tag{211}$$

$$\overline{V}_r = \overline{V}_r - \overline{I_r Z} \tag{212}$$

where

$$I_r = \frac{(P_L) - j(Q_L)}{V_r^*} \quad (213)$$

Now, with DG, the current I_r concerning the active power and reactive power support are expressed as:

$$I_r = \frac{(P_L - P_G) - j(Q_L - Q_G)}{V_r^*} \quad (214)$$

Combining Equations (212) and (214), and writing real and imaginary terms separately, we obtain:

$$P_L - P_G = \frac{|V_r||V_s|}{V_r^*} cos(\theta - \delta_s + \delta_r) - \frac{|V_r|^2}{Z} cos(\theta) \quad (215)$$

$$Q_L - Q_G = \frac{|V_r||V_s|}{V_r^*} sin(\theta - \delta_s + \delta_r) - \frac{|V_r|^2}{Z} sin(\theta) \quad (216)$$

Rearranging Equation (215), we obtain:

$$|V_r|^2 - \frac{|V_r||V_s| cos(\theta - \delta)}{cos(\theta)} + \frac{Z(P_L - P_G)}{cos(\theta)} = 0 \quad (217)$$

Hence, Equation (217) is in a quadratic form in terms of V_r. When stable, the node voltage must contain real roots, i.e., the discriminant is more significant than zero, and the index derives as:

$$PSI = \frac{4R_{ij}(P_L - P_G)}{[|V_i| cos(\theta - \delta)]^2} \leq 1 \quad (218)$$

If the PSI value is less than 1, the system is stable; if the value is nearer to zero, the system is more stable. Hence, the PSI determines the optimal location of the DG as the calculated value of the PSI for each line patterned from maximum to minimum value. The optimal location of the DG is at the line-end with a maximum value of the PSI.

2.2.10. Voltage Deviation Index (VDI)

The Voltage Deviation Index proposed in [37] determines the voltage deviation's real value compared to 1 per unit. Assuming an N-bus system, the total VDI is the addition of all N buses' voltage deviation index.

$$VDI_j = |1 - V_j| \quad (219)$$

$$VDI_T = \sum_{j=1}^{N} |1 - V_j| \quad (220)$$

2.2.11. Simplified Voltage Stability Index (SVSI)

An advanced SVSI proposes [38] to evaluate the system's stability. This SVSI depends on the Relative Electrical Distance (RED) concept, which considers identifying the convenient generator to a particular load bus; moreover, the relation of electrical parameters for enhancing the performances.

- Relative electrical distance (RED)

The concerned system, the interrelation among the current (I), Generator bus voltage phasor (V), and the load buses (L) represented in the form of an admittance matrix can be shown as:

$$\begin{bmatrix} I_G \\ I_L \end{bmatrix} = \begin{bmatrix} Y_{GG} & Y_{GL} \\ Y_{LG} & Y_{LL} \end{bmatrix} \begin{bmatrix} V_G \\ V_L \end{bmatrix} \quad (221)$$

Rearranging Equation (221), we obtain:

$$\begin{bmatrix} V_L \\ I_G \end{bmatrix} = \begin{bmatrix} Z_{LL} & F_{GL} \\ K_{GL} & Y_{GG} \end{bmatrix} \begin{bmatrix} I_L \\ V_G \end{bmatrix} \tag{222}$$

$F_{LG} = -\mid Y_{LL} \mid^{-1} \mid Y_{LG} \mid$ represents a composite matrix, which indicates the association between the voltages of the source and load buses. The REDs determined from the matrix F_{LG} are shown as:

$$R_{LG} = [A] - abs[F_{LG}] = [A] - abs\left(|Y_{LL}|^{-1}|Y_{LG}|\right) \tag{223}$$

Here, the size of matrix A is (n-g) X g, the total number of buses is represented by n, and the total generator count is denoted by g. Matrix A is a particular matrix in that all elements are 1, i.e., a matrix of ones. The data obtained from matrix R_{LG} are considered rather than path algorithms for determining the electrical distance from the generator bus to the load bus. Subsequently, after identifying the shortest distance of the generator to a particular load with the R_{LG} matrix, the Thevenin impedance voltage drop ΔV_i can be estimated with:

$$\Delta V_i = \sum_{b=1}^{n_j-1} \left|\vec{V}_b - \vec{V}_{b+1}\right| \cong \left|\vec{V}_g - \vec{V}_i\right| \tag{224}$$

V_l and V_g are the phasor voltages at the load bus and closest generator. While an increase in demand for power reaches the maximum loading point in one particular bus or various buses in the system, some buses may experience high voltage drops. This process illustrates the potential of voltage collapse and comprises the formation of the $SVSI$ in terms of a correction factor β.

$$\beta = 1 - (max(|V_m| - |V_l|))^2 \tag{225}$$

The derived correction factor is proportionate with higher variations between the voltage magnitude of bus m and bus l, which can be determined directly by measuring the PMU during the power system analysis, considering particular operating conditions. Thus, the SVSI is as follows:

$$SVSI_i = \frac{\Delta V_i}{\beta \times V_i} \tag{226}$$

Suppose the obtained value of the proposed index is near the unity. In that case, the system considers the voltage instability on these terms: the Thevenin impedance voltage drop equals the voltage of the load bus.

2.2.12. P-Index

The P-index [39] evaluates the distance to collapse and measures the load shedding. The P-index is derived from a standard two-bus system, shown in Figure 1. The load at the receiving end bus is $P_r + jQ_r$, and V_r is the voltage magnitude. The equivalent load admittance is $G_r - jB_r$,

$$G_r = \frac{P_r}{V_r^2}, B_r = \frac{Q_r}{V_r^2} \tag{227}$$

Let ΔP_r, ΔQ_r be an incremental increase in load without varying the power factor. The equivalent increment in admittance is ΔG_r, ΔB_r. The voltage magnitude changed by an amount of ΔV_r, is typically negative; therefore, the updated bus voltage is $V_r + \Delta V_r$. The corresponding active power variation is as follows:

$$\begin{aligned} \Delta P_r &= (V_r + \Delta V_r)^2 (G_r + \Delta G_r) - V_r^2 G_r \\ &= (V_r + \Delta V_r)^2 \Delta G_r + (2V_r + \Delta V_r) G_r \Delta V_r \end{aligned} \tag{228}$$

The substantial consequences of the two terms in Equation (228) are: the primary term is positive and describes the power attained because of the additional load ΔG_r, while the secondary term is negative and is the power wasted on the original load G_r because of the voltage drop ΔV. Because of these opposite terms, the overall active power gained is balanced. At the stability limiting point, these two terms diminish, and the overall increment in power is zero.

The P-index bus stability index depends on the ratio of two terms in Equation (228), i.e., the ratio of power lost to power gained. When ΔV is negative for positive ΔG_r, a minus sign is added to make the index positive.

$$\text{P-index} = -\frac{(2V_r + \Delta V_r)G_r}{(V_r + \Delta V_r)^2}.\frac{\Delta V_r}{\Delta G_r} \tag{229}$$

In the limiting case, such as ΔG_r, $\Delta V \rightarrow 0$,

$$\text{P-index} = -\frac{2G_r}{V_r}.\frac{dV_r}{dG_r} \tag{230}$$

Although dV_r/dG_r is often not defined in terms of network terminology, it is simple to do so in terms of the power and voltage sensitivity of the system. Provided that dV_r/dG_r:

$$\frac{dV_r}{dG_r} = \frac{dV_r}{dP_r}.\frac{dP_r}{dG_r} \tag{231}$$

Thus, differentiating the power, $P_r = V_r{}^2\ G_r$ is as follows:

$$dP_r = V_r^2 dG_r + 2V_r^2 G_r dV_r \tag{232}$$

Or:

$$\frac{dP_r}{dG_r} = V_r^2 + 2V_r G_r \frac{dV_r}{dG_r} \tag{233}$$

Substituting Equation (233) in Equation (231), we obtain:

$$\frac{dV_r}{dG_r} = \frac{dV_r}{dP_r}\left(V_r^2 + 2V_r G_r \frac{dV_r}{dG_r}\right) \tag{234}$$

Further simplifying Equation (234), it is reduced as:

$$\frac{dV_r}{dG_r} = \frac{V_r^2 \frac{dV_r}{dP_r}}{1 - 2V_r G_r \frac{dV_r}{dP_r}} \tag{235}$$

Substituting Equation (235) in the P-index Equation (230), we obtain:

$$\text{P-index} = \frac{-2V_r G_r \frac{dV_r}{dP_r}}{1 - 2V_r G_r \frac{dV_r}{dP_r}} \tag{236}$$

Expressing Equation (236) in terms of active power, we obtain:

$$\text{P-index} = \frac{-2\frac{P_r}{V_r} \times \frac{dV_r}{dP_r}}{1 - 2\frac{P_r}{V_r} \times \frac{dV_r}{dP_r}} \tag{237}$$

Therefore, the P-index is expressed in terms of power and voltage sensitivities. When $dV/dP_r = \infty$, the theoretical value of the stability index is 1.0 at the stability limit. The stability limit varies from 0 (no load) to 1.0 at the collapse point of the system.

3. Summary

The Voltage Stability Indices that are preferred in power system analysis are derived in detail in this paper. Our review of stability indices examined the overall parameters of voltage stability analysis, including the voltage stability margin, stability indices, identifying voltage collapse proximity, and the maximum loadability. This comprehensive review features the significance of definite stability assessment and specifies the effectiveness and vulnerability of various bus and line Voltage Stability Indices.

The line stability indices, like the VSLI, L_{mn}, LQP, LP, FVSI, NLSI, VQI, and NVSI, and the bus stability indices like the L-Index, Voltage Instability, and VCPI, are derived from the power flow in a standard two-bus system. In contrast, the stability indices like the VCPI, PTSI, VSI_1, VSM, SDC, and PSI are determined by considering the maximum power transfer and maximum power losses. At the same time, stability indices like the SI, LCPI, LVSI, and BVSI derive from the quadratic voltage equation—the overall objectives of the listed stability indices are expressed in Figure 7.

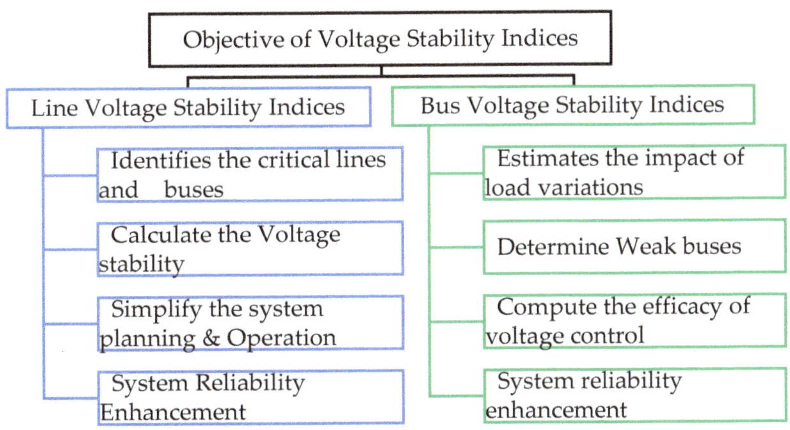

Figure 7. The objective of Voltage Stability Indices.

The VSLI stability index depends on the load reactive power demand ratio to the available reactive power. It helps to determine the critical points and identifies the load bus nearer to the collapse point. The L_{mn} stability index is online monitoring that predicts the voltage collapse; the reliability test on the IEEE 24 bus system results in an accurate voltage collapse prediction. It also identifies the stressed lines and weak areas that are exposed to voltage collapse; moreover, it allows for flexible operation. The LQP is a static indicator that performs faster, evaluates stability, and determines the origin of voltage collapse.

The VCPI is a proximity index that determines the range of the collapse point; this index depends upon the systems' generation and load characteristics and identifies exposure to voltage collapse. The VCPI is a method for the real-time prediction of voltage collapse. This proximity index illustrates the collapse position and is an alarming tool for preventing system collapse. LP performs voltage stability assessment in a radial distribution system. It predicts instabilities easily when changing the load. The FVSI can determine the collapse point, weakest bus, critical line, and maximum loadability. The VSLBI is used for local monitoring of threshold voltage collapse and emergency management during voltage-sensitive loads. The VSLBI operates within a small stability margin and maintains reactive reserve power. The VSMI presents data about weak locations, evaluates the voltage stability margin, and maintains stable voltage levels during load variations. Moreover, the VSMI depends on the relation between voltage stability and angular difference between the receiving- and sending-end buses. VCPI_1 is an online voltage stability evaluation index that depends on power transmission lines and identifies the weakest transmission lines by distinguishing the minor power outage buses, including the electrical distance.

The Critical Voltage (V_{cr}) specifies the minimum voltage where the system performs away from voltage collapse; the system is unstable if the Jacobian power flow matrix is singular. In V_{cr}, the Jacobian power matrix is a static voltage stability index specifying the range of the voltage stability limit. The PTSI is an essential technique that forecasts dynamic voltage collapse, calculates the effect of adding additional equipment like generators or transmission lines, and determines the weak points in the system. VSI_1 is a novel online stability index for predicting steady-state voltage stability. Coordinated with the network simplification method, it presents the stability margin of every load bus, sequentially determining the buses which are vulnerable to voltage collapse.

The NLSI performs by varying the active and reactive power and evaluates the voltage stability more accurately than other indices that vary only the reactive power. Moreover, the NLSI evaluates the voltage collapse point, specifies the order of transmission lines based on loadability, and considers the consistency of wind and solar power output and the response time of the system. The SI identifies the most vulnerable bus that is exposed to voltage collapse in the radial distribution system, is easily implemented for load flow analysis, and does not need extra parameters. The VSM is the variation between the minimal permissible voltage and absolute voltage at a particular bus; it determines the loss of voltage contribution during the voltage collapse. The VSM can efficiently prevent voltage collapse from voltage crisis, and can be measured by PV and QV curves. The VQI is an effective technique that determines the critical bus, detects instability in large-scale systems, and determines the distance of the collapse point.

The LCPI combines the influence of the proportional flow of active and reactive power with ABCD parameters, determines the consequences of different power system arrangements, and specifies the most sensitive lines. The NVSI evaluates the voltage stability effectively by varying the active and reactive power, determining the weakest bus and sensitive line, and utilizes the Reactive Power Sensitivity theory. The ITLTI was first derived for radial transmission networks and subsequently adapted for larger systems. The ITLTI contains extended integrated significance; moreover, it is probably a better resolution for online stability assessment and specifies the weakest line and the specific data about a critical voltage instability range. The CBI operates with the Lagrange Constant Computational Method (LCM); the main motive of this index is to determine the critical boundaries and voltage stability with lesser parameters. Varying the active and reactive power predicts voltage stability margin precisely with better accuracy. The LVSI determines the stability margin by considering the transmission line (ABCD) parameters and is expressed in terms of MVA. The LVSI can determine the voltage collapse point and critical lines, and involves only the information on the phasor bus voltage. The BVSI forecasts the voltage security level, considering various operating conditions and identifies the sensitive lines that are nearer to stability. This index can identify the capability of the bus to place a single DG and size effectively; moreover, it can accurately identify the critical lines and weakest bus.

The L-index in online testing depends on the eigenvalues of the Jacobian matrix of a network; these eigenvalues determine the gap between the voltage collapse point and the current operating point. Moreover, the L-index measures stability linearly, and non-linear behavior may not be measured thoroughly. The VIPI determines very sensitive buses that are exposed to voltage instability and finds the possible precautions for preventing voltage collapse; the VIPI depends on the relation between the instability and various load flow solutions. The VCPI is a real-time prediction that determines the system's proximity to voltage collapse, which is evaluated by analyzing the present voltage and voltage levels at the point of collapse and utilizing the magnitude of voltage and angle; the admittance matrix forecasts the voltage collapse. The SDC's utilization in the protection algorithm for voltage collapse depends on the local bus phasor of the voltage and current at every line's relay point. The relay operation is local to voltage collapse, and the overall rise in apparent power is because of transmission losses to the supply.

The ISI calculates the system's stability; this index illustrates the ratio of variation in system impedance to load impedance. The ISI is a local stability index that is derived using Tellegen's theorem and adjoint network. The VSIBUS is a simple index which identifies the distance to the collapse point with the help of local data like the magnitude of voltage and load current, and thoroughly examines the influence of line tripping, reactive power limits, and non-linear variation in the load. The Z_L/Z_S measures the derivation of apparent power concerning admittance, quickly calculating both the power and admittance results in switching the impedance load tap-changing of a transformer. The SVSI utilizes the data of present operating conditions combined with phasor voltage measurement. The SVSI is easily accessible and less computationally derived.

P-index calculation is simple. It involves the system model, nodal power, and only the single datum of system states. The only computation performed is Jacobian matrix inversion, which can be attained smoothly. Therefore, the P-index is appropriate for online voltage stability analysis. Mainly, the load shedding is evaluated by the P-index. Instead of system-wide shedding, the method makes use of single-load shedding.

The Stability index's concept, assumptions considered during derivation, mathematical equation, condition for index stability, objective, and some references are represented in Tables 1 and 2 of the line Voltage Stability Indices and bus Voltage Stability Indices, respectively.

Figure 8 represents the number of articles published on the line and bus Voltage Stability Indices in the last decade. It shows an eventual rise in research in Voltage Stability Indices to enhance voltage stability and supply the desired power under various load conditions. Table 3 represents some distinct characteristics of Voltage Stability Indices that are the optimal location of the DG and sizing of the DG, as well as dependency on VSI.

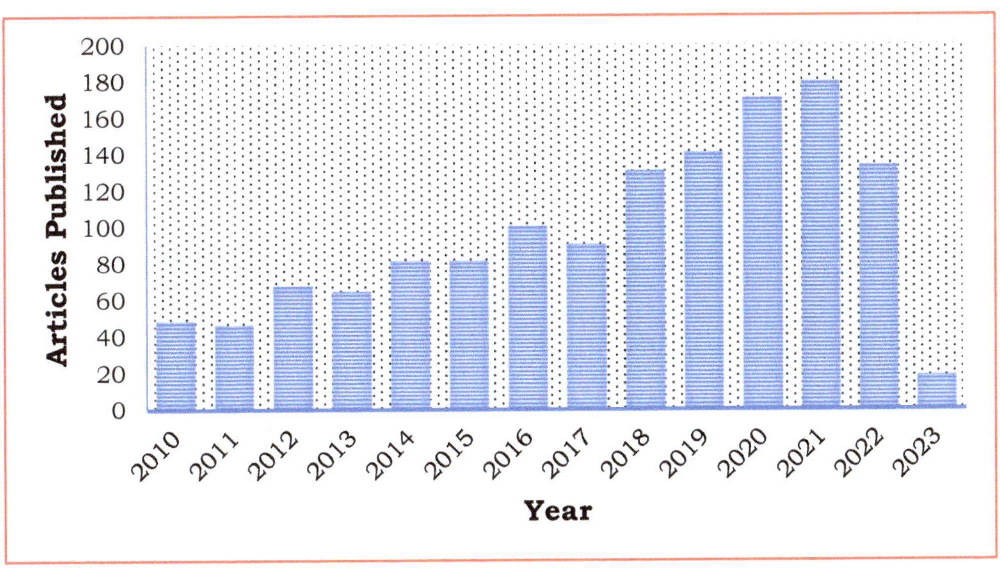

Figure 8. Number of voltage stability indices articles published.

Table 1. Line Voltage Stability Indices.

VSI	Concept	Assumption	Equation	Condition for Stability	Objective	References
VSLI (1995)	Power flow in a transmission line of a two-bus system.	Shunt admittance is ignored.	$VSLI = \frac{4[V_s V_r \cos(\delta) - V_r^2 \cos(\delta)^2]}{V_s^2}$	Stable: VLSI < 1 Unstable: VLSI > 1	Identify the critical buses near voltage collapse.	[40–42]
L_{mn} (1998)	Power flow in a transmission line of a two-bus system.	The effect of active power is ignored, and the shunt admittance is approximately zero.	$L_{mn} = \frac{4xQ_r}{[V_s sin(\theta - \delta)]^2}$	Stable: $L_{mn} < 1$ Unstable: $L_{mn} > 1$	Online monitoring that predicts the voltage collapse and identifies the stressed condition.	[43–45]
LQP (1998)	Power flow in a transmission line of a two-bus system.	$Y \approx 0$ & $R/Z << 1$	$LQP = 4\left(\frac{X}{V_s^2}\right)\left(Q_r + X\frac{P_s^2}{V_s^2}\right)$	Stable: LQP < 1 Unstable: LQP > 1	Performs faster, evaluates the static voltage collapse.	[46–48]
VCPI (1998)	Maximum Power loss & Maximum Power transfer in a Transmission Line.	Constant Power Factor Constant Shunt admittance is ignored.	$VCPI\ (1) = \frac{P_r}{P_{r(max)}}$ $P_{r(max)} = \frac{V_s^2}{Z_s}\frac{\cos\varnothing}{4\cos^2\frac{(\theta-\varnothing)}{2}}$ $VCPI\ (2) = \frac{Q_r}{Q_{r(max)}}$ $Q_{r(max)} = \frac{V_s^2}{Z_s}\frac{\sin\varnothing}{4\cos^2\frac{(\theta-\varnothing)}{2}}$ $VCPI\ (3) = \frac{P_l}{P_{l(max)}}$ $P_{l(max)} = \frac{V_s^2}{Z_s}\frac{\cos\theta}{4\cos^2\frac{(\theta-\varnothing)}{2}}$ $VCPI\ (4) = \frac{Q_r}{Q_{r(max)}}$ $Q_{l(max)} = \frac{V_s^2}{Z_s}\frac{\sin\theta}{4\cos^2\frac{(\theta-\varnothing)}{2}}$	Stable: VCPI < 1 Unstable: VCPI > 1	Determines the range of the collapse point, which depends upon the systems' generation load characteristics.	[49–52]
L_p (2001)	Power flow in a transmission line of a two-bus system.	Shunt admittance is ignored, and the effect of reactive power.	$L_p = \frac{4RP_r}{[V_s \cos(\theta - \delta)]^2}$	Stable: $L_p < 1$ Unstable$L_p > 1$	Voltage stability assessment in a radial distribution. Simple index with higher accuracy.	[53–56]
FVSI (2002)	Power flow in a transmission line of a two-bus system.	Shunt admittance is ignored and $\sin\delta \approx 0$, $\cos\delta \approx 1$, $R\sin\delta \approx 0$, $X\cos\delta \approx X$.	$FVSI = \frac{4Z^2 Q_r}{V_s^2 X}$	Stable: FVSI < 1 Unstable: FVSI > 1	Determines collapse point, weakest bus, critical line, and maximum loadability.	[57–60]
VSLBI (2003)	During the maximum power conditions, the voltage drop in impedance equals load bus voltage.	Thevenin equivalent impedance connected to the sending end bus is ignored.	$VSLBI = \frac{V_r}{\Delta V}$	Stable: VSLBI > 1 Unstable: VLSBI < 1	The Access to voltage collapse is local monitoring, watchful, and emergency control during voltage-sensitive load.	[61,62]
VSMI (2004)	VSMI considers the relation between the angular difference of voltage and maximum power transfer.	The shunt admittance neglected	$VSMI = \frac{\delta_{rmax} - \delta}{\delta_{rmax}}$	Stable: VSMI > 0 Unstable: VLMI < 0	Evaluate stability margin and determine weak locations.	[63–66]

Table 1. *Cont.*

VSI	Concept	Assumption	Equation	Condition for Stability	Objective	References
VCPI_1 (2005)	The voltage drop across the Thevenin impedance is equal to the load; at the collapse point.	Thevenin equivalent impedance connected to the sending end bus is ignored.	$VCPI\_1 = V_r cos\delta - 0.5V_s$	VCPI_1 $\geq$ 0: Stable VCPI_1 < 0: Unstable	Online evaluation index. Identify the weakest lines by distinguishing the minor power outage buses with distance.	[67,68]
V_{cr} (2006)	The load flow equations and Eigenvalue theorem.	Constant Power factor and load.	$V_{cr} = \frac{E}{\sqrt{2(1+cos(\theta-\varnothing))}}$ $V_{cr} = \frac{E}{2cos\theta}$	The system is unstable if the Jacobian power matrix is singular	Specifies the minimum voltage where the system performs away from collapse.	[69,70]
PTSI (2006)	Maximum Power loss and maximum power transferable through a line are limited.	Shunt admittance is ignored.	$PTSI = \frac{2S_L Z(1+cos(\theta-\varphi))}{V_s^2}$	Stable: PTSI < 1 Unstable: PTSI > 1	Predicts the dynamic voltage collapse and calculates the effect of adding additional equipment.	[71–74]
VSI_1 (2006)	Maximum Power loss and maximum power transferable through a line are limited.	The resistance of line and shunt admittance is ignored.	$VSI = min\left(\frac{P_{margin}}{P_{max}}, \frac{Q_{margin}}{Q_{max}}, \frac{S_{margin}}{S_{max}}\right)$	Stable: VSI_1 > 0 Unstable: VSI_1 < 0	Predicts steady-state voltage stability, determines the stability margin of every load bus.	[75,76]
NLSI (2007)	Power flow in a transmission line of a two-bus system, "Critical Clearing Time" (CCT).	The minimal angular difference between receiving and sending voltage and the shunt admittance is ignored.	$NLSI = \frac{R_{sr}P_r + X_{sr}Q_r}{0.25V_s^2}$	Stable: NLSI < 1 Unstable: NLSI > 1	By varying active and reactive power, evaluating the collapse point, Rank transmission lines.	[42,77,78]
SI (2007)	Voltage Quadratic Equation.	The shunt admittance is ignored.	$SI(r) = 2V_s^2V_r^2 - V_r^4 - 2V_r^2(PR + QX) - \lvert Z\rvert^2(P^2 + Q^2)$	Stable: SI $\neq$ 0 Unstable: SI = 0	Predict the most vulnerable bus exposed to collapse in the radial distribution system.	[79–81]
VSM (2009)	Maximum Power loss and maximum power transfer in a transmission line.	Constant power factor and shunt admittance is ignored.	$VSM_Z = \frac{Z_r - Z_0}{Z_0}$ $S_{cr} = \frac{V_s^2}{2Z_0\lvert 1+cos(\theta_0-\theta_r)\rvert}$	Stable: VSM > 0 Unstable: VSM < 0	Determining the loss of voltage, VSM can efficiently prevent voltage collapse.	[82–85]
VQI (2010)	Power flow in a transmission line of a two-bus system.	Zero angular difference and shut admittance are ignored.	$VQI = \frac{4Q_r}{\lvert B_{sr}\rvert\lvert V_r\rvert^2} \leq 1$	Stable: VQI < 1 Unstable: VQI > 1	Determines the critical bus, detects the instability in large-scale systems, and the distance of the collapse point.	[86–89]
LCPI (2012)	Voltage Quadratic Equation.	The transmission lines model is like π model.	$LCPI = \frac{4Acos\alpha(P_rBcos\beta + Q_rBsin\beta)}{(V_s cos\delta)^2}$	Stable: LCPI < 1 Unstable: LCPI > 1	Combines the influence of the relative flow of reactive and active power flow with ABCD.	[90–93]

Table 1. *Cont.*

VSI	Concept	Assumption	Equation	Condition for Stability	Objective	References
NVSI (2013)	Power flow in a transmission line. Reactive Power Sensitivity (RPS).	The Shunt admittance and line resistance are ignored.	$NVSI = \frac{2X\sqrt{(P_r^2+Q_r^2)}}{4Q_rX-V_s^2}$	Stable: NVSI < 1 Unstable: NVSI > 1	Evaluate the voltage stability effectively by varying the P & Q, determining the weak bus and sensitive line.	[94–97]
ITLTI (2016)	ABCD parameters, the power factor of the receiving end, and the power angle between the receiving and sending ends.	Two power circles with two distinct centers but identical radius.	$S_{r\_index} = \frac{sin(\theta_r'+\delta')sin\delta'}{\left(\cos\frac{\theta_r'}{2}\right)^2}$	Stable: ITLTI < 1 Unstable: ITLTI $\geq$ 1	For radial transmission networks and subsequently adapted for larger systems, the Weakest line.	[4,98]
CBI (2018)	Active and Reactive power changes.	Negligible system impedance, linearized power flow model.	$CBI_{sr} = \sqrt{\Delta P_{sr}^2 + \Delta Q_{sr}^2}$	Stable: CBI > 1 Unstable: CBI = 0	Operates with Lagrange Constant Computational Method (LCM); Determines the critical boundaries and voltage stability with lesser parameters.	[99–102]
LVSI (2018)	Voltage Quadratic Equation, ABCD Parameters.	line's resistance and charging capacitance are ignored.	$LVSI = \frac{2V_rA\cos(\beta-\alpha)}{V_s\cos(\beta-\delta)}$	Stable: LVSI > 1 Unstable: LVSI < 1	Evaluate the stability margin considering the ABCD parameters expressed in MVA.	[103–105]
BVSI (2022)	Voltage Quadratic Equation.	Line shunt admittance and the reactive power's effects are ignored.	$BVSI = \frac{4RP_rZ^2}{(V_s(Rcos\delta+Xsin\delta))}$	Stable: BVSI < 1 Unstable: BVSI > 1	Optimal location and sizing of distributed generations.	[27]

Table 2. Bus Voltage Stability Indices.

VSI	Concept	Assumption	Equation	Condition for Stability	Objective	References
L–Index (1986)	Power flow equation solution, Eigen Values.	All generator voltages remain constant.	$L = \underset{j\in\alpha_L}{\mathrm{MAX}}\left\lvert 1 - \frac{\Sigma_{i\in\alpha_G} F_{ji}V_i}{\underline{V}_j}\right\rvert$	Stable: L-Index < 1 Unstable: L-Index > 1	Identify the critical points of the system.	[106–108]
VIPI (1989)	Power flow equation solution.	steady-state condition, system impedance is negligible.	$VIPI = \theta = \cos^{-1}\frac{Y_s^TY(a)}{\lVert Y_s\rVert.\lVert Y(a)\rVert}$	The operational solutions estimate critical points and a proximate fictitious solution.	The potential voltage instability problems and the efficient control approach for avoiding instability.	[109–112]
VCPI (2004)	Power Flow Equation.	It utilizes offline and online measurement.	$VCPI_{k^{th}bus} = \left\lvert 1 - \frac{\Sigma_{m=1, m\neq k}^{N}\lvert V_m'\rvert}{V_k}\right\rvert$	Stable: VCPI < 1 Unstable: VCPI > 1	The system's proximity to voltage collapse.	[113–116]
SDC (2004)	Maximum Power Transfer Theorem.	Minimal values, such as a change in voltage at receiving, are ignored.	$SDC = \left\lvert 1 + \frac{\Delta \underline{V}_j^{(k+1)}\underline{I}_{ji}^{(k)*}}{\underline{V}_j^{(k)}\Delta\underline{I}_{ji}^{(k+1)*}}\right\rvert$	SDC > 0: Stable SDC < 0: Unstable	Used to protect voltage collapse depends on local bus phasor current and voltage at every line's relay point.	[117–119]

Table 2. *Cont.*

VSI	Concept	Assumption	Equation	Condition for Stability	Objective	References
ISI (2006)	Maximum transferred when Thevenin impedance's magnitude equals the load impedance's amplitude.	Constant system topology.	$ISI = (\hat{Z}_r - \hat{Z}_{th})/\hat{Z}_r$ $ISI = 1 - \lvert \underline{I}_r \Delta \underline{V}_r \rvert / \lvert \underline{V}_r \Delta \underline{I}_r \rvert$	ISI > 0: Stable ISI < 0: Unstable	Calculates the stability of the system.	[120–123]
VSI_{BUS} (2007)	The rise in sending apparent power no longer yields a rise in receiving line power.	Minimal values, such as a change in voltage at receiving, are ignored.	$VSI_i = \left[1 + \left(\frac{I_i}{V_i}\right)\left(\frac{\Delta V_i}{\Delta I_i}\right)\right]^{\alpha}$	The VSIBUS value lies between unity and zero	Identifying the distance to the collapse point with the help of local voltage.	[124,125]
Z_L/Z_S Ratio (2007)	Maximum transferred when the magnitude of Thevenin impedance is equal to the amplitude of the load impedance.	$730 \leq \phi s \leq 870$	$\frac{Z_L}{Z_S} = \frac{M+1}{-M\cos\beta + \left[(M\cos\beta)^2 - M^2 + 1\right]^{0.5}}$ $M = \left(\frac{dS^*}{dY^*}\right) = \frac{(S_2 - S_1)(Y_2 + Y_1)}{(S_2 + S_1)(Y_2 - Y_1)}$	Stable : $\frac{Z_L}{Z_S}$ Ratio > 1 Unstable : $\frac{Z_L}{Z_S}$ Ratio < 1	ENCVI is accurate in design and measuring and simple in real-time implementation.	[110,126,127]
ENVCI (2009)	Equivalent system model (ESM) & Equivalent local network model (ELNM).	Consider the effects of the local network and the system outside the local network.	$ENVCI = 2(e_k e_n + f_k f_n) - (e_k^2 + f_k^2)$	Stable: ENVCI > 0 Unstable: ENVCI < 0	Optimal placement depends on ESM and considers only local voltage phasors.	[4,128–130]
PSI (2012)	Maximum power transfer theorem.	DG depends on the most critical bus.	$PSI = \frac{4R_{ij}(P_L - P_G)}{[\lvert V_i \rvert \cos(\theta - \delta)]^2}$	Stable: PSI < 1 Unstable: PSI > 1	DG and sizing for distribution networks.	[110,131,132]
VDI (2012)	Negligible phase angle deviations.	The real value of the deviation of the bus voltage.	$VDI_T = \sum_{j=1}^{N} \lvert 1 - V_j \rvert$	0: Perfect voltage regulation 1: worst case voltage regulation.	Optimal placement of DG and sizing for distribution network.	[133–135]
SVSI (2014)	During the maximum Power, the load bus voltage equals the impedance voltage drop across the line.	The voltage at the nearest generator to the load bus equals the Thevenin load voltage.	$SVSI_i = \frac{\Delta V_i}{\beta \times V_i}$	Stable: SVSI < 1 Unstable: SVSI > 1	Uses the data of present operating conditions combined with phasor voltage measurement.	[136–138]
P-Index (2017)	The ratio of power loss to power gained.	For incremental increase, the power factor is unchanged.	$\text{P-index} = \frac{-2\frac{P_r}{V_r}\frac{dV_r}{dP_r}}{1 - 2\frac{P_r}{V_r}\frac{dV_r}{dP_r}}$	At stability limit when $dV_r/dP_r = \infty$, the value would be 1	Dynamic voltage stability assessment and load shedding purposes.	[39]

Table 3. Characteristics of Stability Indices.

Characteristic	Stability Indices
Optimal placement of DG & DG sizing	Line VSIs, Bus VSIs except for SDC, VSI_{bus}, ISI, and Z_L/Z_S ratio
Impedance dependent VSI	VSLI, L VCPI_1, VSLBI, ISI, SDC, VSI_{bus}, Z_L/Z_S ratio
Independent VSI	VSMI, SI, LCPI, $VCPI_{BUS}$, NLSI, VCPI, NVSI, SVSI, FVSI, L_{mn}, LQP, L_p, VIPI
Reduce Power Losses	Line VSIs, Bus VSIs

4. Conclusions

Voltage stability indices are essential numerical parameters for measuring and evaluating stability points in the power system network. These indices' significant roles help detect instability points early, preventing systems from voltage collapse and power outages, minimizing the actual power losses, and enhancing the system's overall efficiency. The evaluation of these indices indicates the power system operators' performances of a transmission system. Further, the stability of a system can improve through proper design, reactive power compensation, and optimal power flow. This article discusses the mathematical modeling of 34 line and bus stability indices. The significance of these indices for different power system problems with suitable objectives is also discussed with the help of suitable works from the literature, around 138. This article may be attractive to a researcher who has started researching power system stability analysis, the placement of reactive power compensators, and DG placement for stability enhancement. The main intention of this review is to summarize and highlight how crucial it is to choose suitable stability indices to preserve systems' reliability and stability. Developing various stability indices for effective utilization in sustaining the instabilities across various operational scenarios has been undertaken.

The VSIs review various elements and perspectives and derive from the concepts of maximum power transfer through a line (VCPI, PTSI, VSI_1, VSMI), solution of voltage equation (VQI_{line}, VSLI, L-index, SI, $VCPI_{bus}$ LCPI, FVSI, L_{mn}, Lp, LQP, NVSI, NLSI), P-V Curve (SDC, SI, VIPI), and maximum power transfer theorem (SDC, VSI_{bus}, ISI, Z_L/Z_S, VCPI_1, VSLBI, SVSI). Most VSIs which correspond to the voltage equation concept should have a solution. A few VSIs neglect the sending-end bus Thevenin impedance (VCPI_1, VSLBI), and few VSIs consider bus Thevenin impedance to be extremely sensitive for a minute variation in parameters of two consecutive measurements (VSI_{bus}, SDC, Z_L/Z_S ratio, and SDC).

In the procedure to enhance the efficiency and minimize the operational time of DG placement and sizing issues, the primary phase considers the simple line and bus VSIs, and the secondary phase considers the bus and overall VSIs operating with improved efficiency. This review article mainly focuses and presents information on the significant measurement of line and bus stability indices for maintaining the voltage and reactive power in a power system network.

Furthermore, this review can assist in choosing the desired stability index for a given scenario. The main findings of this paper can be highlighted as follows:

- The power system stability analysis;
- A comprehensive review of 34 Voltage Stability Indices derived mathematically;
- Voltage Stability Indices evaluated the sizing and placement of distributed energy sources;
- Various power system issues and their corresponding application of Voltage Stability Indices were presented;
- The corresponding data: name, mathematical calculation, concept, assumptions, condition for stability, and objective for each VSI are listed;
- This review article supports researchers, power system operators, and engineers regarding stability indices.

Funding: This work was supported by the Basic Science Research Program through the National Research Foundation of Korea (NRF), which is funded by the Ministry of Education under Grant NRF-2021R1I1A3053429.

Data Availability Statement: Not Applicable.

Conflicts of Interest: The authors declare no conflict of interest.

References

1. Eremia, M.; Shahidehpour, M. *Handbook of Electrical Power System Dynamics: Modeling, Stability, and Control*; Wiley: Hoboken, NJ, USA, 2013.
2. Kundur, P.; Paserba, J.; Ajjarapu, V.; Andersson, G.; Bose, A.; Canizares, C. Definition and classification of power system stability IEEE/CIGRE joint task force on stability terms and definitions. *IEEE Trans. Power Syst.* **2004**, *19*, 1387–1401.
3. Modarresi, J.; Gholipour, E.; Khodabakhshian, A. A comprehensive review of the voltage stability indices. *Renew. Sustain. Energy Rev.* **2016**, *63*, 1–12. [CrossRef]
4. Danish, M.S.S.; Senjyu, T.; Danish, S.M.S.; Sabory, N.R.; Narayanan, K.; Mandal, P. A recap of voltage stability indices in the past three decades. *Energies* **2019**, *12*, 1544. [CrossRef]
5. Lim, Z.J.; Mustafa, M.W.; Muda, Z.B. Evaluation of the effectiveness of voltage stability indices on different loadings. In Proceedings of the International Power Engineering and Optimization Conference, Melaka, Malaysia, 6–7 June 2012.
6. Rahman, T.K.A.; Jasmon, G.B. A new technique for voltage stability analysis in a power system and improved load flow algorithm for the distribution network. In Proceedings of the International Conference on Energy Management and Power Delivery EMPD, Singapore, 21–23 November 1995.
7. Moghavvemi, M.; Omar, F.M. Technique for contingency monitoring and voltage collapse prediction. *IEE Proc. Gener. Transm. Distrib.* **1998**, *145*, 634–640. [CrossRef]
8. Mohamed, A.; Jasmon, G.B.; Yusof, S. A static voltage collapse indicator using line stability factors. *J. Ind. Technol.* **1998**, *7*, 73–85.
9. Moghavvemi, M.; Faruque, O. Real-time contingency evaluation and ranking technique. *IEE Proc.—Gener. Transm. Distrib.* **1998**, *145*, 517–524. [CrossRef]
10. Moghavvemi, M.; Faruque, M.O. Technique for assessment of voltage stability in the ill-conditioned radial distribution network. *IEEE Power Eng. Rev.* **2001**, *21*, 58–60. [CrossRef]
11. Musirin, I.; Khawa, T.; Rahman, A. Novel fast voltage stability index (FVSI) for voltage stability analysis in a power transmission system. In Proceedings of the Student Conference on Research and Development Proceedings, Shah Alam, Malaysia, 17 July 2002.
12. Milosevic, B.; Begovic, M. Voltage-stability protection and control using a wide-area network of phasor measurements. *IEEE Trans. Power Syst.* **2003**, *18*, 121–127. [CrossRef]
13. He, T.; Kolluri, S.; Mandal, S.; Galvan, F.; Rastgoufard, P. Identification of weak locations in bulk transmission systems using voltage stability margin index. In Proceedings of the IEEE Power Engineering Society General Meeting, Denver, CO, USA, 6–10 June 2004.
14. Wang, L.; Liu, Y.; Luan, Z. Power transmission paths-based voltage stability assessment. In Proceedings of the IEEE/PES Transmission and Distribution Conference and Exhibition: Asia and Pacific, Dalian, China, 18 August 2005.
15. Sun, H.; Zhou, X.; Li, R. Accuracy analysis of static voltage stability indices based on power flow model. In Proceedings of the 2005 IEEE/PES Transmission and Distribution Conference and Exposition: Asia and Pacific, Dalian, China, 18 August 2005.
16. Nizam, M.; Mohamed, A.; Hussain, A. Performance evaluation of voltage stability indices for dynamic voltage collapse prediction. *Appl. Sci.* **2006**, *6*, 1104–1113. [CrossRef]
17. Gong, Y.; Schulz, N.; Guzman, A. Synchro phasor-based real-time voltage stability index. In Proceedings of the PSCE Conference, Atlanta, GA, USA, 29 October–1 November 2006.
18. Yazdanpanah-Goharrizi, A.; Asghari, R. A novel line stability index (NLSI) for voltage stability assessment of power systems. In Proceedings of the 7th WSEAS International Conference on Power Systems, Beijing, China, 15–17 September 2007.
19. Eminoglu, U.; Hocaoglu, M.H. A voltage stability index for radial distribution networks. In Proceedings of the 42nd International Universities Power Engineering Conference, Brighton, UK, 4–6 September 2007.
20. Guiping, D.; Yuanzhang, S.; Jian, X. A new index of voltage stability considering distribution network. In Proceedings of the Asia-Pacific Power and Energy Engineering Conference, Wuhan, China, 27–31 March 2009.
21. Althowibi, F.A.; Mustafa, M.W. Line voltage stability calculations in power systems. In Proceedings of the IEEE International Conference on Power and Energy, Kuala Lumpur, Malaysia, 29 November–1 December 2010.
22. Tiwari, R.; Niazi, K.R.; Gupta, V. Line collapse proximity index for prediction of voltage collapse in power systems. *Int. J. Electr. Power Energy Syst.* **2012**, *41*, 105–111. [CrossRef]
23. Kanimozhi, R.; Selvi, K. A novel line stability index for voltage stability analysis and contingency ranking in power system using fuzzy based load flow. *J. Electr. Eng. Technol.* **2013**, *8*, 694–703. [CrossRef]
24. Chuang, S.J.; Hong, C.M.; Chen, C.H. Improvement of integrated transmission line transfer index for power system voltage stability. *Int. J. Electr. Power Energy Syst.* **2016**, *78*, 830–836. [CrossRef]
25. Furukakoi, M.; Adewuyi, O.B.; Danish, M.S.S.; Howlader, A.M.; Senjyu, T.; Funabashi, T. Critical Boundary Index (CBI) based on active and reactive power deviations. *Int. J. Electr. Power Energy Syst.* **2018**, *100*, 50–57. [CrossRef]
26. Ratra, S.; Tiwari, R.; Niazi, K.R. Voltage stability assessment in power systems using line voltage stability index. *Comput. Electr. Eng.* **2018**, *70*, 199–211. [CrossRef]
27. Ismail, B.; Wahab, N.I.A.; Othman, M.L.; Radzi, M.A.M. New Line Voltage Stability Index (BVSI) for Voltage Stability Assessment in Power System: The Comparative Studies. *IEEE Access* **2022**, *10*, 103906–103931. [CrossRef]
28. Kessel, P.; Glavitsch, H. Estimating the voltage stability of a power system. *IEEE Trans. Power Deliv.* **1986**, *1*, 346–354. [CrossRef]

29. Tamura, Y.; Sakamoto, K.; Tayama, Y. Voltage instability proximity index (VIPI) based on multiple load flow solutions in ill-conditioned power systems. In Proceedings of the 27th IEEE Conference on Decision and Control, IEEE, Austin, TX, USA, 7–9 December 1988.
30. Balamourougan, V.; Sidhu, T.S.; Sachdev, M.S. Technique for online prediction of voltage collapse. *IEEE Gener. Transm. Distrib.* **2004**, *151*, 453–460. [CrossRef]
31. Verbic, G.; Gubina, F. A new concept of voltage-collapse protection based on local phasors. *IEEE Trans. Power Deliv.* **2004**, *19*, 576–581. [CrossRef]
32. Smon, I.; Verbic, G.; Gubina, F. Local voltage-stability index using Tellegen's theorem. *IEEE Trans. Power Syst.* **2006**, *21*, 1267–1275. [CrossRef]
33. Haque, M.H. Use of local information to determine the distance to voltage collapse. In Proceedings of the International Power Engineering Conference, Singapore, 3–6 December 2007.
34. Wiszniewski, A. New criteria of voltage stability margin for the purpose of load shedding. *IEEE Trans. Power Deliv.* **2007**, *22*, 1367–1371. [CrossRef]
35. Wang, Y.; Li, W.; Lu, J. A new node voltage stability index based on local voltage phasors. *Electr. Power Syst. Res.* **2009**, *79*, 265–271. [CrossRef]
36. Aman, M.M.; Jasmon, G.B.; Mokhlis, H.; Bakar, A.H.A. Optimal placement and sizing of a DG based on a new power stability index and line losses. *Int. J. Electr. Power Energy Syst.* **2012**, *43*, 1296–1304. [CrossRef]
37. Yang, C.F.; Lai, G.G.; Lee, C.H.; Su, C.T.; Chang, G.W. Optimal setting of reactive compensation devices with an improved voltage stability index for voltage stability enhancement. *Int. J. Electr. Power Energy Syst.* **2012**, *37*, 50–57. [CrossRef]
38. Perez-Londōno, S.; Rodríguez, L.F.; Olivar, G. A Simplified Voltage Stability Index (SVSI). *Int. J. Electr. Power Energy Syst.* **2014**, *63*, 806–813. [CrossRef]
39. Kamel, M.; Karrar, A.A.; Eltom, A.H. Development and Application of a New Voltage Stability Index for On-Line Monitoring and Shedding. *IEEE Trans. Power Syst.* **2018**, *33*, 1231–1241. [CrossRef]
40. Makasa, K.J.; Venayagamoorthy, G.K. Online voltage stability load index estimation based on PMU measurements. In Proceedings of the 2011 IEEE Power and Energy Society General Meeting, Detroit, MI, USA, 24–28 July 2011.
41. Alias, M.R.N.M.; Verayiah, R.; Yin, L.J. Investigation of Power Flow Tracing Strategy Using an Index in a Transmission System with Solar PV for Congestion Management. In Proceedings of the 11th International Conference on Power and Energy Systems (ICPES), Shanghai, China, 18–20 December 2021.
42. Oukennou, A.; Sandali, A. Analysis and Comparison of Line Voltage Stability Indices. In Proceedings of the 6th International Renewable and Sustainable Energy Conference (IRSEC), Rabat, Morocco, 5–8 December 2018.
43. Salah, M.; Wang, Y.; Chen, T. A novel collapse prediction index for voltage stability analysis and contingency ranking in power systems. *Prot. Control. Mod. Power Syst.* **2023**, *8*, 7.
44. Shekhawat, N.; Gupta, A.K.; Sharma, A.K. Voltage Stability Assessment Using Line Stability Indices. In Proceedings of the 3rd International Conference and Workshops on Recent Advances and Innovations in Engineering (ICRAIE), Jaipur, India, 22–25 November 2018.
45. Calma, E.R.B.; Pacis, M.C. Artificial Neural Network-based Voltage Stability Analysis of Power Transmission Networks with Distributed Generation Using Phasor Measurement Unit Synthetic Data. In Proceedings of the IEEE 12th Control and System Graduate Research Colloquium (ICSGRC), Shah Alam, Malaysia, 6 August 2021.
46. Singh, P.; Parida, S.K.; Chauhan, B.; Choudhary, N. Online Voltage Stability Assessment Using Artificial Neural Network Considering Voltage Stability Indices. In Proceedings of the 21st National Power Systems Conference (NPSC), Gandhinagar, India, 17–19 December 2020.
47. Chintakindi, R.; Mitra, A. Detecting Weak Areas in a Real-time Power System on a Static Load Model using Standard Line Voltage Stability Indices. In Proceedings of the IEEE 2nd International Conference on Smart Technologies for Power, Energy, and Control (STPEC), Chhattisgarh, India, 12 November 2021.
48. Verayiah, R.; Abidin, I.Z. A Study on static voltage collapse proximity indicators. In Proceedings of the IEEE 2nd International Power and Energy Conference, Johor Bahru, Malaysia, 1–3 December 2008.
49. Isaiah, A.; Yanxia, S. New performance indices for voltage stability analysis in a power system. *Energies* **2017**, *10*, 2042.
50. Salama, M.M.; Saied, E.M.; Abou-Elsaad, M.M.; Ghariany, E.F. Estimating the voltage collapse proximity indicator using an artificial neural network. *Energy Convers. Manag.* **2001**, *42*, 69–79. [CrossRef]
51. Chaithra, A.; Modi, S. Power System Vulnerability Assessment using Voltage Collapse Proximity Index. In Proceedings of the 2nd IEEE International Conference on Power Electronics, Intelligent Control and Energy Systems (ICPEICES), Delhi, India, 22–24 October 2018.
52. Chaithra, A.; Modi, S. Evaluation of Power System Stability using Voltage Proximity Index. In Proceedings of the International Conference on Recent Innovations in Electrical, Electronics & Communication Engineering (ICRIEECE), Bhubaneswar, India, 27–28 July 2018.
53. Khunkitti, S.; Premrudeepreechacharn, S. Voltage Stability Improvement Using Voltage Stability Index Optimization. In Proceedings of the 2020 International Conference on Power, Energy, and Innovations (ICPEI), Chiangmai, Thailand, 14–16 October 2020.

54. Xi, Z.; Kong, W.C. Comparison of voltage stability indexes considering dynamic load. In Proceedings of the 2011 IEEE Electrical Power and Energy Conference, Winnipeg, MB, Canada, 3 October 2011.
55. Soyemi, A.; Misra, S.; Oluranti, J.; Ahuja, R. Evaluation of Voltage Stability Indices. In *Information Systems and Management Science*; Springer: Cham, Switzerland, 2022; Volume 303.
56. Hosseinzadeh, N.; Aziz, A.; Mahmud, A.; Gargoom, A.; Rabbani, M. Voltage Stability of Power Systems with Renewable-Energy Inverter-Based Generators: A Review. *Electronics* **2021**, *10*, 115. [CrossRef]
57. Jirjees, M.A.; Al-Nimma, D.A.; Al-Hafidh, M.S.M. Voltage Stability Enhancement Based on Voltage Stability Indices Using FACTS Controllers. In Proceedings of the 2018 International Conference on Engineering Technology and their Applications (IICETA), Al-Najaf, Iraq, 8–9 May 2018.
58. Pacis, M.C.; Antonio, I.J.M.; Bañaga, I.J.T. Under Voltage Load Shedding Algorithm using Fast Voltage Stability Index (FVSI) and Line Stability Index (LSI). In Proceedings of the IEEE 13th International Conference on Humanoid, Nanotechnology, Information Technology, Communication and Control, Environment, and Management, Manila, Philippines, 28–30 November 2021.
59. Ali, M.B.; Kazmi, S.A.A.; Khan, Z.A.; Altamimi, A.; Alghassab, M.A.; Alojaiman, B. Voltage Profile Improvement by Integrating Renewable Resources with Utility Grid. *Energies* **2022**, *15*, 8561. [CrossRef]
60. Meena, G.; Verma, K.; Mathur, A. FVSI-Based Meta-heuristic Algorithm for Optimal Load Shedding to Improve Voltage Stability. In Proceedings of the IEEE 10th Power India International Conference (PIICON), New Delhi, India, 25–27 November 2022.
61. Ramírez-Perdomo, S.L.; Lozano, C.A. Evaluation of indices for voltage stability monitoring using PMU measurements. *Ing. E Investig.* **2014**, *34*, 44–49. [CrossRef]
62. Arbelaez Ramirez Marcela, L. Monitoring Voltage Stability Margin across Several Transmission Lines Using Synchro Phasor Measurements. Ph.D. Thesis, Iowa State University, Ames, Iowa, 2016.
63. Sajan, K.S.; Kumar, V.; Tyagi, B. Genetic algorithm-based support vector machine for online voltage stability monitoring. *Int. J. Electr. Power Energy Syst.* **2015**, *73*, 200–208. [CrossRef]
64. Naganathan, G.S.; Babulal, C.K. Optimization of support vector machine parameters for voltage stability margin assessment in the deregulated power system. *Soft Comput.* **2019**, *23*, 10495–10507. [CrossRef]
65. Reddy, M.V.; Murthy, V.S.K. Improvement of voltage stability based on static and dynamic criteria. In Proceedings of the 16th National Power Systems Conference, Hyderabad, India, 15–17 December 2010.
66. Yunhwan, L.; Hwachang, S. Decentralized Load Shedding Method Based on Voltage Stability Margin Index Using Synchrophasor Measurement Technology. *Electronics* **2018**, *7*, 277.
67. Dharmapala, K.D.; Rajapakse, A.; Narendra, K.; Zhang, Y. Machine Learning Based Real-Time Monitoring of Long-Term Voltage Stability Using Voltage Stability Indices. *IEEE Access* **2020**, *8*, 222544–222555. [CrossRef]
68. Mesgarnejad, H.; Shahrtash, S.M. Power system voltage stability assessment employing phasor measurement units. In Proceedings of the 2nd International Conference on Computer and Automation Engineering (ICCAE), Singapore, 26–28 February 2010.
69. Ankit, S.; Akash, S. Application of supervised learning for voltage stability assessment using GVSM. *Electr. Electron. Technol. Open Access J.* **2018**, *2*, 281–287.
70. Hameedullah, Z.; Shah, D.M.S.; Tomonobu, S.; Mikaeel, A.; Abdul, N.; Mohebullah, W.; Mahdi, K.; Paras, M. A Contemporary Novel Classification of Voltage Stability Indices. *Appl. Sci.* **2020**, *10*, 1639.
71. Ghassan, S.H.; Husham, I. Assessment of voltage stability based on power transfer stability index using com-putational intelligence models. *Int. J. Electr. Comput. Eng.* **2021**, *11*, 2790–2797.
72. Izzri, N.; Mehdi, O.H.; Abdalla, A.N.; Jaber, A.S.; Shalash, N.A.; Noori, Y.L. Fast Prediction of Power Transfer Stability Index Based on Radial Basis Function Neural Network. *Int. J. Phys. Sci.* **2011**, *6*, 7978–7984.
73. Nizam, M.; Mohamed, A.; Hussain, A. Dynamic Voltage Collapse Prediction on a Practical Power System Using Power Transfer Stability Index. In Proceedings of the 5th Student Conference on Research and Development, Selangor, Malaysia, 11–12 December 2007; pp. 1–6. [CrossRef]
74. Woldu, T.A.; Ziegler, C.; Wolter, M. A New Method for Prediction of Static and Dynamic Voltage Collapse Using Node Parameters in Large Power Networks. In Proceedings of the IEEE PES Innovative Smart Grid Technologies Europe (ISGT-Europe), The Hague, The Netherlands, 26–28 October 2020.
75. Jamroen, C.; Dechanupaprittha, S. Synchrophasor-based early-warning voltage instability index. In Proceedings of the IEEE Manchester PowerTech, Manchester, UK, 18–22 June 2017.
76. Rohikaa, R.M.; Lakshmi, R.; Sunitha, R.; Ashok, S. Assessment of voltage stability in microgrid. In Proceedings of the International Conference on Electrical, Electronics, and Optimization Techniques (ICEEOT), Chennai, India, 3–5 March 2016.
77. Oukennou, A.; Sandali, A. Assessment and analysis of Voltage Stability Indices in electrical networks using PSAT Software. In Proceedings of the Eighteenth International Middle East Power Systems Conference (MEPCON), Cairo, Egypt, 27–29 December 2016.
78. Mahiraj, R.S.; Vadhera, S. Voltage Stability Assessment Techniques for Modern Power Systems. In *Novel Advancements in Electrical Power Planning and Performance*; IGI Global: Hershey, PA, USA, 2020; pp. 128–176.
79. Gupta, A.R. Effect of optimal allocation of multiple DG and D-STATCOM in the radial distribution system for minimizing losses and THD. In Proceedings of the 7th International Symposium on Embedded Computing and System Design (ISED), Durgapur, India, 18–20 December 2017.

80. Narayan, K.S.; Kumar, A. Probabilistic analysis of radial distribution network performance with varying wind speed levels. In Proceedings of the 6th IEEE Power India International Conference (PIICON), Delhi, India, 5–7 December 2014.
81. Pattabhi, M.B.; Lakshmikantha, B.R.; Sundar, K.S. A Novel Method for Contingency Ranking Based on Voltage Stability Criteria in Radial Distribution Systems. *Technol. Econ. Smart Grids Sustain. Energy* **2022**, *7*, 9. [CrossRef]
82. Sagwal, R.; Kumar, A. Voltage Stability Margin, based congestion management for a hybrid system, considering secure bilateral trading: MINLP approach. In Proceedings of the IEEE 6th International Conference on Power Systems (ICPS), New Delhi, India, 4–6 March 2016.
83. Rao, A.R.N.; Vijaya, P.; Kowsalya, M. Voltage stability indices for stability assessment: A review. *Int. J. Ambient Energy* **2021**, *42*, 829–845.
84. Hauqe, M.H. A linear static voltage stability margin for radial distribution systems. In Proceedings of the IEEE Power Engineering Society General Meeting, Montreal, QC, Canada, 18–22 June 2006.
85. Ahmad, N.; Mohamad, F.; Sulaiman, M.; Fazliana, A.; Kadir, A.; Bin, R.O. Classifications of voltage stability margin (VSM) and load power margin (LPM) using probabilistic neural network (PNN). *ARPN J. Eng. Appl. Sci.* **2017**, *12*, 5591–5596.
86. Merlyn, M.; Suresh, D.; Venkataramu, P.S.; Nagaraj, M.S. Effect of TCSC on Line Voltage Stability Indices under Single Line Outage Condition. *Int. J. Innov. Res. Electr. Electron. Instrum. Control Eng.* **2015**, *3*, 101–105.
87. Singh, B.; Sharma, P.A.K. Voltage Stability Calculations in Power Transmission Lines: Indications and Allocations (IEEE 30 BUS SYSTEM). *Int. J. Nov. Res. Electr. Mech. Eng.* **2015**, *2*, 63–66.
88. Mathew, M.; Gosh, S.; Suresh, D.; Ansari, A.A. An Assessment of Voltage Stability Based on Line Voltage Stability Indices and its Enhancement Using TCSC. *IOSR J. Electr. Electron. Eng.* **2015**, *10*, 81–88.
89. Zhang, Y.; Huang, L.; Chen, X. An Enhanced Voltage-Reactive Power Index Considering Power Injection Limitations for Power System Stability Assessment. *IEEE Trans. Power Syst.* **2021**, *36*, 262–273.
90. Kumar Kavuturu, K.V.; Sai Tejaswi, K.N.V.; Janamala, V. Performance and security enhancement using generalized optimal unified power flow controller under contingency conditions and renewable energy penetrations. *J. Electr. Syst. Inf. Technol.* **2022**, *9*, 18. [CrossRef]
91. Alluri, A.; Lanka, R.S.; Rayapudi, S.R. Optimal Location for FACTS Devices: Hybrid Algorithm to Improve Power Flow and System Security. *Cybern. Syst.* **2022**, *54*, 938–960. [CrossRef]
92. Ismail, B.; Wahab, N.I.A.; Othman, M.L.; Radzi, M.A.M.; Vijyakumar, K.N.; Naain, M.N.M. A Comprehensive Review of Optimal Location and Sizing of Reactive Power Compensation Using Hybrid-Based Approaches for Power Loss Reduction, Voltage Stability Improvement, Voltage Profile Enhancement, and Loadability Enhancement. *IEEE Access* **2020**, *8*, 222733–222765. [CrossRef]
93. Prabhakar, P. Voltage Stability Indices: Formulation and classification perspectives. *Int. J. Innov. Technol. Explor. Eng.* **2019**, *8*, 5777–5785. [CrossRef]
94. Salah, M.; Yifei, W.; Tiancong, C. Modern voltage stability index for prediction of voltage collapse and estimation of maximum load-ability for weak buses and critical lines identification. *Int. J. Electr. Power Energy Syst.* **2023**, *145*, 1–19.
95. Gitanjali, S.; Kabir, C.; Priyanath, D. Enrichment of voltage stability in power system through novel generalized approximate reasoning based intelligent control with African buffalo optimization approach. *Soft Comput.* **2022**, *27*, 7473–7496.
96. Kushwaha, P.K.; Bhattacharjee, C. Research on a selection of appropriate stability index under adverse system conditions for the assessment of voltage stability of an IEEE 14 bus power system. *Eur. J. Electr. Eng.* **2020**, *22*, 435–446. [CrossRef]
97. Samy, A.; Venkadesan, A. Evaluation of Voltage Stability Margin in a Power System using an Artificial Neural Network. In Proceedings of the IEEE 10th Power India International Conference (PIICON), New Delhi, India, 25–27 November 2022.
98. Ademola, A.S.; Sun, Y. Power system optimization approach to mitigate voltage instability issues: A review. *Cogent Eng.* **2023**, *10*, 2153416.
99. Bode, O.A.; Lotfy, M.E.; Olabisi, B.A.; Rashid, H.O.; Senjyu, H.T.; Narayanan, K. Security-constrained optimal utility-scale solar PV investment planning for weak grids: Short reviews and techno-economic analysis. *Appl. Energy* **2019**, *245*, 16–30.
100. Adewuyi, O.B.; Danish, M.S.S.; Howlader, A.M.; Senjyu, T.; Lotfy, M.E. Network Structure-Based Critical Bus Identification for Power System Considering Line Voltage Stability Margin. *J. Power Energy Eng.* **2018**, *6*, 97–111. [CrossRef]
101. Akito, N.; Miyara, R.; Alharbi, T.; Prabaharan, N. Dynamic Voltage Stability Assessment in Remote Island Power System with Renewable Energy Resources and Virtual Synchronous Generator. *Energies* **2021**, *14*, 5851.
102. Bode, O.A.; Shigenobu, R.; Ooya, K.; Senjyu, T.; Howlader, M.A. Static voltage stability improvement with battery energy storage considering optimal control of active and reactive power injection. *Electr. Power Syst. Res.* **2019**, *172*, 303–312.
103. Ratra, S.; Singh, D.; Bansal, R.C.; Naidoo, R.M. Stochastic Estimation and Enhancement of Voltage Stability Margin considering Load and Wind Power Intermittencies. In Proceedings of the IEEE 6th International Conference on Computing, Communication, and Automation (ICCCA), Arad, Romania, 17–19 December 2021.
104. Saifullah, M.K.; Kabir, M.M.; Mia, N.; Tanvir, A. Optimal Placement of FACTS Devices in the Power System Using ANN to Increase the System Loadability. Electronic Systems and Intelligent Computing. In *Lecture Notes in Electrical Engineering*; Mallick, P.K., Meher, P., Majumder, A., Das, S.K., Eds.; Springer: Singapore, 2015; Volume 686.
105. Kushwaha, P.K.; Bhattacharjee, C. Role of Energy Storage System for improving System Stability considering Customer Behavior Uncertainties. In Proceedings of the International Conference on Contemporary Computing and Applications (IC3A), Lucknow, India, 5–7 February 2020.

106. Adebayo, I.G.; Sun, Y. Voltage Stability Enhancement Capabilities of LTCT and STATCOM in a Power System. In Proceedings of the IEEE PES/IAS PowerAfrica, Cape Town, South Africa, 28–29 June 2018.
107. Shuai, C.; Deyou, Y.; Weichun, G.; Chuang, L.; Guowei, C.; Lei, K. Global sensitivity analysis of voltage stability in the power system with correlated renewable energy. *Electr. Power Syst. Res.* **2021**, *192*, 106916. [CrossRef]
108. Adebayo, I.G.; Jimoh, A.A.; Yusuff, A.A.; Sun, Y. An alternative method for the identification of critical nodes leading to voltage instability in a power system. *Afr. J. Sci. Technol. Innov. Dev.* **2018**, *10*, 323–333. [CrossRef]
109. Lakshmi, C.; Manjula, S.S. Detection of Proximity to Voltage Collapse Indicator Using Artificial Neural Networks. *Int. J. Sci. Eng. Res.* **2019**, *10*, 59–63.
110. Salama, H.S.; Vokony, I. Voltage stability indices–A comparison and a review. *Comput. Electr. Eng.* **2022**, *98*, 107743. [CrossRef]
111. Nanba, M.; Huang, Y.; Kai, T.; Iwamoto, S. Studies on VIPI-based control methods for improving voltage stability. *Int. J. Electr. Power Energy Syst.* **1997**, *20*, 141–146. [CrossRef]
112. Takehara, A.; Tanaka, Y.; Iwamoto, S. Voltage stability preventive and emergency-preventive control using VIPI sensitivity. *Electr. Eng. Jpn.* **2002**, *143*, 22–30. [CrossRef]
113. Gubina, F.; Strmcnik, B. Voltage collapse proximity index determination using voltage phasors approach. *IEEE Trans. Power Syst.* **1995**, *10*, 788–794. [CrossRef]
114. Adebayo, I.G.; Jimoh, A.A.; Yusuff, A.A. Prediction of voltage collapse through voltage collapse proximity index and inherent structural characteristics of a power system. In Proceedings of the IEEE PES Asia-Pacific Power and Energy Engineering Conference (APPEEC), Brisbane, QLD, Australia, 15–18 November 2015.
115. Venu, Y.; Kanthala, S. Optimal placement and Sizing of Compensator for Voltage Stability improvement Using Voltage Collapse Proximity Indicator. In Proceedings of the 3rd International Conference on Electronics, Communication and Aerospace Technology (ICECA), Coimbatore, India, 12–14 June 2019.
116. Adebayo, I.G.; Jimoh, A.A.; Yusuff, A.A. Analysis of Voltage Collapse in a Power System Using Voltage Stability Indices. *J. Energy Technol. Policy* **2018**, *8*, 40–45.
117. Šmon, I.; Pantoš, M.; Gubina, F. An improved voltage-collapse protection algorithm based on local phasors. *Electr. Power Syst. Res.* **2008**, *78*, 434–440. [CrossRef]
118. Maheshkumar, M. Online voltage stability analysis using synchrophasor technology. *Int. J. Sci. Res. Dev.* **2014**, *2*, 273–275.
119. Amroune, M.; Bouktir, T.; Musirin, I. Power system voltage instability risk mitigation via emergency demand response-based whale optimization algorithm. *Prot Control Mod Power Syst.* **2019**, *4*, 1–14. [CrossRef]
120. Gümüş, T.E.; Emiroglu, S.; Yalcin, M.A. Optimal DG allocation and sizing in distribution systems with Thevenin-based impedance stability index. *Int. J. Electr. Power Energy Syst.* **2023**, *144*, 108555. [CrossRef]
121. Cuffe, P.; Milano, F. Validating Two Novel Equivalent Impedance Estimator. *IEEE Trans. Power Syst.* **2018**, *33*, 1151–1152. [CrossRef]
122. Sadeghi, S.E.; Akbari Foroud, A. A new approach for static voltage stability assessment in distribution networks. *Int. Trans. Electr. Energy Syst.* **2019**, *30*, 1–21. [CrossRef]
123. Stefan, P.; Herwig, R. Voltage stability monitoring methods for distribution grids using the Thevenin impedance. *CIRED—Open Access Proc. J.* **2017**, *2017*, 1535–1539.
124. Zhou, Q.; Guan, Y.; Liu, Y.; Zhao, Z. A novel online local voltage stability monitoring approach based on clustering analysis. *Int. J. Electr. Power Energy Syst.* **2021**, *132*, 106807.
125. Pan, J.; Dong, A.; Fan, J.; Li, Y. Online Static Voltage Stability Monitoring for Power Systems Using PMU Data. *Math. Probl. Eng.* **2020**, *2020*, 1–8. [CrossRef]
126. Alzaareer, K.; Saad, M. Real-Time Voltage Stability Monitoring in Smart Distribution Grids. In Proceedings of the International Conference on Renewable Energy and Power Engineering (REPE), Toronto, ON, Canada, 24–26 November 2018.
127. Lee, Y.; Han, S. Real-Time Voltage Stability Assessment Method for the Korean Power System Based on Estimation of Thévenin Equivalent Impedance. *Appl. Sci.* **2019**, *9*, 1671. [CrossRef]
128. Wang, Y.; Wang, C.; Lin, F.; Li, W.; Wang, L.Y.; Zhao, J. A new transfer impedance-based system equivalent model for voltage stability analysis. *Int. J. Electr. Power Energy Syst.* **2014**, *62*, 38–44. [CrossRef]
129. Akhil, P.; Atul, M.; Prabhakar, K.; Kothari, D.P. Assessment on various node voltage stability indices—A review. In Proceedings of the International Conference on Smart Technologies for Smart Nation (SmartTechCon), Bengaluru, India, 17–19 August 2017.
130. Nagendra, P.; Datta, T.; Halder, S.; Paul, S. Power System Voltage Stability Assessment Using Network Equivalents-A Review. *J. Appl. Sci.* **2010**, *10*, 2147–2153. [CrossRef]
131. Xinglong, W.; Zheng, X.; Zheren, Z. Power Stability Analysis and Evaluation Criteria of Dual-Infeed HVDC with LCC-HVDC and VSC-HVDC. *Appl. Sci.* **2021**, *11*, 5847.
132. Xu, Z.; Xu, Z.; Xiao, L. Analysis and assessment standards of power stability of multi-send HVDC systems. *J. Eng.* **2018**, *2019*, 748–753. [CrossRef]
133. Alam, M. Determination of Power System Contingency Ranking Using Novel Indices. Recent Advances in Power Systems. In *Lecture Notes in Electrical Engineering*; Gupta, O.H., Sood, V.K., Malik, O.P., Eds.; Springer: Singapore, 2022; Volume 812.
134. Mehebub, A.; Shubhrajyoti, K.; Sankar, T.S.; Sumit, B. Determination of Critical Contingency Based on L-Index and Impact Assessment on Power System. In *Smart Energy and Advancement in Power Technologies*; Springer: Berlin, Germany, 2022; pp. 583–596.

135. Khunkitti, S.; Boonluk, P.; Siritaratiwat, A. Optimal Location and Sizing of BESS for Performance Improvement of Distribution Systems with High DG Penetration. *Int. Trans. Electr. Energy Syst.* **2022**, *2022*, 6361243. [CrossRef]
136. Rodríguez Lopez, D.K.; Perez-Londoño, S.; Rodriguez-Garcia, L. Optimal under voltage load shedding based on voltage stability index. *Ing. E Investig.* **2016**, *36*, 43–50.
137. Shabir, M.; Nawaz, S.; Vijayvargiya, A. Voltage Stability Enhancement Using SVC in PSCAD Software. In *Innovations in Electrical and Electronic Engineering. Lecture Notes in Electrical Engineering*; Favorskaya, M.N., Mekhilef, S., Pandey, R.K., Singh, N., Eds.; Springer: Singapore, 2021; Volume 661.
138. Akbarzadeh Aghdam, P.; Khoshkhoo, H. Voltage stability assessment algorithm to predict power system loadability margin. *IET Gener. Transm. Distrib.* **2020**, *14*, 1816–1828. [CrossRef]

energies MDPI

Article

Real-Time Control of a Battery Energy Storage System Using a Reconfigurable Synchrophasor-Based Control System †

Prottay M. Adhikari ‡, Luigi Vanfretti *, Hao Chang and Koushik Kar

Rensselaer Polytechnic Institute, Electrical, Computer and Systems Engineering Department, Troy, NY 12180, USA; prottaymondaladhikari@gmail.com (P.M.A.)
* Correspondence: vanfrl@rpi.edu
† This paper is an extended version of our paper published in 2022 International Conference on Smart Grid Synchronized Measurements and Analytics (SGSMA), Split, Croatia, 24–26 May 2022; pp. 1–6.
‡ Current address: Eaton Corporation, Moon Twp, PA 15108, USA.

Abstract: Synchrophasor-driven smart grid applications aiming to orchestrate a diverse set of Distributed Energy Resources (DERs) require extensive infrastructure including substantial instrumentation hardware, communication network extensions and controller installations for coordinated operation. This can make the overall installation expensive. Additionally, due to the computational complexity and data-intensive nature of the PDC functionality, most of the existing PDC implementations are on a purely software level, making them unsuitable for the real-time applications. To address this, the current paper proposes an alternate architecture for the real-time synchrophasor-based control of DER applications (e.g., microgrids) incorporating a centralized synchronization hardware designed to replace aggregation Phasor Data Concentrators (PDCs) and supplementary control algorithms into a singular reconfigurable hardware. This particular hardware is termed a Synchrophasor Synchronization Gateway and Controller (SSGC). The robustness of the proposed architecture is tested by using real-time (RT) Controller Hardware-In-the-Loop (CHIL) simulation-based experiments by manipulating the communication network that connects the SSGC with multiple Phasor Measurement Unit (PMU) streams broadcasting data through the IEEE C37.118.2 protocol in real time. These PMU streams were generated by using a real-time microgrid model running on a Typhoon HIL 604 simulator. To manipulate the communication interface between the proposed SSGC hardware and the PMU streams, a configurable Wide Area Network (WAN) emulator and communication network impairment appliance deployed in the Candela Technologies CT910 external hardware was utilized. The real-time control system was expanded by incorporating a low-pass filter to eliminate the potential overswitching of a Battery Energy Storage System (BESS). The proposed architecture demonstrated a reliable performance under ideal to moderately tampered communication networks. However, under a significantly corrupted network, the performance of this architecture is acutely affected.

Keywords: synchrophasor; Typhoon HIL 604; IEEE C37.118; DER; communication protocols; PMU; PMU-based control; CHIL; HIL; BESS

Citation: Adhikari, P.M.; Vanfretti, L.; Chang, H.; Kar, K. Real-Time Control of a Battery Energy Storage System Using a Reconfigurable Synchrophasor-Based Control System. *Energies* **2023**, *16*, 6909. https://doi.org/10.3390/en16196909

Academic Editor: Ying-Yi Hong

Received: 3 September 2023
Revised: 22 September 2023
Accepted: 26 September 2023
Published: 30 September 2023

1. Introduction

1.1. Motivation

In modern interconnected power systems deploying synchrophasor technology, Phasor Data Concentrators (PDCs) are expected to receive, parse, align, store and publish the measurement data received from Phasor Measurement Units (PMUs). As networked devices, PDCs should be compliant with synchronized data transmission standards, i.e., IEEE C37.118.2 [1] While maintaining this compatibility, the existing PDC hardware is inadequate to comply with hard **real-time** control requirements. Currently, most state-of-the-art *real-time compliant* PDC implementations are developed entirely at the software level.

On the other end, existing industrial PDCs [2–4] are **not** real-time compliant as they are implemented in nondeterministic operating systems (e.g., Windows 10/11 and similar). This gap in engineering development has been studied and reported by the authors of [5–7].

As illustrated in Figure 1, this makes the application of PDCs in the real-time networked control of power systems a challenging problem, with the major drawback of latency build-up if the current approach for real-time synchrophasor control is adopted. In [8], a hardware platform for wide-area control system (WACS) applications that can only function upon a single incoming PMU stream was proposed. To the best of the authors' knowledge, currently, there is no real-time compliant real-time PDC hardware architecture that can operate on multiple PMU streams. This becomes a bottleneck in implementing synchrophasor-driven real-time control systems for microgrid applications that require measurements from diverse Distributed Energy Resources (DERs). To bridge this gap, as illustrated in Figure 2, this paper proposes an alternative real-time control architecture which replaces aggregator PDCs and secondary control functionalities into a singular reconfigurable hardware and demonstrates a microgrid control system featuring the proposed hardware.

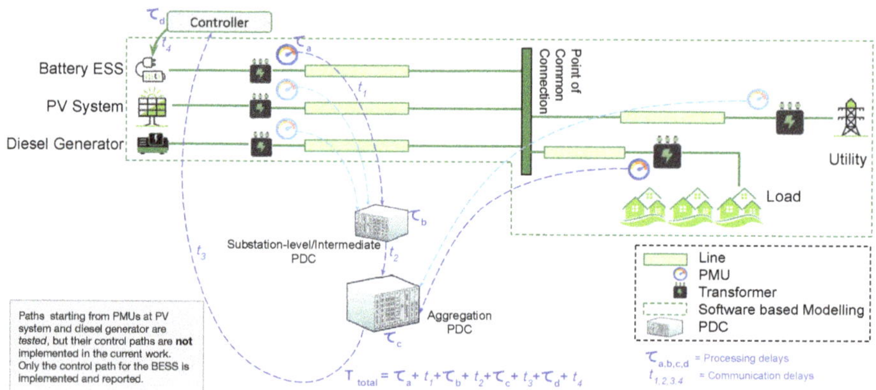

Figure 1. Typical infrastructure for real-time control using networked PMUs and PDCs.

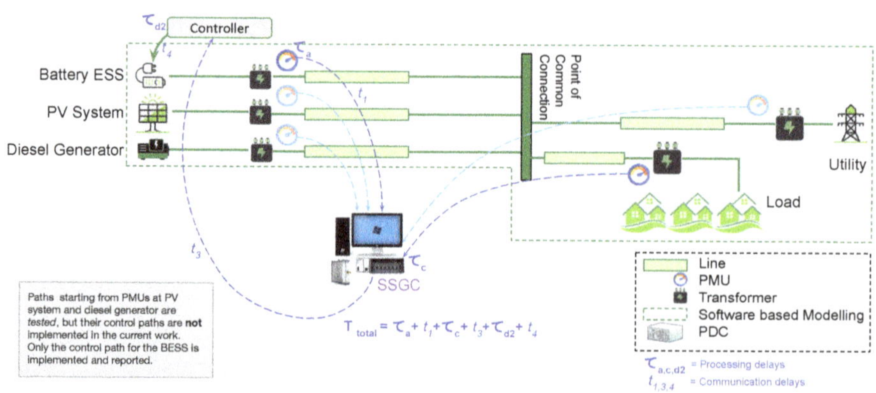

Figure 2. Proposed approach for real-time DER control with networked PMUs and SSGC.

1.2. Related Works

Synchrophasor-based networked control architectures for power systems is an emerging field of research. Due to the latency build-up within the network, the computational limitations of various edge devices, and the strict real-time compliance requirement required for designing a control system, the deployment of a fully functional real-time synchrophasor-based networked control architecture becomes extremely challenging. These challenges have been summarized in [9].

Most of the existing research in this domain proposes and utilizes very specific edge devices and software tools to implement such a control system, which is summarized in Table 1. The authors of [10] developed custom PMU hardware (MDPMU) to deploy their proposed fast-load-control system based on synchrophasor measurements retrieved from those custom PMUs. Similar custom PMUs (both on hardware and software levels) were developed by the authors of [11–13] while deploying their custom FNET/GridEye wide-area synchrophasor network. However, these implementations focused on the monitoring and event-detection applications only, leaving open questions related to real-time control. Meanwhile, industrial research laboratories (e.g., SEL) have published hardware developments for PMU-based control applications in [14]. However, these developments did not explore the utilization of concurrent PMU streams coming from different locations and focused on local control application development from local PMU streams. Thus, this development was not tested for robustness under corrupted network conditions, which take into account incoming measurement data from remote PMUs.

Table 1. Summary of existing research related to PMU/PDC-based real-time control.

References	Hardware/ Software	RT Compliant	Suitable for Control	Available at Production Scale	Comments
[10]	Hardware	Yes	Yes	No	Cannot utilize existing PMUs in the network
[11–13]	Both	Yes	No	Yes	Control applications not tested
[14]	Hardware	Yes	No	Yes	Tested only for local control
[15,16]	Software	No	Yes	No	No major hardware development
[17]	Software	Yes	No	No	Proposed a PDC placement scheme for control

The decentralized nature of DERs coupled with the hierarchical character of the microgrid control system (reported in [18–21]) makes it possible to break up the control system into different networked hardware according to the transient requirements of the control action. Based on the survey presented in [21], it can be hypothesized that the *secondary* and *tertiary* control actions for microgrids can be put into networked devices because these control actions are relatively slow in operation. The current paper only focuses on implementing *secondary* control actions, more specifically the control of active and reactive power flow. The research reported in [22,23] established the theoretical background for controlling BESS in real time.

The authors of [24] reported the architecture for the Synchrophasor Synchronization Gateway (SSG) and introduced the hardware and software associated with the implementation. On the software level, the SSG uses the C-based *Khorjin* library reported in [25] to parse multiple and concurrent PMU data streams. The SSG's Graphical User Interface (GUI) was designed by using LabVIEW; it can be configured to accommodate additional incoming PMU/PDC streams and to apply modifications in communication-network specifications. Because some of the functionalities provided by the proposed hardware are similar to the functionalities offered by a PDC, it is important to take note of the existing standardization efforts in the domains of PDC hardware implementation. The authors of [6] summarized the standard functional blocks and communication interfaces associated with the PDC architecture. A similar study was reported in [26] that also explored the communication protocol between the PMUs and PDCs.

Unlike a conventional PDC, the SSGC enables real-time control functions in addition to the aforementioned PDC functions, and some of the control functions that can be implemented by using synchrophasor-based control were reviewed as part of the literature survey. In the domain of control system design for microgrids, the authors of [21,27] surveyed and classified the existing control strategies into three different classes depending on the priorities, time scales and required speed of the various control actions. The primary control class consists of the fastest control actions including voltage and current control algorithms for the individual DERs. The secondary control class evaluates the power flows to and from the different existing DERs and helps the microgrid navigate between the islanded and grid-tied modes. The secondary control class tackles slower dynamic responses (e.g., power flow) compared to the primary control class. Finally, the tertiary control class consists of supplementary control algorithms sitting on top of both primary and secondary classes of control and enables the microgrid to operate in an economically optimized fashion. The research reported in [28] demonstrated significant efforts of standardization across these three classes of control systems in microgrids. The authors of [29] explored the utilization of synchrophasor data to monitor microgrids and to increase the reliability of measurement data. To this end, this research proposed an Advanced Phasor Data Concentrator (APDC) hardware that is capable of operating under a tampered network and estimating missing data points in the synchrophasor streams. However, this hardware was not time synchronized, and the reported experiments were performed by a programmable voltage source, lacking the testing rigor that can be brought by using Hardware-In-the-Loop experimental testing techniques [30]. The present manuscript addresses both gaps, i.e., time synchronization, and it exploits RT Controller Hardware-In-the-Loop real-time simulations with microgrid models. The experiments reported in [31] illustrated a synchrophasor-based control architecture for microgrids, where the synchrophasor data are used to formulate reduced-order dynamic models for the DERs within the microgrid, and used those models to seamlessly navigate the microgrid between the islanded mode and the grid-tied mode. Meanwhile, Ref. [32] demonstrated the utilization of adaptive network-management tools within the PDC to compensate the network delays between the PDC and the individual PMUs.

In [24], the authors proposed a framework where multiple concurrent instances of the same dynamically linked library (DLL) were used to parse multiple PMU streams in real time. This procedure includes both deciphering the measurement data and extracting the synchrophasor configuration information. The communication link requires bidirectional capabilities as the PMUs send header information, measurement data and configuration information; they also receive a command message to enable or disable the data transmission. The research presented in [24,33] utilizes multiple instances of this *Khorjin* thread to facilitate the real-time reception and parsing of synchrophasor data streamed from multiple sources. Finally, it is important to note that while the current research utilized the same Khorjin-based SSG architecture reported in [24,33], the major contribution of the current paper lies in the extension of the SSG architecture into the domains of secondary control actions and their validation by using real-time CHIL experiments by using a microgrid model with DERs, where the secondary controller drives a Battery Energy Storage System (BESS).

1.3. Contributions

The current paper proposes the following contributions:

- The capabilities of the *Khorjin* library are extended to develop a Synchrophasor Synchronization Gateway and Control (SSGC) architecture, consisting of a synchronization layer and a control layer.
- A new PMU-based approach for control of the DERs is introduced and demonstrated by using real-time Controller Hardware-In-the-Loop (CHIL) including a microgrid model. This approach exploits the proposed SSGC architecture.
- The performance of the proposed SSGC architecture is analyzed under varying communication network conditions (including randomized configurations) with multiple PMU streams.
- The proposed control system is further enhanced by applying Low Pass Filters (LPFs) in the real-time controller's control error signal to mitigate the overswitching of the primary controllers inside a Battery Energy Storage System (BESS).
- The real-time performance of the SSGC is demonstrated under the presence of faster transients in the system when the control signal is passed through the enhanced controller (with LPFs).
- The SSGC architecture was validated by deploying a second type of secondary controller. The controller utilizes the phase-angle difference between two PMU locations to control the active power output.

1.4. Paper Organization

The remainder of this paper is organized as follows. Section 2 gives an overview of the real-time controller's architecture and the experimental setup developed to validate the proposed approach in this paper. Section 3 presents the results of the performance of the SSGC under both ideal and impaired communication network conditions and the enhancement of the Battery Energy Storage System (BESS) response by enhancing the control function through LPFs, limiting unnecessary BESS switching. Finally, Section 4 outlines future work and Section 5 presents this paper's major conclusions.

2. Architecture and Experimental Setup

2.1. Controller Architecture

The authors of [25] presented the *Khorjin* library, which receives, extracts and parses a C37.118.2 data stream before feeding the information to subscriber applications. The authors of [24,33] demonstrated the data unwrapping and time synchronization functionalities on the real-time embedded system (host), utilizing multiple instances of *Khorjin* extracting data from multiple incoming concurrent PMU streams in parallel.

In this paper, the 'virtual instrument code' running on the host side (in an RT OS) is extended and improved to incorporate crucial secondary control functionalities targeted to control DERs in a microgrid, whose one-line diagram and simplified view is shown in Figure 3. The real-time control system utilizes PI controllers that calculate their set point and process variables by unwrapping the current and voltage phasors received from the concurrent *Khorjin* instances. A simplified view of the system is demonstrated in Figure 4, which identifies both the electrical connections and communication interfaces of the setup. The detailed experimental setup is illustrated in Figure 5.

A *Khorjin* thread requires bidirectional data transmission capabilities for the communication link. After securing the TCP/IP connection, the SSG sends a `'turn off transmission'` command and a `'send CFG-2'` command. The PMU responds by sending the `CFG-2` frame. At this point, the SSG has received all the required information from this frame to initiate the data transmission. Hence, the SSG now requests synchrophasor measurement data, and the PMU begins data transmission. This exact protocol is implemented in all the concurrent *Khorjin* instances. This process is graphically illustrated in Figure 6.

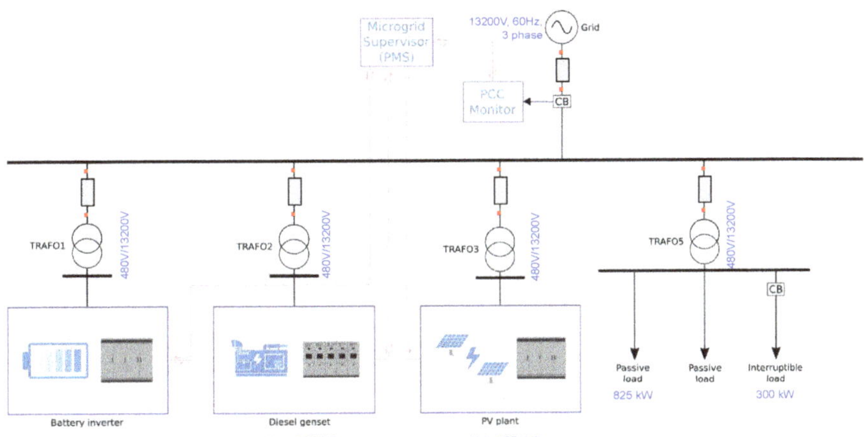

Figure 3. Simplified real-time microgrid model implemented by using Typhoon HIL 604.

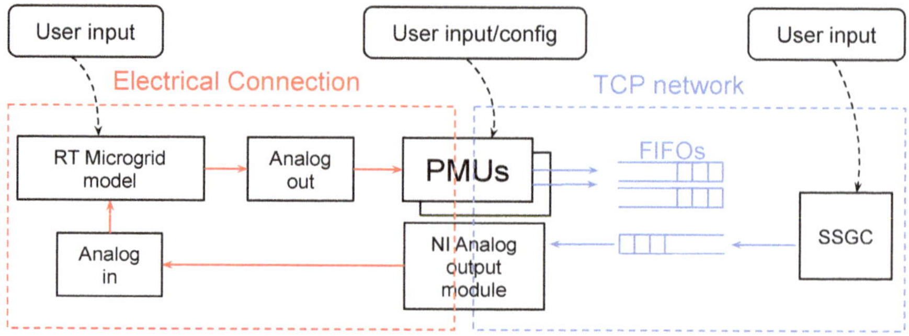

Figure 4. Simplified block diagram of the experimental setup.

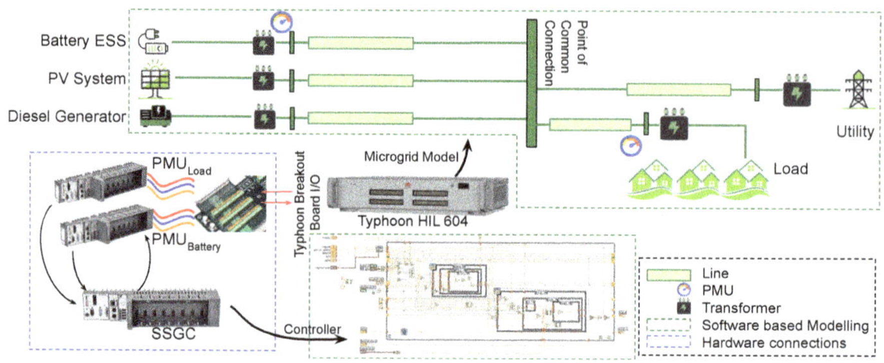

Figure 5. Detailed architecture of the experimental setup demonstrating the software and hardware components.

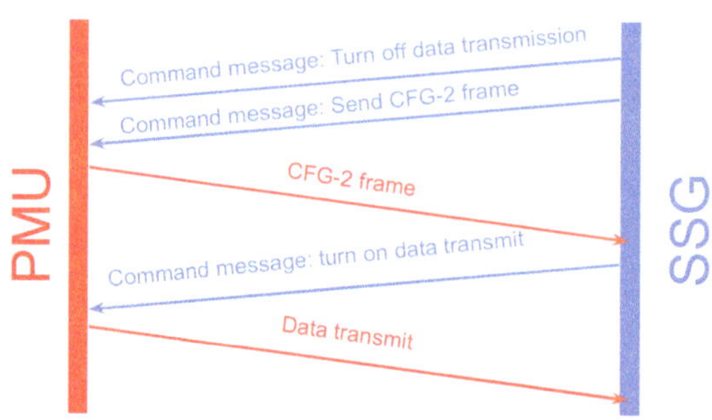

Figure 6. PMU to SSG/SSGC data transmission protocol by utilizing Khorjin instance.

As can be observed in Figures 5 and 7, the main goal of the real-time controller is to maintain the power at the Point of Common Connection (PCC) constant from the point of view of the utility, meaning that the microgrid self-balances by taking advantage of measurements at the PCC and other locations (e.g., at the BEES terminals). Because the goal of the controller is to redispatch the battery to meet load/generation changes within the microgrid, it is important to mitigate the BESS response in specific time scales. To achieve this goal, the real-time control system was enhanced by incorporating LPFs, as shown in Figure 8, which would prevent the controller from responding to possible high-frequency transients on the load side (e.g., switching transients).

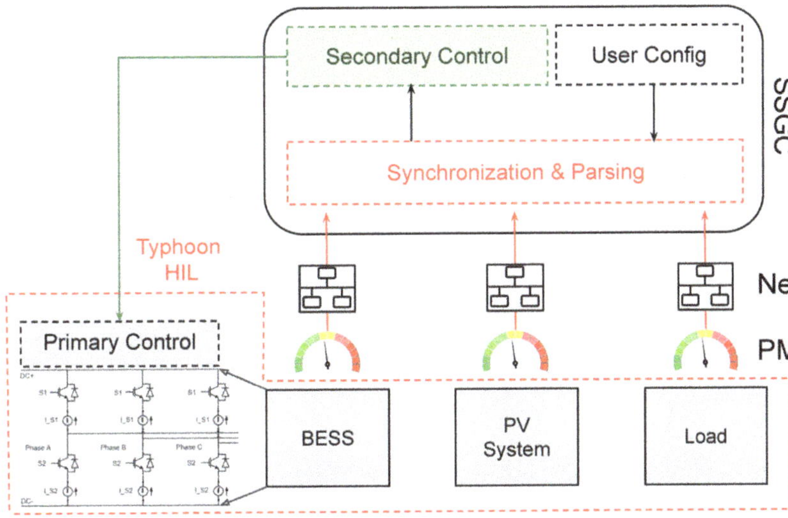

Figure 7. SSGC's utilization within hierarchical control system for microgrid.

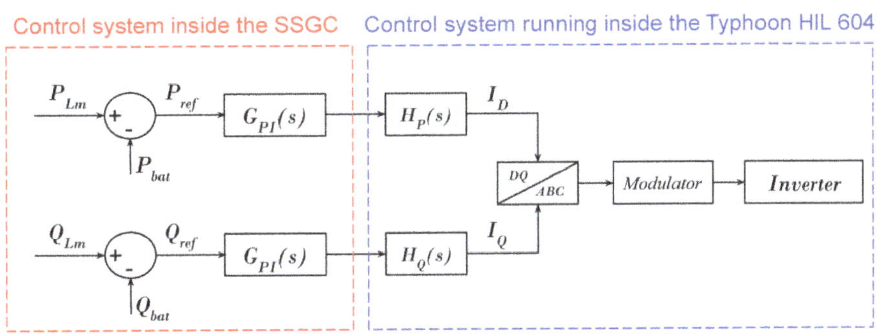

Figure 8. Microgrid controller implementation utilized in the experimental setup.

The current work demonstrates a scenario where the proposed SSGC architecture is configured to regulate the power output from a Battery Energy Storage System (BESS)-based DER. As shown in Figure 8, the PI controller coded inside the SSGC hardware (situated at the remote location) determines the set point to be utilized at the local (primary) controller inside the BESS. The individual parameters for the PI controller were based on the specifications provided in the model illustrated in [34]. The PI controller was followed by a saturation block (with the upper and lower limits set at 1 and −1, respectively) to ensure stability. The value of K_p and K_i were taken from the Typhoon specifications directly (for this particular application, K_p = 0.5 and K_i = 3.0). This control algorithm computes P and Q in real time from the voltage and current phasors it receives from the multiple incoming PMU streams. The controller uses P and Q as the control variable. It is also possible to utilize the voltage magnitude and phase-angle differences ($\Delta\delta$) as the control input variables [35–38]. The controller output is fed back to the real-time simulator's analog input channels and used as a control input for the *primary controller* of the Li-Ion BESS, as shown in Figure 7. The proposed control architecture can take advantage of the hierarchical structure of the control system for microgrids, secondary control in particular, as summarized in [21].

It must be noted that, as shown in Figures 3 and 5, the microgrid RT simulation model comprises a diesel generator, a PV system and an external grid alongside the BESS. All these energy resources work in synchronism to supply a configurable load. These models were based on the existing components available in the Typhoon HIL's model library [34,39].

2.2. Experimental Setup

This paper reports a reconfigurable SSGC hardware that can suitably incorporate secondary and tertiary control functionalities. As proof of concept, the authors aim to demonstrate its functionalities through a real-time secondary controller to regulate the power output of the BESS implemented on the SSGC in this work. This is illustrated in the block diagram of Figure 8.

2.2.1. Scope of the Experiments Supported by the Experimental Setup

As shown in Figure 8, the proposed secondary control action utilizes a set point, which is computed in real time from the two parameters P_{Lm}, which represents the measurement from the load side, and P_{bat}, which represents the BESS output power. The experiments in this work will assume that the P_{PV} (active power output from the PV system), P_{Uti} (active power dispatch from the utility) and P_{DSG} (active power output from the diesel generator) are constant. Under this condition, the parameter P_{ref} (reference power) will be a product of the changes in the total load consumption P_{load} only. For this part of the experiment, the two PMUs were placed near the load and near the BESS as the current measurements at these locations were expected to experience changes along with changes in the load and battery dispatch. As an extension, a further experiment was performed by taking the phase-angle difference ($\delta_{PCC} - \delta_{Load}$) as the control variable to regulate the active power output from the BESS instead of using the active powers directly. For this part of the experiment, the phase angles were measured at the point of common coupling (PCC) and the load. Thus, the PMUs were placed accordingly. The reprogrammable and reconfigurable nature of the proposed SSGC hardware can be exploited to adapt the hardware to deploy different control schemes and control of different types of DERs. However, those experiments are considered to be beyond the scope of the current paper. For demonstration, only real-time synchrophasor-based P-Q control of the BESS is presented in this paper. Hence, this paper primarily focuses on the implementation of the SSGC architecture and its validation in the context of simple control-system development targeted for DERs. We also demonstrate how the control performance can be improved by implementing a Low Pass Filter (LPF) ($LPF(s) = \frac{1}{1+\tau s}$, where $\tau = \frac{1}{\omega_c}$) to make the BESS operate safely under a condition where switching transients are present in the load.

2.2.2. Real-Time Communications

It has been discussed that the SSGC requires real-time synchrophasor data (i.e., following the IEEE C37.118.2 std.) to be streamed at its input. Modern RT simulators have the capability to stream synchrophasor data directly from the RT model being simulated, without requiring external PMU hardware. However, the present work was completed by using *Typhoon HIL control center*'s 2020.4 release, where the IEEE C37.118 library was primitive and unstable (as discovered through experimentation) when connected to external hardware such as a PDC or SSGC (note that the 2022.4 and 2023.1 releases of the *Typhoon HIL control center* toolkit have stable and dedicated library components for streaming C37.118 data, and the authors of this paper chose not to adjust the hardware configuration and keep the PMUs as external hardware to facilitate impairment). This was chosen to facilitate tampering between the communications network between the PMUs and the SSGC hardware. With the introduction of external PMU hardware based on [40,41], it is simple to pick and choose individual PMUs and corrupt the communication network between any single PMU and the SSGC hardware.

2.3. Hardware Integration

The aforementioned external PMUs (the hardware used is that reported in [42] and extended in [43]) were connected to the low-voltage analog outputs of the real-time simulator. It is crucial to note that these PMU designs feature both current and voltage inputs to function as PMUs whereas Typhoon HIL 604 hardware **only** provides analog voltage outputs. Thus, an external *burden circuit* was introduced in the experimental setup to perform the voltage-to-current conversion, as demonstrated in Figure 9c. The connection between Typhoon HIL 604 (the hosting microgrid model) and the PMUs (streaming synchrophasor data to SSGC) is shown in Figure 9a. Figure 9b demonstrates how the PMUs receive GPS signals to obtain precision timing, and Figure 9d shows the SSGC deployed in a remote location, communicating to the rest of the experimental setup through a standard Ethernet-based communication network.

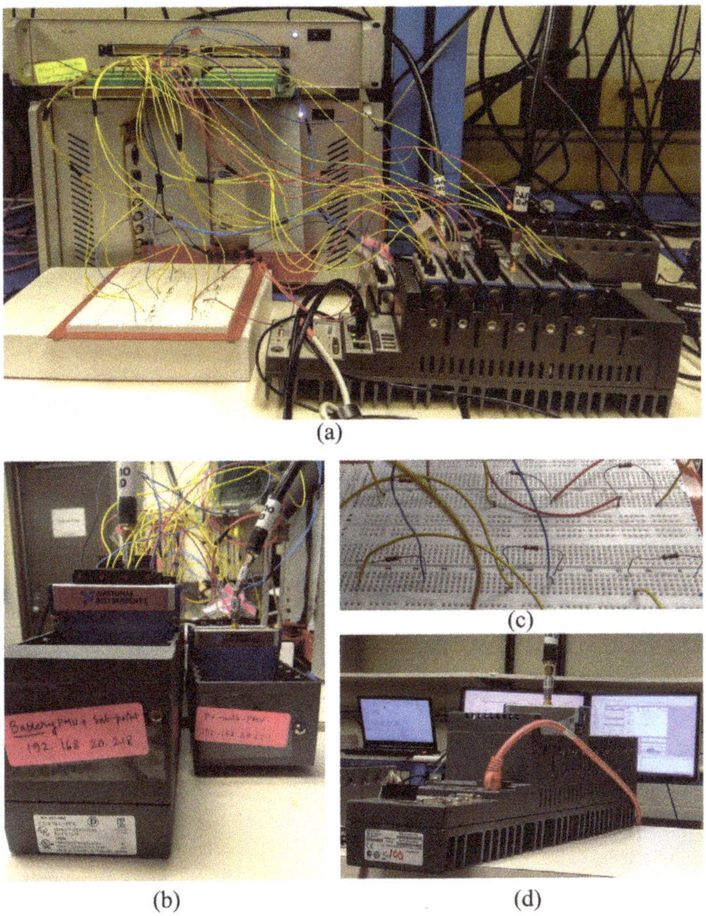

Figure 9. Hardware arrangement for RT-HIL-testing the SSGC hardware: (**a**) connection between the microgrid and the PMUs, (**b**) PMUs receiving timing information, (**c**) conversion of RT low-voltage signals into current signals and (**d**) SSGC connected remotely to the microgrid RT model.

SSGC Configuration and Integration in the Experimental Setup

Figure 10 shows how the LabVIEW GUI interacts with the physical inputs and outputs of the Typhoon HIL 604 and how the analog signal levels are scaled to obtain the measurements of actual voltages and currents. This technique enables the SSGC operation without any high external amplifiers that are typically required for HIL experiments [30], such as [44], making the experimental environment safer.

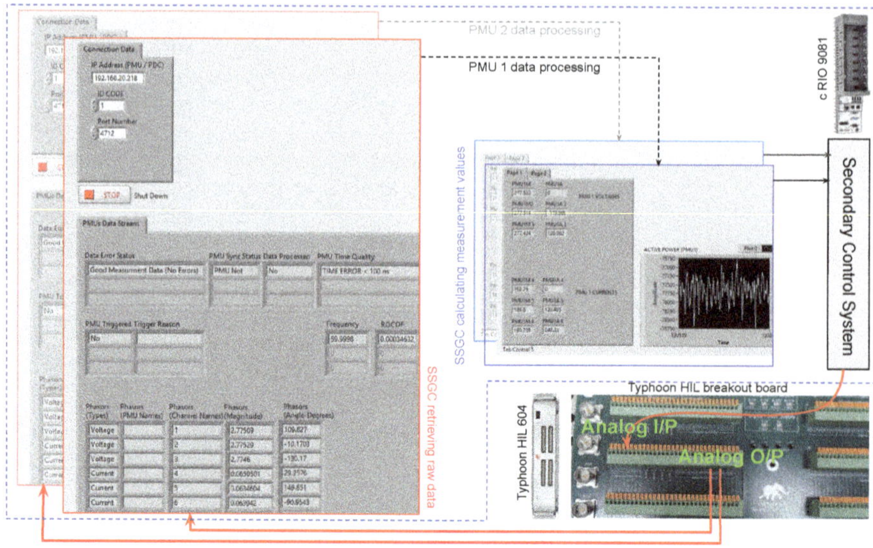

Figure 10. Details of the control system implemented inside the SSGC.

The communication network between the SSGC hardware and the PMUs was intentionally impaired during experimentation in order to test the robustness of the proposed control architecture. To perform these network impairments, the external *Wide-Area Network emulator and impairment appliance hardware CandelaTech CT910* [45] was connected inside the network, which would tamper the communication link between PMUs and the SSGC. Using CT910, it is possible to introduce custom delays, data drops, jitters and bandwidth limitations within the network. The values of these parameters can be reprogrammed every second, and this procedure can be automated.

3. Results

3.1. SSGC Performance under Ideal Network Conditions

In this experiment, the total load is increased in a step by turning on the *configurable load*, as shown in Figure 3. Initially, a fixed load of 825 kW was supplied by the PV unit (125 kW), diesel generator (500 kW) and the utility (100 kW). This makes the initial dispatch for the BESS inverter to be fixed at 100 kW. With the system running in this "steady state", a step increase of 300 kW in load was triggered externally.

Under this scenario, the control system must respond to this change in load. Hence, the BESS inverter and its dispatch increases from 100 kW to 400 kW. It is important to note that the portion of the controller (in Figure 8) within the dashed red box is implemented within the SSGC hardware. This portion is capable of utilizing synchrophasor measurement data obtained from the PMUs placed at the load and at the BESS. The PMU data are utilized to compute the active and reactive powers, which are then used to calculate a new set point (by using the PI controller block G_{PI}) to operate the BESS. This set point is then utilized by the internal control algorithm (implemented locally inside the BESS) to control the individual current and voltage output of the inverter inside the BESS. This portion of the control system was implemented locally within the BESS model of the real-time simulator because it generates high-frequency switching sequences for the individual semiconductor switches in the inverters that are modeled within the simulator.

Figure 11a shows the 300 kW manual load injection in the system. The SSGC incorporates a PI controller onboard. The output of this PI controller modulates the set point for the local BESS controller to operate, as shown in Figure 11b. This measurement is taken from the SSGC side. Meanwhile, in Figure 11c, the BESS power output response is shown. It can be observed from this figure that the power output of the battery increases from

100 kW to 400 kW to cover for the step increase in the load. Figure 11a–c demonstrate the SSGC's performance under ideal conditions of the communication network while there are no external communication disturbances.

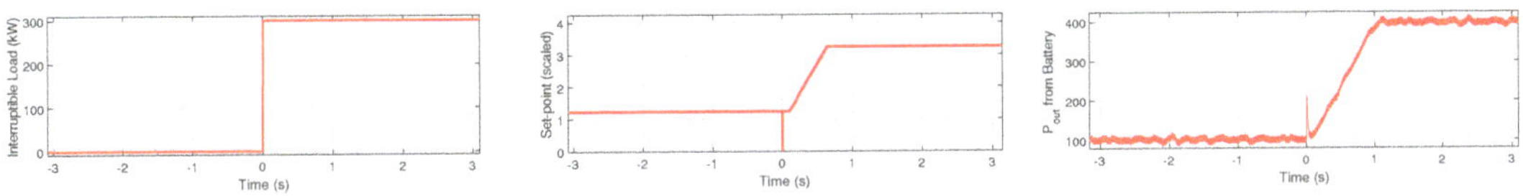

(**a**) (**b**) (**c**)

Figure 11. Control system performance for the BESS under ideal network conditions: (**a**) 300 kW load injection by switching the interruptible load, (**b**) response at the output of the PI controller inside SSGC and (**c**) active power output from the Battery Energy Storage System (BESS).

3.2. *SSGC Performance under Nonideal Communication Network Conditions*

Even though the SSGC hardware can be configured to control different DERs and employ sophisticated control algorithms, such demonstrations are considered to be beyond the scope of the current paper. For experimentation purposes, only the real-time synchrophasor-based control of the BESS is presented in this paper. Hence, the primary focus of this paper is the implementation of the SSGC architecture and its validation in the context of the P-Q control of the BESS-based DER.

Because this controller architecture relies on real-time synchrophasor communications, it is important to assess the performance of the proposed architecture under nonideal communication network conditions. Therefore, to test the robustness of the SSGC architecture, the communication network between the SSGC hardware and the PMUs was tampered with, as shown in Figure 12. As mentioned earlier, to perform experiments impairing the communications, an additional external hardware *CandelaTech CT910* [45] was connected between the SSGC and the rest of the communication network system. This hardware enables the user to introduce custom delays and data drops within the network through a GUI or command line and is used for the experiments in the sequel.

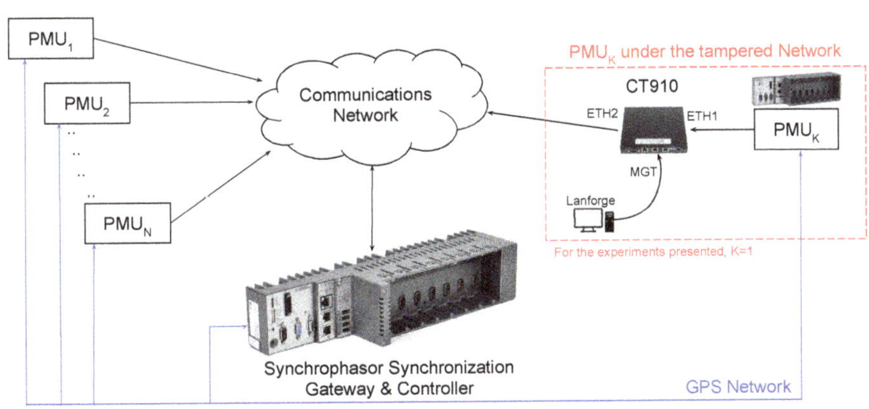

Figure 12. Block diagram of a system comprising SSGC and N PMUs; the network between the SSGC and the kth PMU is impaired, while all the hardware receive timing information from the same time base.

3.2.1. Network Tampering by Introducing Fixed Delay

In this particular experiment, artificial delays were introduced in between PMU_1 and the SSG prototype. To this end, a network-traffic generator within the CT910 was reconfigured to introduce artificial delays and thereby tamper with the network between the two ports where the PMU_1 and SSGC prototype were connected.

Figure 12 shows a representative connection where the CT910 device is utilized to tamper with the communication link between PMU_1 and the SSGC. The delays are introduced by CT910 (CT910 uses the *LanForge 5.4.6* software library, which provides crucial command-line utilities and a stable GUI to interact with the communication link in real time). The delay between the time when a synchrophasor packet is generated and when that particular packet is received at the SSGC end can be computed by sharing the respective time stamps. This procedure is explained in detail by the authors of [24].

As a simple representative experiment, the network was impaired by injecting an additional 100 ms delay, and the network delay was computed in real time by time-stamp sharing. This network delay is plotted in Figure 13, which shows a window of 10 s, during which the communication link was tampered with an additional 100 ms delay.

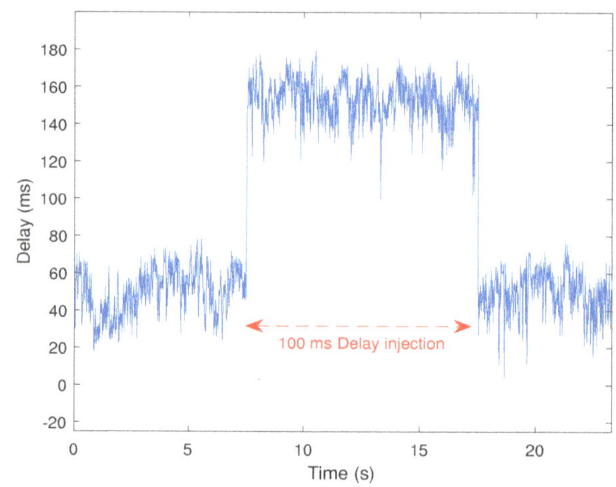

Figure 13. Measured delay at the SSGC due to a 100 ms delay injection from the CT910.

3.2.2. Network Tampering by Introducing Varying Delay

In this experiment, the communication link between the PMU and the SSGC is subjected to a network tampering that would randomize the injected delay every second. The hardware restricts the frequency of the network reconfiguration and dictates that, at most, the network can be reconfigured **once every second**. To achieve this, a custom script was executed on the CT910 hardware that would reconfigure its network interface with a random amount of delay every second. A pregenerated data file contained the amount of the injected delay in milliseconds and was used by the script to reconfigure the hardware with varying amounts of delay every second. The overall execution is shown in the flowchart of Figure 14a.

By following the same methodology to compute the network delay as described in Section 3.2.1, the network delay of this experiment is computed in real time. These results are shown in Figure 14b. It can be seen that the network reconfigures itself with a new value of the injected delay every second. The data file containing the raw values of the *network delays* was generated from a *random variable* following a **uniform distribution** between 0 and 200 ms.

Additionally, for the second part of this experiment, *random* delays were also generated to compute the network delay in real time by using a *random variable* following a **normal distribution** with a fixed mean and standard deviation (all the negative values generated were discarded because network delay cannot be negative). The experiment is carried out for 10 min (with the network reconfiguring the delay every 1 s) and repeated 10 times. If the network sustains itself (i.e., the controller keeps operating after 10 min of real-time simulation), the experiment run is termed as *successful*.

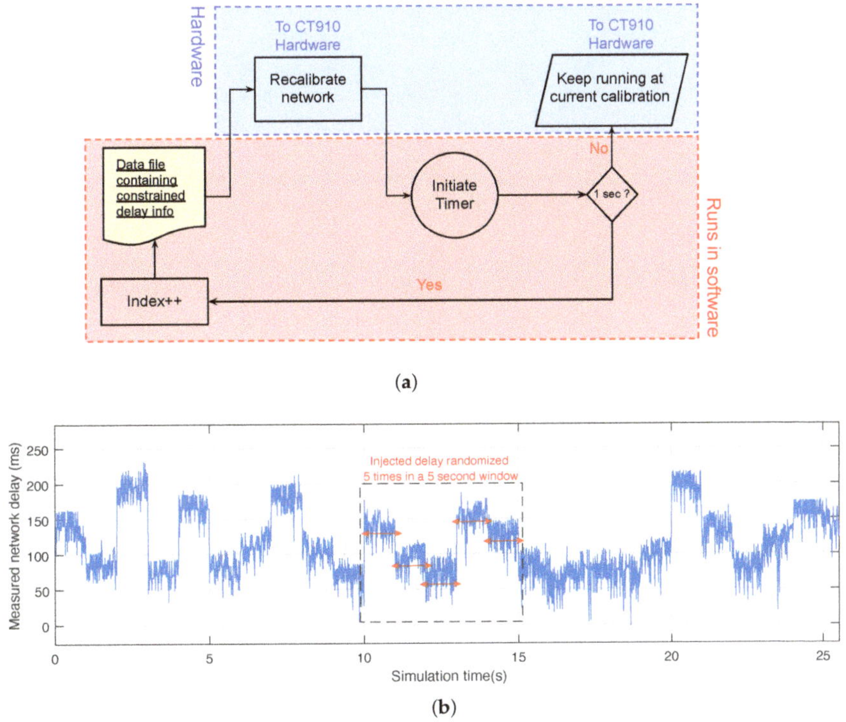

Figure 14. Randomized delay injection: (**a**) algorithm for random delay injection, (**b**) real-time delay measurement after randomized delay injection.

This batch of trials (10 runs, each for a duration of 10 min) was performed with different values of the mean and standard deviation of the injected delay, and whether or not the 10 individual runs were *successful* is summarized in Table 2. The mean delay was varied from 50 ms to 750 ms, and the standard deviation was varied from 5 ms to 200 ms. It can be seen that the performance progressively deteriorates at higher values of the mean delay. It can also be seen that when the mean delay is 500 ms, a small increase in the standard deviation leads to a significant number of *unsuccessful* trials. It is interesting to note that the *unsuccessful* runs failed because the First-In First-Out (FIFO) instances, which transmit data from the PMUs to the SSGC, gradually filled up as the PMUs keep streaming data into a slow network. This shows that in real-time applications under nonideal communication interfaces, the adequate sizing of elastic storage elements in the embedded systems is crucial for the control architecture to function.

Table 2. SSGC performance under varying random network delays.

σ (ms) / μ (ms)	5	10	25	50	100	200
50	10/10	10/10	10/10	10/10	10/10	10/10
100	10/10	10/10	10/10	10/10	10/10	10/10
200	10/10	10/10	10/10	10/10	10/10	9/10
500	10/10	10/10	5/10	5/10	4/10	3/10
750	0/10	0/10	0/10	0/10	0/10	1/10

Note: the color scale is used to denote from green (100% successful) to red (unsuccessful) experiments.

3.2.3. Effect of Data Drop in the Communication Network

In this experiment, the quality of a controller's output signal is analyzed for varying network delays and varying data-drop rates injected simultaneously in the communication network. The controller regulates the power output of the inverter inside the BESS, as demonstrated in Section 3.1. However, its performance is expected to deteriorate under stressed communication network conditions (i.e., higher delay and higher data-drop rates). The results of these tests are summarized in Figure 15. It can be observed that the analog output of the remote controller loses the control signal resolution under higher network delays and higher data-drop rates. However, as discussed earlier in Section 3.1, the control objective of the SSGC does not involve any management of fast system dynamics. Thus, in the short term (up to 30 s), the SSGC-driven control architecture can sustain itself even while operating within a tampered network under drastic delay and data-drop conditions.

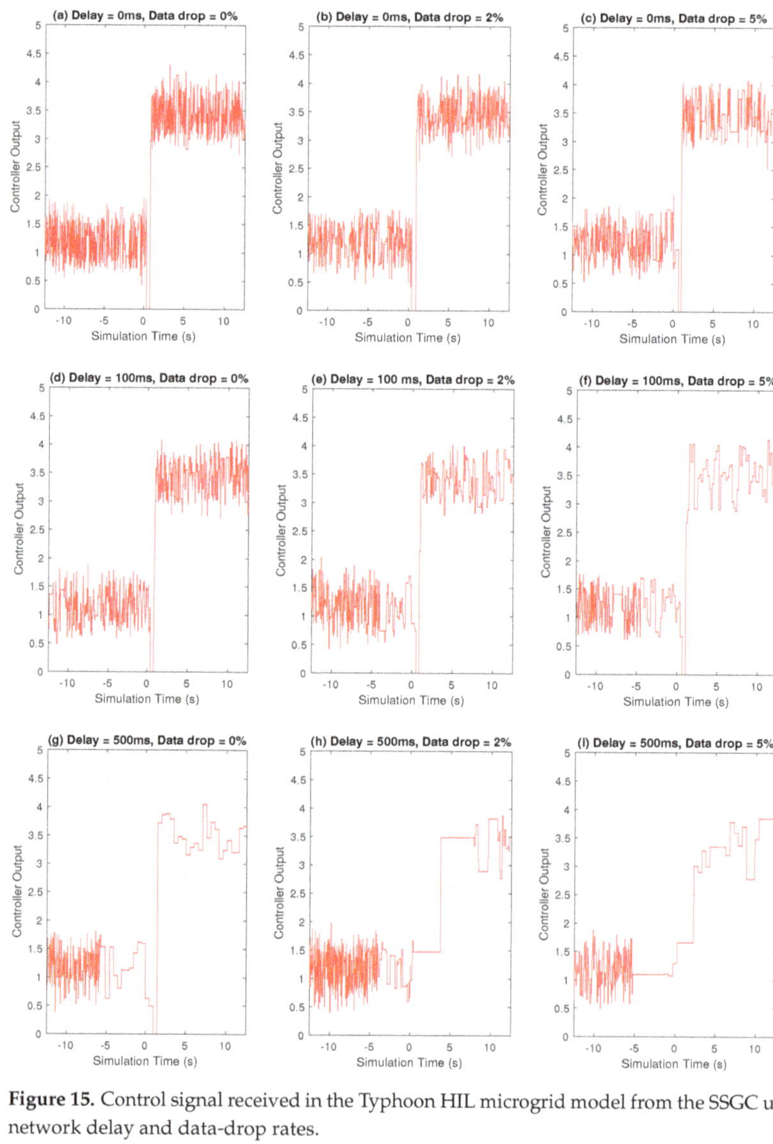

Figure 15. Control signal received in the Typhoon HIL microgrid model from the SSGC under varying network delay and data-drop rates.

An additional set of trials were conducted for this experiment. In these trials, the SSGC was set to operate freely under tampered network conditions, and experiments were run to test whether or not it can sustain itself for **longer periods** of time. The network was tampered with by introducing network delays and data drop. Under these conditions, the network was kept running for 10 min. After 10 min, it was determined if the SSGC was still receiving all the PMU streams successfully and if the real-time simulator was still receiving the controller's output. Observations were taken 10 times for each communication network condition. The summary of these results is shown in Table 3. As can be observed, the network delay and data drop can both adversely effect the robustness of the SSGC. In fact, in a situation where the SSGC is subjected to both a high network delay coupled with a high data-drop rate, the SSGC is almost certain to be unable to sustain itself for a long period of time. However, for lower delays and lower data-drop rates, the SSGC is proven to be reliable. For instance, consider the case for a 0.5% data drop and 50 ms delay: 10 out of 10 runs were sustained, while for a 5% data drop and 200 ms delay, only 2 out 10 runs were sustained.

Table 3. SSGC performance under varying communication network conditions.

Delay / Drop Rate (%)	0%	0.5%	1%	2%	5%	10%
0 ms	10/10	10/10	10/10	10/10	10/10	3/10
50 ms	10/10	10/10	10/10	10/10	9/10	0/10
100 ms	10/10	10/10	10/10	9/10	7/10	0/10
200 ms	10/10	10/10	10/10	6/10	2/10	0/10
500 ms	10/10	10/10	5/10	0/10	0/10	0/10
750 ms	1/10	0/10	0/10	0/10	0/10	0/10

Note: the color scale is used to denote from green (100% successful) to red (unsuccessful) experiments.

3.2.4. Effect of Jitter on the Communication Network

Jitter is defined as the variation in the time delay between when a signal is transmitted and when it is received over a network connection. The CT910 hardware has the provision of incorporating two parameters: *jitter amount* and *jitter frequency* in the network configuration. The parameter *jitter amount* defines the maximum amount of delay mismatch introduced in the network, and the *jitter frequency* determines what percent of data packets would experience the jitter. For this experiment, a jitter frequency of 10% and a maximum jitter delay of 500 ms was introduced in the network. This test was repeated with and without the jitter present. Figure 16 shows the effect of jitter on the control signal being transmitted by SSGC. The jitter delay was maxed at 500 ms, which meant that only a few packets would be experiencing a jitter delay around that range. Experimentally, it was observed that the effect of jitter is visible intermittently for some brief windows (denoted by red boxes in Figure 16), during which the control signal slows down its update rate significantly (thus dropping the high-frequency components). However, these windows are short and are only observed while the applied jitter delay is close to its maximum permissible limit of 500 ms.

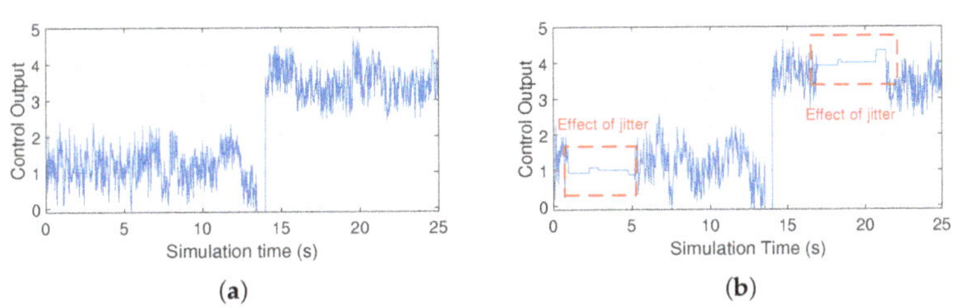

Figure 16. Control signal received in the Typhoon HIL microgrid model from the SSGC with (**a**) and without (**b**) jitter present in the network.

3.3. Control Performance Enhancement through LPF Functions

It needs to be noted that only one scenario for controlling the DERs within a microgrid is demonstrated in this paper. In that scenario, the battery is covering for a step increase in the load. While this experiment is an important 'proof of concept', additional experimentation is crucial before implementing the proposed architecture to control real-world DERs.

In real-world applications, the behavior of a BESS-based DER in a power system is more complex than what simulation models can capture, and more importantly, the operation of the controllers are restricted by the physics of the battery. In fact, the batteries utilized in energy storage systems are limited by their response times, which are different during charging and discharging operations in addition to their power availability. The authors of [46] reported that the average response time for a Battery Energy Storage System during charging is about 2.2 s and during discharging is about 0.6 s. Keeping these numbers in mind, the battery cannot be subjected to faster changes than those allowed by the charge and discharge rates. Hence, when aiming to control power flows to balance load/generation changes, any frequency variations, if kept unfiltered, would rapidly increase the switching and would generate excessive heat, thus compromising the health of the batteries.

To address this issue, a practical approach is adopted to restrict the response bandwidth of the proposed controller. To this end, a low-pass filter was introduced as an SSG function, with the goal to eliminate the high-frequency variations in the control signal. In addition, it may also slow down the response of the overall control response. Hence, the LPF needs to be adequately tuned. To illustrate, experiments are conducted by varying the cut-off frequency of this LPF from 4 Hz to 0.25 Hz. The control responses of each of these cases are demonstrated in Figure 17. It can be seen from the results that at a lower cut-off frequency ($\leq$0.5 Hz), the control system is free of high-frequency components, and the control signal can be safely employed in the BESS without exposing the BESS to overswitching. These filters can reliably bring the response of the SSGC-based control system within the permissible limit for battery operations, as reported in [46].

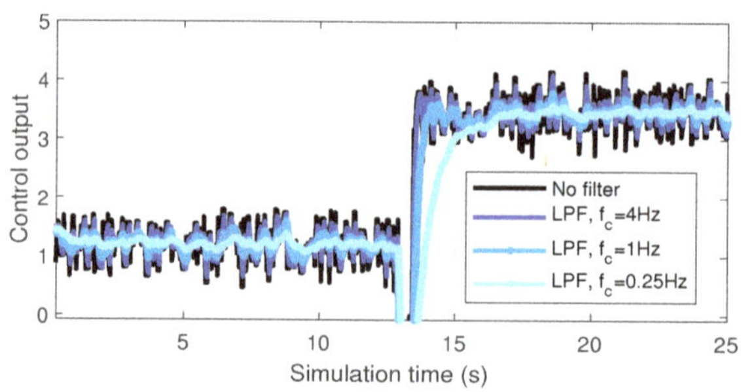

Figure 17. Application of LPFs to reduce overswitching of the BESS unit being controlled by the SSGC hardware.

3.4. Control System Redesign to Utilize Phase-Angle Measurements

As a 'proof of concept' that the SSGC can support different control functions, in this section, the control system is redesigned to utilize the phase-angle difference as the controller input, and the power flow is regulated the same way as before. The usage of the phase angle to control the power flow is a tried and tested technique which has been proposed, theorized and archived in [47–49]. Synchrophasor technology is capable of reporting accurate phase-angle measurements in real time [35–38]. This makes this technology suitable for phase-angle-based power-flow control. In this paper, the phase-angle difference between the load and the point of common coupling is used as the control variable to direct the power flow from the BESS. It is to be noted that, for distribution systems, the angle difference between the buses can be very small. Hence, the controller needs to be tuned with precision. Thus, even though the control system architecture is independent of the control variable (the active power or phase angle), the internal PI controller requires significant retuning.

Experiments were performed with the BESS power output being controlled by the phase-angle difference $\delta_{PCC} - \delta_{load}$. The results of this experiment are summarized in Figure 18a–c. It can be observed from Figure 18c that the active power output increases from $\simeq$100 kW to $\simeq$400 kW. This response is consistent with the results demonstrated previously in Figure 11c.

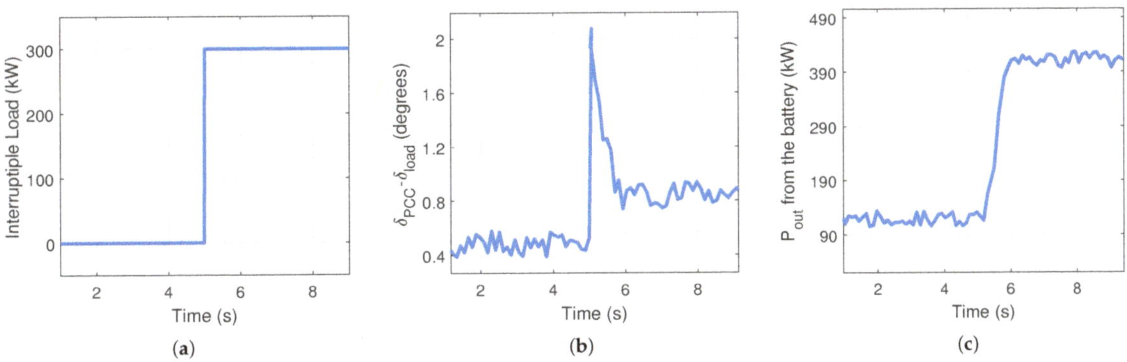

Figure 18. Control system performance for the BESS under ideal network conditions with phase angle utilized as the control signal: (**a**) 300 kW load injection by switching load, (**b**) response of the control variable $\delta_{PCC} - \delta_{load}$ to the load change and (**c**) active power output from the Battery Energy Storage System (BESS) equipped with phase-angle-based active power control.

4. Future Works

The current paper introduces a new architecture for control system implementation for DERs in microgrids by utilizing synchrophasor measurements. As a 'proof of concept', one specific case study was tested where the BESS output is increased to match the step increase in the load. Different cases featuring the control of a full range of DERs and a full range of behaviors on the load side, microgrid conditions, etc., must be performed in order to demonstrate the resilience and robustness of the proposed architecture. This work is outside the scope of this paper and is left as future work.

For BESSs, the state of charge (SOC) of the battery is a crucial parameter. In all realistic BESS control systems, based on the SOC, the battery is often put into either charging or discharging mode. The current paper assumes the SOC to be high enough so that the battery can reliably operate in discharging mode; i.e., it can feed active loads. For real systems, this will not be the case. To address this, an additional control loop must be designed to utilize the SOC of the battery and ensure the safe and reliable operation of the BESS. This requires coordination and information exchange with Battery Management Systems (BMSs) and will be the subject of future research.

5. Conclusions

The proposed Synchrophasor Synchronization Gateway and Control system can reduce the complexity and latency of the synchrophasor system and communication network when compared to traditional WAMS/WAMPAC systems by exploiting and expanding upon the existing *Khorjin* library. This architecture was utilized successfully to implement a networked control system for a real-time microgrid model.

A scenario for controlling the DERs within the microgrid was used for experimentation in this paper. In that scenario, the BESS is capable of using synchrophasor data from different locations to respond to a step increase in the load. It was also demonstrated that this architecture can be successful irrespective of the actual control algorithm being employed within it.

Experimentally, the SSGC architecture was tested under different network conditions to establish its validity and robustness. More specifically, the network was subjected to delay, jitter and data drops through external hardware. The performance of the controller was evaluated under those different network conditions. It was observed that for lower values of the data-drop rate ($\leq$2%) and lower values of the network delay ($\leq$100 ms), the SSGC-based control system is reliable. It was also observed that the introduction of jitter makes the controller lose a significant amount of high-frequency components. However, up to a maximum jitter occurrence of 10% and maximum jitter delay of 500 ms, the SSGC sustains itself successfully even for longer runs.

Finally, an LPF was designed to prevent the BESS from overswitching. This filter would ensure that the battery operations are within the physical limits of the battery and thus increase the lifetime of the battery units within the BESS.

The main limitation of the implementation presented in this work is the cost associated with the National Instruments' cRIO hardware. While such hardware was ideal for prototyping and testing, it is cost prohibitive for product development. For the real-world deployment of the proposed architecture, it would be crucial to utilize a low-cost hardware platform to make it economically viable. Consequently, migrating the *Khorjin* dynamically linked library (DLL) to a different platform is the first step towards the authors' future work.

Author Contributions: Conceptualization, P.M.A., L.V. and K.K.; Methodology, P.M.A., L.V. and K.K.; Software, P.M.A. and H.C.; Validation, P.M.A.; Investigation, P.M.A., L.V., H.C. and K.K.; Resources, L.V.; Writing—original draft, P.M.A.; Writing—review & editing, L.V.; Supervision, L.V. and K.K.; Project administration, L.V.; Funding acquisition, L.V. All authors have read and agreed to the published version of the manuscript.

Funding: This research was funded in part by the New York State Energy Research and Development Authority (NYSERDA) under agreement number 137948.

Data Availability Statement: The The data presented in this study are available on request from the corresponding author.

Conflicts of Interest: The funders had no role in the design of the study; in the collection, analyses, or interpretation of data; in the writing of the manuscript; or in the decision to publish the results.

Abbreviations

BESS	Battery Energy Storage System
DER	Distributed Energy Resources
HIL	Hardware-In-the-Loop
CHIL	Controller Hardware-In-the-Loop
LPF	Low Pass Filter
PCC	Point of common coupling
PMU	Phasor Measurement Unit
PDC	Phasor Data Concentrator
PV	Photovoltaic
SSG	Synchrophasor Synchronization Gateway
SSGC	Synchrophasor Synchronization Gateway and Controller
WACS	Wide-area control system
WAMPAC	Wide-Area Monitoring Protection and Control System

References

1. *IEEE Std C37.118.2-2011*; IEEE Standard for Synchrophasor Data Transfer for Power Systems. Revision of IEEE Std C37.118-2005. IEEE: New York, NY, USA, 2011; pp. 1–53. [CrossRef]
2. Available online: https://www.gegridsolutions.com/multilin/catalog/p30.htm (accessed on 25 September 2023).
3. Available online: https://selinc.com/products/3573/ (accessed on 25 September 2023).
4. Available online: https://nrec.com/en/index.php/product/productInfo/571.html (accessed on 25 September 2023).
5. Castello, P.; Muscas, C.; Pegoraro, P.A.; Sulis, S. Low-cost implementation of an active phasor data concentrator for smart grid. In Proceedings of the 2018 Workshop on Metrology for Industry 4.0 and IoT, Brescia, Italy, 16–18 April 2018; pp. 78–82. [CrossRef]

6. Adamiak, M.G.; Kanabar, M.; Rodriquez, J.; Zadeh, M.D. Design and implementation of a synchrophasor data concentrator. In Proceedings of the 2011 IEEE PES Conference on Innovative Smart Grid Technologies—Middle East, Jeddah, Saudi Arabia, 17–20 December 2011; pp. 1–5. [CrossRef]
7. Retty, H.; Delport, J.; Centeno, V. Development of tests and procedures for evaluating phasor data concentrators. In Proceedings of the 2013 IEEE Grenoble Conference, Grenoble, France, 16–20 June 2013; pp. 1–5. [CrossRef]
8. Vanfretti, L.; Jónsdóttir, G.M.; Almas, M.S.; Rebello, E.; Firouzi, S.R.; Baudette, M. Audur—A platform for synchrophasor-based power system wide-area control system implementation. *SoftwareX* **2018**, *7*, 294–301. [CrossRef]
9. Danielson, C.F.M.; Vanfretti, L.; Almas, M.S.; Choompoobutrgool, Y.; Gjerde, J.O. Analysis of communication network challenges for synchrophasor-based wide-area applications. In Proceedings of the 2013 IREP Symposium Bulk Power System Dynamics and Control—IX Optimization, Security and Control of the Emerging Power Grid, Rethymno, Greece, 25–30 August 2013; pp. 1–13. [CrossRef]
10. Yao, W.; You, S.; Wang, W.; Deng, X.; Li, Y.; Zhan, L.; Liu, Y. A Fast Load Control System Based on Mobile Distribution-Level Phasor Measurement Unit. *IEEE Trans. Smart Grid* **2020**, *11*, 895–904. [CrossRef]
11. Liu, Y.; You, S.; Yao, W.; Cui, Y.; Wu, L.; Zhou, D.; Zhao, J.; Liu, H.; Liu, Y. A Distribution Level Wide Area Monitoring System for the Electric Power Grid–FNET/GridEye. *IEEE Access* **2017**, *5*, 2329–2338. [CrossRef]
12. You, S.; Zhao, J.; Yao, W.; Liu, Y.; Cui, Y.; Wu, L.; Guo, J.; Liu, Y. FNET/GridEye for Future High Renewable Power Grids—Applications Overview. In Proceedings of the 2018 IEEE PES Transmission & Distribution Conference and Exhibition—Latin America (T & D-LA), Lima, Peru, 18–21 September 2018; pp. 1–5. [CrossRef]
13. Liu, Y.; Yao, W.; Zhou, D.; Wu, L.; You, S.; Liu, H.; Zhan, L.; Zhao, J.; Lu, H.; Gao, W.; et al. Recent developments of FNET/GridEye—A situational awareness tool for smart grid. *CSEE J. Power Energy Syst.* **2016**, *2*, 19–27. [CrossRef]
14. Schweitzer, E.O.; Whitehead, D.; Zweigle, G.; Ravikumar, K.G. Synchrophasor-based power system protection and control applications. In Proceedings of the 2010 63rd Annual Conference for Protective Relay Engineers, College Station, TX, USA, 29 March–1 April 2010; pp. 1–10.
15. Rodrigues, Y.R.; Abdelaziz, M.; Wang, L. D-PMU Based Secondary Frequency Control for Islanded Microgrids. *IEEE Trans. Smart Grid* **2020**, *11*, 857–872. [CrossRef]
16. Zenelis, I.; Wang, X.; Kamwa, I. Online PMU-Based Wide-Area Damping Control for Multiple Inter-Area Modes. *IEEE Trans. Smart Grid* **2020**, *11*, 5451–5461. [CrossRef]
17. Chenine, M.; Nordström, L. Investigation of communication delays and data incompleteness in multi-PMU Wide Area Monitoring and Control Systems. In Proceedings of the 2009 International Conference on Electric Power and Energy Conversion Systems, (EPECS), Sharjah, United Arab Emirates, 10–12 November 2009; pp. 1–6.
18. Microgrid Knowledge. Available online: https://microgridknowledge.com/microgrids-navigant/ (accessed on 25 September 2023).
19. Lasseter, R.H. MicroGrids. In Proceedings of the 2001 IEEE Power Engineering Society Winter Meeting, Conference Proceedings (Cat. No.01CH37194), Columbus, OH, USA, 28 January–1 February 2001; Volume 1, pp. 146–149. [CrossRef]
20. Hatziargyriou, N.; Asano, H.; Iravani, R.; Marnay, C. Microgrids. *IEEE Power Energy Mag.* **2007**, *5*, 78–94. [CrossRef]
21. Bidram, A.; Davoudi, A. Hierarchical Structure of Microgrids Control System. *IEEE Trans. Smart Grid* **2012**, *3*, 1963–1976. [CrossRef]
22. Canciello, G.; Cavallo, A.; Guida, B. Control of Energy Storage Systems for Aeronautic Applications. *J. Control Sci. Eng.* **2017**, *2017*, 2458590. [CrossRef]
23. Cavallo, A.; Canciello, G.; Guida, B. Energy Storage System Control for Energy Management in Advanced Aeronautic Applications. *Math. Probl. Eng.* **2017**, *2017*, 4083132. [CrossRef]
24. Adhikari, P.M.; Vanfretti, L. Delay Analysis of a Real-Time Hard Reconfigurable Synchrophasor Synchronization Gateway. In Proceedings of the International Conference on Control and Optimization of Renewable Energy Systems (CORES), Anaheim, CA, USA, 6–7 December 2019. Available online: https://www.actapress.com/PaperInfo.aspx?paperId=456528 (accessed on 25 September 2023).
25. Firouzi, S.R.; Vanfretti, L.; Ruiz-Alvarez, A.; Hooshyar, H.; Mahmood, F. Interpreting and implementing IEC 61850-90-5 Routed-Sampled Value and Routed-GOOSE protocols for IEEE C37.118.2 compliant wide-area synchrophasor data transfer. *Electr. Power Syst. Res.* **2017**, *144*, 255–267. ISSN 0378-7796. [CrossRef]
26. Buttayak, S.; Wornpuen, A.; Promparn, N.; Charbkaew, N.; Bunyagul, T. Design of Phasor Data Concentrator for phasor monitoring system. In Proceedings of the 2012 IEEE Conference on Sustainable Utilization and Development in Engineering and Technology (STUDENT), Kuala Lumpur, Malaysia, 6–9 October 2012; pp. 102–107. [CrossRef]
27. Bihari, S.P.; Sadhu, P.K.; Sarita, K.; Khan, B.; Arya, L.D.; Saket, R.K.; Kothari, D.P. A Comprehensive Review of Microgrid Control Mechanism and Impact Assessment for Hybrid Renewable Energy Integration. *IEEE Access* **2021**, *9*, 88942–88958. [CrossRef]
28. Guerrero, J.M.; Vasquez, J.C.; Matas, J.; de Vicuna, L.G.; Castilla, M. Hierarchical Control of Droop-Controlled AC and DC Microgrids—A General Approach Toward Standardization. *IEEE Trans. Ind. Electron.* **2011**, *58*, 158–172. [CrossRef]
29. Pourramezan, R.; Seyedi, Y.; Karimi, H.; Zhu, G.; Mont-Briant, M. Design of an advanced phasor data concentrator for monitoring of distributed energy resources in smart microgrids. *IEEE Trans. Ind. Inform.* **2017**, *13*, 3027–3036. [CrossRef]

30. Guillaud, X.; Faruque, M.O.; Teninge, A.; Hariri, A.H.; Vanfretti, L.; Paolone, M.; Dinavahi, V.; Mitra, P.; Lauss, G.; Dufour, C.; et al. Applications of Real-Time Simulation Technologies in Power and Energy Systems. *IEEE Power Energy Technol. Syst. J.* **2015**, *2*, 103–115. [CrossRef]
31. Konakalla, S.A.R.; Valibeygi, A.; de Callafon, R.A. Microgrid dynamic modeling and islanding control with synchrophasor data. *IEEE Trans. Smart Grid* **2020**, *11*, 905–915. [CrossRef]
32. Castello, P.; Muscas, C.; Pegoraro, P.A.; Sulis, S. Active phasor data concentrator performing adaptive management of latency. *Sustain. Energy Grids Netw.* **2018**, *16*, 270–277. [CrossRef]
33. Adhikari, P.M.; Vanfretti, L.; Mishra, C.; Jones, K.D. A Reconfigurable Synchrophasor Synchronization Gateway & Controller Architecture for DERs. In Proceedings of the 2022 International Conference on Smart Grid Synchronized Measurements and Analytics (SGSMA), Split, Croatia, 24–26 May 2022; pp. 1–6. [CrossRef]
34. Microgrids. Available online: https://bit.ly/TyphoonHILMG (accessed on 25 September 2023).
35. Almas, M.S.; Baudette, M.; Vanfretti, L. Utilizing synchrophasor-based supplementary damping control signals in conventional generator excitation systems. *Electr. Power Syst. Res.* **2018**, *157*, 157–167. [CrossRef]
36. Díez-Maroto, L.; Vanfretti, L.; Almas, M.S.; Jónsdóttir, G.M.; Rouco, L. A WACS exploiting generator Excitation Boosters for power system transient stability enhancement. *Electr. Power Syst. Res.* **2017**, *148*, 245–253. [CrossRef]
37. Rebello, E.; Vanfretti, L.; Almas, M.S. Experimental Testing of a Real-Time Implementation of a PMU-Based Wide-Area Damping Control System. *IEEE Access* **2020**, *8*, 25800–25810. [CrossRef]
38. Chompoobutrgool, Y.; Vanfretti, L. Using PMU signals from dominant paths in power system wide-area damping control. *Sustain. Energy Grids Netw.* **2015**, *4*, 16–28. [CrossRef]
39. Plug-and-Play Microgrid Library and Testing of Microgrid Controller. Available online: https://bit.ly/TyphoonControl (accessed on 25 September 2023).
40. NI Expertise: Power Quality Analysis. Available online: https://bit.ly/NIPQAnalysis (accessed on 25 September 2023).
41. Derviškadić, A.; Romano, P.; Pignati, M.; Paolone, M. Architecture and experimental validation of a low-latency phasor data concentrator. *IEEE Trans. Smart Grid* **2018**, *9*, 2885–2893. [CrossRef]
42. LabVIEW Electrical Power Toolkit API Reference. Available online: https://bit.ly/NIPMUvis (accessed on 25 September 2023).
43. Adhikari, P.M.; Hooshyar, H.; Vanfretti, L. Experimental Quantification of Hardware Requirements for FPGA-Based Reconfigurable PMUs. *IEEE Access* **2019**, *7*, 57527–57538. [CrossRef]
44. Available online: https://www.typhoon-hil.com/products/hil-connect/ (accessed on 25 September 2023).
45. CT 910. Available online: https://www.candelatech.com/ct910_product.php (accessed on 25 September 2023).
46. Song, L.; Wang, Z.; Luo, W. Analysis on the Response Time of the Battery Energy Storage System. *Mater. Sci. Energy Eng. (CMSEE 2014)* **2015**, 553–560. [CrossRef]
47. Edris, A.; Mehraban, A.S.; Rahman, M.; Gyugyi, L.; Arabi, S.; Reitman, T. Controlling the flow of real and reactive power. *IEEE Comput. Appl. Power* **1998**, *11*, 20–25. [CrossRef]
48. Pierce, R.E.; Hamilton, B.W. Phase-angle control of system interconnections. *Electr. Eng.* **1939**, *58*, 83–92. [CrossRef]
49. Jolissaint, C.H.; Arvanitidis, N.V.; Luenberger, D.G. Decomposition of Real and Reactive Power Flows: A Method Suited for On-Line Applications. *IEEE Trans. Power Appar. Syst.* **1972**, *PAS-91*, 661–670. [CrossRef]

energies

MDPI

Article

Detection of False Data Injection Attacks in a Smart Grid Based on WLS and an Adaptive Interpolation Extended Kalman Filter

Guoqing Zhang [1,2], Wengen Gao [1,2,*], Yunfei Li [1,2], Xinxin Guo [1,2], Pengfei Hu [1,2] and Jiaming Zhu [1,2]

1 School of Electrical Engineering, Anhui Polytechnic University, Wuhu 241000, China; gq17556964576@163.com (G.Z.); lyf@mail.ahpu.edu.cn (Y.L.); guoxinxin@ahpu.edu.cn (X.G.); h382024188@163.com (P.H.); m18315378156@163.com (J.Z.)

2 Key Laboratory of Advanced Perception and Intelligent Control of High-End Equipment, Chinese Ministry of Education, Wuhu 241000, China

* Correspondence: ahpuchina@ahpu.edu.cn

Abstract: An accurate power state is the basis of the normal functioning of the smart grid. However, false data injection attacks (FDIAs) take advantage of the vulnerability in the bad data detection mechanism of the power system to manipulate the process of state estimation. By attacking the measurements, then affecting the estimated state, FDIAs have become a serious hidden danger that affects the security and stable operation of the power system. To address the bad data detection vulnerability, in this paper, a false data attack detection method based on weighted least squares (WLS) and an adaptive interpolation extended Kalman filter (AIEKF) is proposed. On the basis of applying WLS and AIEKF, the Euclidean distance is used to calculate the deviation values of the two-state estimations to determine whether the current moment is subjected to a false data injection attack in the power system. Extensive experiments were conducted to simulate an IEEE-14-bus power system, showing that the adaptive interpolation extended Kalman filter can compensate for the deficiency in the bad data detection mechanism and successfully detect FDIAs.

Keywords: false data injection attacks; adaptive interpolation extended Kalman filter; state estimation; Euclidean distance; smart grid

Citation: Zhang, G.; Gao, W.; Li, Y.; Guo, X.; Hu, P.; Zhu, J. Detection of False Data Injection Attacks in a Smart Grid Based on WLS and an Adaptive Interpolation Extended Kalman Filter. *Energies* **2023**, *16*, 7203. https://doi.org/10.3390/en16207203

Academic Editors: Ying-Yi Hong, Javier Contreras and Michael Negnevitsky

Received: 8 August 2023
Revised: 10 October 2023
Accepted: 20 October 2023
Published: 23 October 2023

1. Introduction

The modern power system is developing towards intelligence, and plenty of intelligent devices, such as smart meters and sensors, promote the transformation of the power system in the modes of power generation, transformation, transmission, and distribution, which makes the smart grid a typical cyber–physical system (CPS) [1,2]. In a smart grid, the supervisory control and data acquisition (SCADA) system collects and analyzes real-time data from field devices across the network. Finally, the SCADA reports back to the control center, which then makes adjustments to the power generation and distribution of the grid based on this information [3].

The susceptibility of the power cyber–physical system (CPS) to cyber attacks is a result of the unpredictable nature of sensor data in the perception layer and the unrestricted communication channels for data exchange [4,5]. Among the many types of cyber attacks, attacks against smart grids and industrial control systems are the most common; the damage caused to the system cannot be underestimated, seriously affecting the normal production activities of society. For example, in 2010, the "Stuxnet" virus attack on a Belarusian enterprise, which caused anomalies in uranium enrichment centrifuges and generators at the Iranian nuclear power plant, resulted in damage to many pieces of equipment [6]. In 2015, Black Energy, a cyber virus targeting the power grid, caused power outages at some Ukrainian power plants, disrupting the power supply to many factories in the Ivano-Frankivsk region and affecting production [7]. The investigation revealed that the incident resulted in the malicious deletion of historical grid measurements stored in the SCADA, which made recovery extremely difficult.

A false data injection attack (FDIA) is a novel attack method specifically targeting the integrity of state estimation data in the power CPS [8,9]. The attackers inject false data, which affects the power flow calculation, control decisions, etc., through smart grid sensors, controllers, and remote control units to tamper with the original data of the grid. This situation can potentially result in the malfunction of grid equipment and, in severe cases, the complete paralysis of the power network, which not only poses a significant threat to grid security but also carries the potential for substantial economic losses. Figure 1 shows the structure of a smart grid system and an illustration of an FDIA.

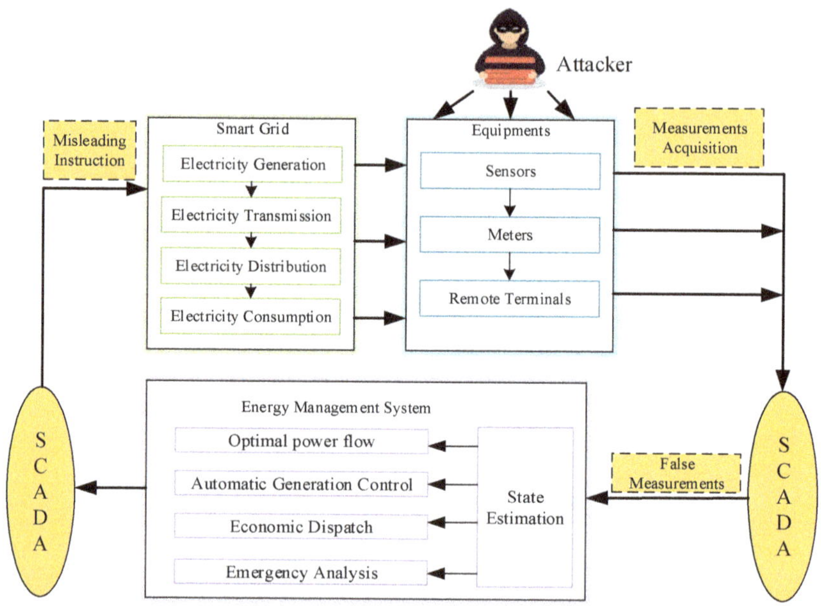

Figure 1. Illustration of an FDIA in a smart grid.

Liu et al. [10] first introduced the topic of FDIAs in the literature, where it was hypothesized that an attacker could access the current configuration information of a smart grid and manipulate meter or sensor measurements. Such an attack could insert false data into specific state variables, avoiding detection by current bad data detection algorithms. Yang et al. [11] delved into the challenge of determining the most effective attack strategy. This strategy, known as an injection attack strategy, involves selecting a specific set of meters to manipulate in a way that maximizes the resulting damage. They not only formalized this problem but also developed efficient algorithms to pinpoint the ideal set of meters for such attacks. It is important to highlight that even if these attacks are isolated to specific devices, their impact on the smart grid can be catastrophic due to the grid's intricate interconnections. As described by He et al. [12], electricity theft by attackers by modifying smart meter data has seriously affected utility security. Therefore, many researchers have devoted themselves to the detection of false data injection attacks in order to safeguard the security of the smart grid.

When the power system is subjected to malicious false data injection, the state estimation result of the WLS under attack is updated in real time by incorporating an estimation algorithm of an extended Kalman filter, which has a hysteresis in its state estimation process, and by observing disparities in the estimation outcomes produced by the two algorithms, making it possible to detect the FDIA. Meanwhile, to improve the accuracy and reduce the linearization error of the EKF, the adaptive interpolation strategy is introduced. Therefore, in this paper, we propose a detection method based on WLS and AIEKF. Considering that the two algorithms have different degrees of correspondence to real-time information, the FDIA can be effectively detected. The main contributions can be summarized as follows:

- Considering the linearization error of the EKF algorithm in state estimation in a power system, the adaptive interpolation strategy is introduced. The pseudomeasurements between two consecutive measurements are inserted by linear interpolation to improve the estimation accuracy.
- We propose a novel FDIA detection method that combines AIEKF and WLS, marking the first instance of their joint application in this context.
- We conduct many experiments on an IEEE-14-bus power system to demonstrate the proposed algorithm's performance in detecting FDIAs. The result shows that the method can effectively detect FDIAs.

The remainder of the paper is structured as follows. Section 2 provides an overview of relevant literature pertaining to the detection of FDIAs. Section 3 outlines the system model employed in this study. In Section 4, we delve into the details of the proposed AIEKF algorithm. In Section 5, the detection principle is described. The experiments and results are presented in Section 6. Finally, in Section 7, we present our concluding remarks and suggest directions for future work. A list of abbreviations and acronyms is provided in Table 1.

Table 1. A list of abbreviations and acronyms.

Abbreviations and Acronyms	Full Name
AIEKF	Adaptive interpolation extended Kalman filter
BDD	Bad data detection
CPS	Cyber–physical system
EMS	Energy management system
FDIAs	False data injection attacks
LNR	Largest normalized residual
PMUs	Phase measurement units
RMSE	Root mean square error
SCADA	Supervisory control and data acquisition
WLS	Weighted least squares

2. Related Work

Since the concept of FDIAs was introduced, the issues related to FDIAs have received a great deal of attention in both academic research and industry. Many scholars have studied bad data in state estimation and proposed corresponding detection methods for FDIAs based on their research. Although FDIA detection algorithms differ from each other, the algorithms can be classified into two categories [13]: model-based algorithms and data-driven algorithms.

2.1. Model-Based Detection Algorithms

Nowadays, with the increasing degree of interconnection of power systems in smart grids, the simple use of weighted least squares is no longer applicable, and many variants have emerged. Moslemi et al. [14] introduced an innovative cyber attack detection technique based on maximum likelihood estimation. This method capitalizes on the near-chordal sparsity (NCS) characteristics of the power grid to establish a robust framework for addressing the corresponding maximum likelihood estimation problem. Furthermore, they broke down this detection approach into a series of localized maximum likelihood estimation problems. This approach not only safeguards privacy but also mitigates potential issues. Chen et al. [15] introduced a novel method for detecting FDIAs using kernel density estimation. This method utilizes the concept of kernel density estimation, which utilizes historical data to estimate the probability distributions of both measurements and control commands. Additionally, it calculates confidence intervals for these estimates at a specified significance level. Zhao et al. [16] proposed a method that employs short-term state prediction in smart grids to detect spurious data injection attacks. This method utilizes real-time and predicted states to pinpoint potential false data injections by analyzing the differences between them. By continuously monitoring and comparing the accuracy of short-term state predictions, this approach enhances the security and resilience of smart grid systems by improving the detection of malicious attacks. Li et al. [17] proposed a detection method based on the watermark embedding technique. The method uses a dynamic watermark embedding technique to embed security-enhancing markers (watermarks) into the grid measurement data to ensure the integrity and authenticity of the data. Then, the data embedded with the watermark are processed and detected using the EKF algorithm to identify possible faulty data insertion attacks. Through the joint application of dynamic watermarking and EKF, this method can improve the detection capability of smart grid systems against FDIAs and provide increased security. The above methods cannot be used alone; they must be used with the help of some determination methods to determine the existence of FDIAs. Some determination methods are available, such as Euclidean distance [16,18], maximum normalized residuals [19,20], chi-square testing [18,21], Cumulative sum testing [22,23], Kullback–Leibler distance [14,24], and cumulative error sum of squares probability density curves [25,26].

2.2. Data-Driven Detection Algorithms

In contrast to detection algorithms that rely on system models, data-driven detection algorithms operate without the need for system parameters and models. The process of detecting FDIAs is independent of these factors, and it relies on a large amount of historical data from the smart grid to speculate on future data [13]. Mahi-Al-Rashid et al. [27] proposed an innovative approach for countering FDIAs. They employed a CNN-LSTM-based self-encoder sequence-to-sequence architecture for prediction-assisted anomaly detection. Additionally, they suggested an adaptive optimal thresholding method based on consumption patterns to identify unusual behaviors. Yu et al. [28] proposed an online method for detecting FDIAs that merges wavelet transform with deep neural networks. This approach involves extracting features from grid sensor data using wavelet transform and subsequently employing a deep neural network for real-time detection. This methodology effectively enhances the security and reliability of the smart grid system by identifying potential attacks as they occur. Wang et al. [29] combined Kalman filtering with recurrent neural networks. The actual state of the grid was first estimated using Kalman filtering. Next, the estimated state was modeled and predicted using recurrent neural networks. By analyzing the disparity between the predicted state and the real measured data, the system can identify the existence of FDIAs. Jorjani et al. [30] introduced a novel algorithm designed for the detection of FDIAs in AC state estimation. This algorithm employs outlier detection techniques to the outcomes of state estimation because the attack may lead to data inconsistencies between buses, resulting in an abnormal graph structure. The above methods can be proven to detect FDIAs.

3. System Model

The state estimation of the power system usually deeply relies on the system model. The selection and establishment of the model have a substantial impact on the results of the system state calculation, which directly lead to the accuracy of the acquired state. State estimation in the power system is a crucial element within EMS, as it provides essential real-time information about the grid's operational status, and it is the basis for other high-level applications to realize the calculation and analysis.

The measurements for power system state estimation are collected from the grid by SCADA or phase measurement units (PMUs). PMUs are able to provide accurate and synchronized phase measurements for geographically dispersed buses in the grid by taking advantage of the high accuracy, sub-microsecond time synchronization, and unprecedented reporting rate [31]. And if the system is completely observable with PMU measurements, the state estimation process is a linear procedure. The proposed algorithm aims at solving the linearization of EKF for state estimation. Therefore, the proposed algorithm can be applied to the mentioned PMU-based state estimation problem by reducing the linearization steps of the AIEKF algorithm. We can discuss a situation in which there are m measurements and n state variables. In an AC power system, the connection between measurements and state variables is characterized by a nonlinear relationship, which can be represented as:

$$z = h(x) + e \tag{1}$$

where $z \in \mathbb{R}^{m\times 1}$ is the measurement vector; $x \in \mathbb{R}^{n\times 1}$ is the state vector, typically bus voltage amplitude and phase; $e \in \mathbb{R}^{m\times 1}$ is the measurement error vector that satisfies $e = (e_1, e_2, \ldots, e_m)^T \sim \mathcal{N}(0_{m\times 1}, R)$; and $h(\cdot)$ represents the nonlinear relationship between the measurement vector (z) and the state vector (x).

To analyze the correlation between the bus voltage, phase angle, and bus current of the grid system and determine the nonlinear relationship $h(\cdot)$, we must streamline the power system branch by representing it through an equivalent circuit, as illustrated in Figure 2. Subsequently, utilizing the AC model of the power system, we establish the connection between the state variables and measurements, which can be formulated as follows:

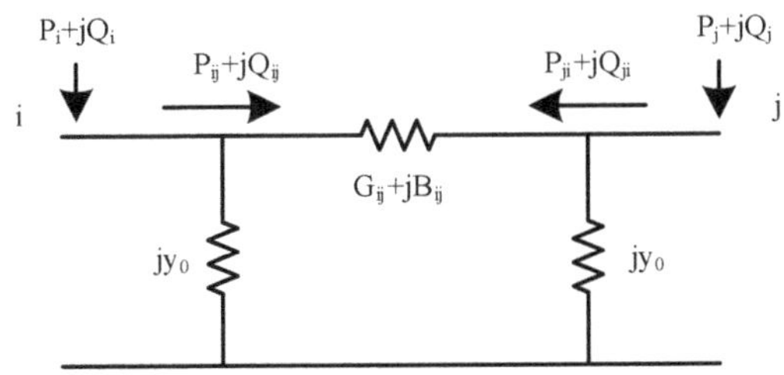

Figure 2. Power system branch equivalent circuits.

$$P_i = V_i \sum_{j\in T} V_j \left[G_{ij}\cos\theta_{ij} + B_{ij}\sin\theta_{ij}\right] \tag{2}$$

$$Q_i = V_i \sum_{j\in T} V_j \left[G_{ij}\sin\theta_{ij} - B_{ij}\cos\theta_{ij}\right] \tag{3}$$

$$P_{ij} = V_i^2 G_{ij} - V_i V_j \left[G_{ij}\cos\theta_{ij} + B_{ij}\sin\theta_{ij}\right] \tag{4}$$

$$Q_{ij} = -V_i^2 B_{ij} - V_i V_j \left[G_{ij}\sin\theta_{ij} - B_{ij}\cos\theta_{ij}\right] \tag{5}$$

where V_i and V_j are the voltage amplitudes at bus i and bus j, respectively; P_i and Q_i represent the active and reactive power injection of bus i, respectively; P_{ij} and Q_{ij} denote active power flow and reactive power flow from bus i to bus j, respectively; G_{ij} and B_{ij} denote the conductance and susceptance of the line from bus i to bus j, respectively; θ_{ij} denotes the phase angle difference of the line voltage from bus i to bus j; and T denotes the set buses adjacent to bus i.

3.1. State Estimation

The most commonly used state estimation in power systems is the weighted least squares method, which is still widely used [32–34]. Under this method, the objective function ($J(\hat{x})$) is the weighted sum of squares of the difference between the measured and estimated values. With the smallest objective function value, the obtained $\boldsymbol{x}$ is the closest approximation to the true state of the system. Based on the weighted least squares method, the objective function ($\boldsymbol{J}(\hat{\boldsymbol{x}})$) can be expressed as:

$$\boldsymbol{J}(\hat{\boldsymbol{x}}) = [\boldsymbol{z} - \boldsymbol{h}(\hat{\boldsymbol{x}})]^T \boldsymbol{R}^{-1}[\boldsymbol{z} - \boldsymbol{h}(\hat{\boldsymbol{x}})]. \tag{6}$$

Then,

$$\hat{\boldsymbol{x}} = \arg\min [\boldsymbol{z} - \boldsymbol{h}(\hat{\boldsymbol{x}})]^T \boldsymbol{R}^{-1}[\boldsymbol{z} - \boldsymbol{h}(\hat{\boldsymbol{x}})] \tag{7}$$

To solve the nonlinear WLS problem, we can linearize the measurement equation around $\boldsymbol{x}$, then apply the linear WLS method. The final result is expressed as:

$$\hat{\boldsymbol{x}}^{k+1} = \hat{\boldsymbol{x}}^k + \left[G\left(\hat{\boldsymbol{x}}^k\right)\right]^{-1} \boldsymbol{H}^T\left(\hat{\boldsymbol{x}}^k\right)\boldsymbol{R}^{-1}\left(\boldsymbol{z} - \boldsymbol{h}\left(\hat{\boldsymbol{x}}^k\right)\right) \tag{8}$$

$$G\left(\hat{\boldsymbol{x}}^k\right) = \boldsymbol{H}^T\left(\hat{\boldsymbol{x}}^k\right)\boldsymbol{R}^{-1}\boldsymbol{H}\left(\hat{\boldsymbol{x}}^k\right) \tag{9}$$

where k is the k-th iteration index, and $\boldsymbol{H} \in \mathbb{R}^{m \times n}$ is the Jacobian matrix of the measurement equation, which can be expressed as:

$$\boldsymbol{H} = \begin{bmatrix} \frac{\partial V_i}{\partial V} & \frac{\partial P_i}{\partial V} & \frac{\partial Q_i}{\partial V} & \frac{\partial P_{ij}}{\partial V} & \frac{\partial Q_{ij}}{\partial V} \\ \frac{\partial V_i}{\partial \theta} & \frac{\partial P_i}{\partial \theta} & \frac{\partial Q_i}{\partial \theta} & \frac{\partial P_{ij}}{\partial \theta} & \frac{\partial Q_{ij}}{\partial \theta} \end{bmatrix}^T \tag{10}$$

3.2. Bad Data Detection

Traditional methods for detecting bad data, like the chi-square test and largest normalized residual (LNR) test, rely on the results obtained from WLS estimation.

By checking the value of the objective function ($\boldsymbol{J}(\hat{\boldsymbol{x}})$), we can determine whether there are bad data in the power system or not. In particular, in the chi-square test, we need to perform null hypothesis testing, which can be expressed as:

$$\begin{cases} \boldsymbol{J}(\hat{\boldsymbol{x}}) > \chi^2_{(m-n),p}, & \text{Reject } \boldsymbol{H}_0 \\ \boldsymbol{J}(\hat{\boldsymbol{x}}) \le \chi^2_{(m-n),p}, & \textit{Accept } \boldsymbol{H}_0 \end{cases} \tag{11}$$

where $\boldsymbol{H}_0$ represents the original hypothesis, i.e., there are no bad data, and $\chi^2_{(m-n),p}$ is the chi-square test threshold with a confidence level of p and a degree of freedom corresponding to $(m-n)$.

The LNR test stands as another commonly employed approach for bad data detection. Its core concept revolves around the normalization of measurement residuals, which can be formulated as follows:

$$r_i = \frac{|z_i - h_i(\hat{x})|}{\sqrt{\Omega_{ii}}} \tag{12}$$

$$\Omega = \left[\boldsymbol{I} - \boldsymbol{H}(\hat{x})\left(\boldsymbol{H}^T(\hat{x})\boldsymbol{R}^{-1}\boldsymbol{H}(\hat{x})\right)^{-1}\boldsymbol{H}^T(\hat{x})\boldsymbol{R}^{-1}\right]\boldsymbol{R} \tag{13}$$

where z_i is the ith measurement, Ω_{ii} is the ith diagonal entry of Ω, and $\boldsymbol{I}$ is the identity matrix. If there exit bad data in the power system, the largest normalized residual is larger than the threshold (ε).

The chi-square test and LNR test are generally effective for detecting natural bad data, which typically induce large measurement residuals [35].

3.3. FDIA Generation

If an attacker possesses precise information regarding real-time state estimation, network topology, and parameters, they can achieve an elaborate FDIA without being detected. When measurement meters are tampered with, the measurement ($\boldsymbol{z}$) changes to $\boldsymbol{z}^f$, and the attacked measurement ($\boldsymbol{z}^f$) changes to:

$$z^f = h(x) + a + e \tag{14}$$

where $a \in \mathbb{R}^{m\times 1}$ is the attacked vector.

As an elaborate FDIA, the attacked vector requires a certain condition, which is expressed as:

$$a = h(\hat{x} + c) - h(\hat{x}) \tag{15}$$

where $c \in \mathbb{R}^{n\times 1}$ is the deviation of the state variable, and $\hat{x}$ is the state-estimated vector without an FDIA.

As indicated by the equation above, FDIAs can lead to an identical measurement residual vector compared to the condition without an attack. To be specific, the measurement residuals between the pre-attack and post-attack states can be described as follows:

$$r = z - h(\hat{x}) \tag{16}$$

$$r^f = z^f - h(\hat{x} + c) = z + a - h(\hat{x}) - a = z - h(\hat{x}) \tag{17}$$

The measurement residual between the pre-attack and post-attack states does not change; hence, an elaborate FDIA is stealthy and can avoid detection by the existing BDD system based on residuals [36].

4. Dynamic State Estimation Model

4.1. Extended Kalman Filter (EKF)

The physical power information system in an AC power system is inherently complex and highly multidimensional and nonlinear. The state and measurement equations for state estimation can be formalized as:

$$x_k = f(x_{k-1}) + \omega_{k-1} \tag{18}$$

$$z_k = h(x_k) + e_k \tag{19}$$

where x_k and z_k denote the state vector and the measurement vector at time k, respectively; $f(.)$ denotes the state transfer equation from $k-1$ to k; $h(.)$ denotes measurement equation; and ω_{k-1} and e_k denote the process and measurement noise, respectively, which are independent of each other.

Since the KF algorithm can only deal with linear system problems, it is not applicable to nonlinear problems such as power systems, so the EKF algorithm is derived. The EKF algorithm first uses Taylor's formula to linearize the nonlinear system, then filters it using the basic formula of the KF algorithm. Specifically, state Equation (18) carries out Taylor series expansion at the state estimation quantity ($\hat{x}_{k-1}$) and ignores items at quadratic levels and higher. Similarly, measurement Equation (19) carries out Taylor series expansion at the state prediction quantity ($\tilde{x}_k$) and ignores items at quadratic levels and higher. The linearization models are expressed as:

$$\begin{aligned} x_k &\approx f(\hat{x}_{k-1}) + \left.\frac{\partial f(\hat{x}_{k-1})}{\partial \hat{x}_{k-1}}\right|_{\hat{x}_{k-1}} (x_{k-1} - \hat{x}_{k-1}) + \omega_{k-1} \\ &= F_{k-1}x_{k-1} + \omega_{k-1} + u_{k-1} \end{aligned} \tag{20}$$

$$\begin{aligned} z_k &\approx h(\tilde{x}_k) + \left.\frac{\partial h(\tilde{x}_k)}{\partial \tilde{x}_k}\right|_{\tilde{x}_k} (x_k - \tilde{x}_k) + e_k \\ &= H_k x_k + e_k + y_k \end{aligned} \tag{21}$$

where $F_{k-1} = \left.\frac{\partial f(\hat{x}_{k-1})}{\partial \hat{x}_{k-1}}\right|_{\hat{x}_{k-1}}$ is the Jacobian matrix of the state equation, $u_{k-1}=f(\hat{x}_{k-1}) - \left.\frac{\partial f(\hat{x}_{k-1})}{\partial \hat{x}_{k-1}}\right|_{\hat{x}_{k-1}} \hat{x}_{k-1}$ is an externality item, $H_k = \left.\frac{\partial h(\tilde{x}_k)}{\partial \tilde{x}_k}\right|_{\tilde{x}_k}$ is the Jacobian matrix of the measurement equation, and $y_k=h(\tilde{x}_k) - \left.\frac{\partial h(\tilde{x}_k)}{\partial \tilde{x}_k}\right|_{\tilde{x}_k} \tilde{x}_k$ is an externality item.

On the basis of Equations (18) and (19), the basic formula of the EKF algorithm is expressed as follows:

(1) Prediction steps:

$$\tilde{x}_{k|k-1} = F_{k-1}\hat{x}_{k-1} \tag{22}$$

$$\tilde{P}_{k|k-1} = F_{k-1}\hat{P}_{k-1}F_{k-1}^{T} + Q_{k-1} \tag{23}$$

(2) Update steps:

$$K_k = \tilde{P}_{k|k-1}{H_k}^{T}(H_k\tilde{P}_{k|k-1}{H_k}^{T} + R_k)^{-1} \tag{24}$$

$$\hat{x}_k = \tilde{x}_{k|k-1} + K_k\left[z_k - H_k\tilde{x}_{k|k-1}\right] \tag{25}$$

$$\hat{P}_k = (I - K_kH_k)\tilde{P}_{k|k-1} \tag{26}$$

where ˜ and ˆ indicate the predicted and estimated quantities, respectively; I is the identity matrix; P is the state covariance matrix; Q and R are the covariance matrices of the process noise and measurement noise error vectors, respectively, which are assumed to be white Gaussian processes; and K is Kalman gain.

The EKF algorithm is extensively employed for dynamic state estimation in power systems due to its straightforward model development and efficient computational performance in practical engineering applications. However, since the EKF algorithm ignores the higher-level items in the linearization process, it results in a large truncation error in power systems with highly nonlinear characteristics, resulting in a decrease in the filtering effect.

4.2. Adaptive Interpolation Strategy

To enhance the dynamic state estimation capabilities of the EKF algorithm in the power system, an adaptive interpolation method is proposed to strike a balance between estimation precision and computational efficiency [37].

Based on Equation (18), we need to quantify the nonlinear index of state function $f(x)$ to obtain η_f, which can be expressed as:

$$\varepsilon_f = x_k - f(\hat{x}_{k-1}) - \left.\frac{\partial f(\hat{x}_{k-1})}{\partial \hat{x}_{k-1}}\right|_{\hat{x}_{k-1}}(x_{k-1} - \hat{x}_{k-1}) \tag{27}$$

$$\eta_f = \varepsilon_f^{T} Q_k^{-1} \varepsilon_f \tag{28}$$

where ε_f is the difference between $f(x)$ and the corresponding linear approximation.

Similarly, based on Equation (19), we can obtain the nonlinear index (η_h) of measurement function $h(x)$, which can be represented as:

$$\varepsilon_h = z_k - h(\tilde{x}_k) - \left.\frac{\partial h(\tilde{x}_k)}{\partial \tilde{x}_k}\right|_{\tilde{x}_k}(x_k - \tilde{x}_k) \tag{29}$$

$$\eta_h = \varepsilon_h^{T} R_k^{-1} \varepsilon_h \tag{30}$$

where ε_h is the difference between $h(x)$ and the corresponding linear approximation.

As shown in Equations (28) and (30), ε_f and ε_h are normalized by Q_k and R_k. Under the process, ε_f and ε_h are numerically non-negative. Hence, if $\varepsilon_f \ll Q_k$ and $\varepsilon_h \ll R_k$, η_f and η_h are both much less than 1, and the system can be considered quasilinear. Otherwise, according to the size of the nonlinearization index, the pseudomeasurements must be added between two consecutive sampling points to increase the sampling rate and reduce the degree of nonlinearity of the system.

The interpolation factor (r) is closely related to the sizes of ε_f and ε_h. The larger nonlinearization indices η_f and η_h are, the larger the interpolation factor (r) is. Conversely, the interpolation factor (r) is smaller. It is important to emphasize that $\varepsilon_f = 0$ and $\varepsilon_h = 0$ in the linear system. Therefore, the system does not interpolate.

The finite state machine model is shown in Figure 3. In practical applications, we can introduce as many states as required to the FSM model to accommodate the nonlinearity indices. There are three parameters in each state (i): the interpolation factor (r_i), the upper threshold (U_i), and the lower threshold (L_i). In addition, as the state (i) changes, the interpolation factor is set to $r_{i+1} > r_i$. The selection of the interpolation factor (r) is shown in Algorithm 1.

The thresholds of each state are different, and they are set depending on different scenarios. When selecting the thresholds, it is necessary to ensure that the upper threshold (U_i) is larger than the lower threshold (L_i). Furthermore, as U_i and L_i become smaller, the interpolation factor (r) and estimation accuracy increase, and the algorithm consumes more time. It is important to highlight that the nonlinear indices can take on discrete values. To maintain small values for both η_f and η_h, here is how the process works: If either η_f or η_h exceeds U_i, r parameter is increased to minimize the nonlinear error. Conversely, if both η_f and η_h are below L_i, r is reduced to lower computational complexity. The specific values of r for each state can be found in Table 2.

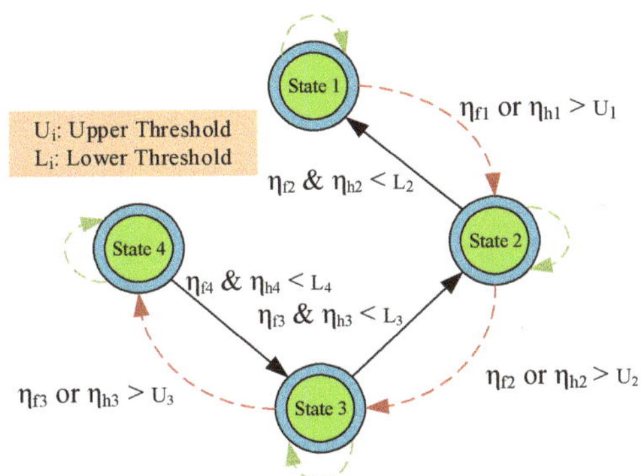

Figure 3. Finite-state machine (four states).

Algorithm 1 Choose the interpolation factor (r).

State $\boldsymbol{i}$ $\quad r = r_i$
if η_{fi} or $\eta_{hi} > U_i$ **then**
  go to the state $\boldsymbol{i+1}$
else
  η_{fi} and $\eta_{hi} < L_i$
  go to the state $\boldsymbol{i-1}$
end if
Stay in the state $\boldsymbol{i}$

Table 2. The size of the interpolation factor (r).

State (i)	Interpolation Factor (r)	Lower Threshold	Upper Threshold
$i = 1$	1	L_1	U_1
$i = 2$	3	L_2	U_2
$i = 3$	7	L_3	U_3
$i = 4$	15	L_4	U_4

4.3. *Adaptive Interpolation EKF (AIEKF)*

Building upon the dynamic model outlined in Section 4.1 and the adaptive interpolation approach discussed in Section 4.2, we introduce the AIEKF algorithm in this section. The AIEKF algorithm effectively strikes a balance between computation time and estimation accuracy, thereby enhancing the performance of the EKF algorithm in power systems.

A flow chart illustrating the AIEKF algorithm is provided in Figure 4, and its detailed steps are as outlined as follows:

(1) Initialization: setting the initial state variable ($\hat{x}_0$) and state error covariance ($\hat{P}_0$).

(2) Adaptive Interpolation: In order to strike a compromise between computational efficiency and estimation precision, the algorithm incorporates an adaptive interpolation strategy, which comprises three key steps. Initially, we calculate the nonlinearity indices of the state transition function and the measurement function (referred to as η_f and η_h, respectively) using Equations (28) and (30), respectively. In the next step, we ascertain the interpolation factor (r) by utilizing a finite-state machine model. Finally, r pseudomeasurements are introduced between two actual measurements through linear interpolation, which is designed to mitigate the adverse impacts of nonlinearity.

(3) EKF: On the basis of determining the number of interpolation factors (r), the power system is estimated using the EKF algorithm. Initially, leveraging the state and its covariance matrix from time $k-1$, we derive a prior estimation at time k in accordance with Equations (22) and (23). Secondly, the correction of the a priori estimation is used to obtain an a posteriori estimation according to Equations (24)–(26). Thirdly, filtering is performed between two consecutive samples based on the size of the interpolation factor. Then, the above steps are repeated until the end of the sampling time.

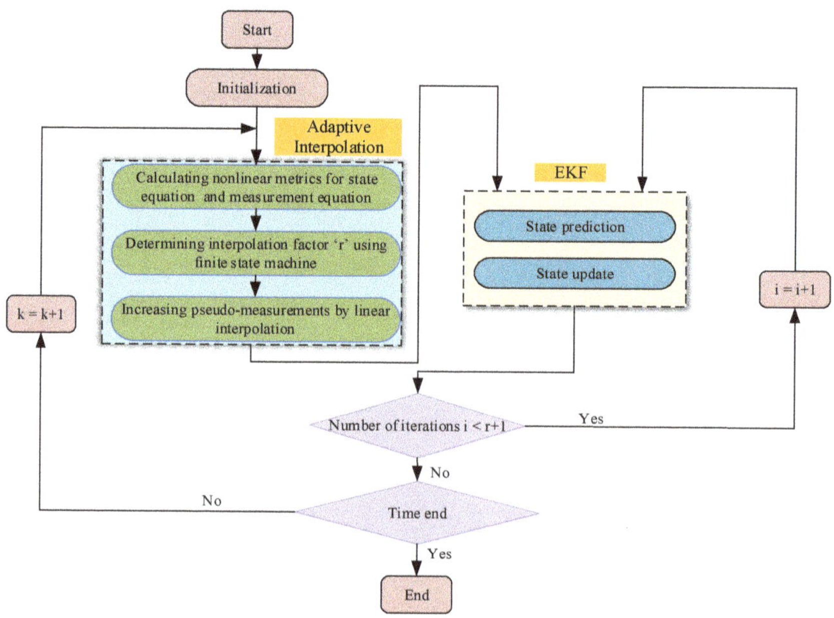

Figure 4. Flow chart of the AIEKF algorithm.

5. Detection of FDIAs

This section proposes a methodology for FDIA detection based on power system state estimation. As a nonlinear system in the smart grid, it is difficult to guarantee the estimation accuracy using traditional state estimation methods. Meanwhile, in order to improve the stability of the detection algorithm, the real-time state information of the grid buses is solved according to the system model equation and AIEKF algorithm.

Once the attacker begins to tamper with the measuring instruments, the result of Equation (25) is different from the previous result and expressed as:

$$\begin{aligned} \hat{x}_k^f &= \tilde{x}_{k|k-1} + K_k[z_k + a_k - H_k\tilde{x}_{k|k-1}] \\ &= \hat{x}_k + K_k a_k \end{aligned} \tag{31}$$

where $\hat{x}_k^f$ is the estimated state after the FDIA. To better facilitate estimation, $c_k = \hat{x}_k^f - \hat{x}_k$ is introduced. Then, for the next time $(k+1)$, it can be represented as:

$$\begin{aligned}\hat{x}_{k+1}^f &= \tilde{x}_{k+1|k}^f + K_{k+1}[z_{k+1}^f - H_{k+1}\tilde{x}_{k+1|k}^f] \\ &= \tilde{x}_{k+1|k}^f + K_{k+1}[z_{k+1} + a_{k+1} - H_{k+1}\tilde{x}_{k+1|k}^f] \\ &= \hat{x}_{k+1} + [I - K_{k+1}H_{k+1}]F_k c_k + K_{k+1}a_{k+1}\end{aligned} \tag{32}$$

The analysis above highlights that the injection bias is influenced by both the currently injected false data and the bias present in the previously estimated state. Over time, this injection bias accumulates and gradually shifts the estimated state closer to the actual system state. When the power system is subjected to an FDIA, the altered measurements make the WLS state estimation results swing towards the new mean. For the AIEKF algorithm, due to the constraints of the state transfer matrix and the fact that its estimation is jointly determined by the predicted and measured values, the state estimation has some hysteresis, and only small oscillations occur.

Based on WLS and AIEKF estimation results, considering the influence of bus states on the system, the Euclidean distance in multidimensional spaces is introduced. The Euclidean distance detection threshold required in FDIA detection is obtained from historical data, and the Euclidean distance between two points estimated by WLS and AIEKF states is calculated online in real time and used as the basis for attack detection. The expression for the Euclidean distance at time k is expressed as follows:

$$d(k) = \sqrt{\sum_{i=1}^{n}\left(\hat{x}_{i,k}^{WLS} - \hat{x}_{i,k}^{AIEKF}\right)^2} \tag{33}$$

where $\hat{x}_{i,k}^{WLS}$ denotes the WLS-based state estimation at time k, $\hat{x}_{i,k}^{AIEKF}$ denotes the AIEKF-based state estimation, and n denotes the system dimension.

In the n-dimensional grid system state space, the Euclidean distance is employed to quantify the spatial separation between two points within the same state space at a given time point. The Euclidean distance of the two state estimation algorithms stabilizes in a certain range during regular power system operation, which provides a basis for false data injection attack detection. The detection threshold is expressed as:

$$\tau_D = \max\{d(1), \ldots, d(n), \ldots\} + \mu \tag{34}$$

where μ is the threshold margin, which is introduced to prevent false alarms triggered by minor data fluctuations while the detection system is operating under normal conditions.

Attack detection is performed by comparing the Euclidean distance between the detection threshold and the two points in the state space, and when $d(k) \geq \tau_D$, it is considered that there exits an FDIA in the power system; otherwise, it is considered that no attack occurs. The relation can be expressed as:

$$\begin{cases} d(k) < \tau_D, & No\ FDIA \\ d(k) \geq \tau_D, & FDIA \end{cases} \tag{35}$$

In order to distinguish between bad data and FDIAs, bad data detection is also required at the end of the above steps. Only if $d(k) \geq \tau_D$ and $J(\hat{x}) < \chi^2_{(m-n),p}$ hold can we conclude that the power system is under FDIAs. The proposed FDIA detection method based on WLS and AIEKF is shown in Algorithm 2.

Algorithm 2 FDIA detection based on WLS and the AIEKF algorithm

1: Initialize state variable $\hat{x}_{k-1}$ and state error covariance $\hat{P}_{k-1}$; the Euclidean distance detection threshold τ_D;
2: Obtain the measurements by SCADA at time k;
3: In traditional static state estimation, WLS is widely used to calculate an estimated state vector $\hat{x}_k^{WLS}$, $\hat{x}_k^{WLS} = \left(H_k^T R_k^{-1} H_k\right)^{-1} H_k^T R_k^{-1} z_k$
4: AIEKF
(1) Calculate calculate the nonlinearity indices of the state transition function and the measurement function (referred to as η_f and η_h) using Equations (28) and (30);
(2) Ascertain the interpolation factor r by utilizing a finite state machine model;
(3) Interpolate r pseudo-measurement between two actual measurements through linear interpolation;
(4) Execute the state prediction step of the EKF by applying Equations (22) and (23);
(5) Conduct the measurement update step of EKF by applying Equations (24)–(26) to calculate estimated state vector $\hat{x}_k^{AIEKF}$, $\hat{x}_k^{AIEKF} = \tilde{x}_{k|k-1} + K_k\left[z_k - H_k\tilde{x}_{k|k-1}\right]$
5: Calculate the Euclidean distance between two points estimated by WLS and AIEKF states;
6: **if** $d(k) \geq \tau_D$ and $J(\hat{x}) < \chi^2_{(m-n),p}$ **then**
7: Exist FDIA and generate early warning;
8: **else**
9: Continue the state estimation process at time $k = k + 1$,
10: **end if**

6. Experiments and Results

This paper introduces a detection approach that relies on state estimation. Considering the effectiveness of the method in real systems, the power standard IEEE-14-bus system shown in Figure 5 is used for MATLAB R2021b simulation. The active and reactive power of each bus are shown in the following Table 3. The data used in this paper come from MATPOWER trend calculation, which is used to obtain the bus voltage magnitude and phase-angle truth values, superimposed with zero-mean Gaussian white noise as the measurements. Furthermore, the estimation computations occur at one-minute intervals, which aligns with the anticipated average sampling frequency for utilities equipped with contemporary Energy management systems (EMS).

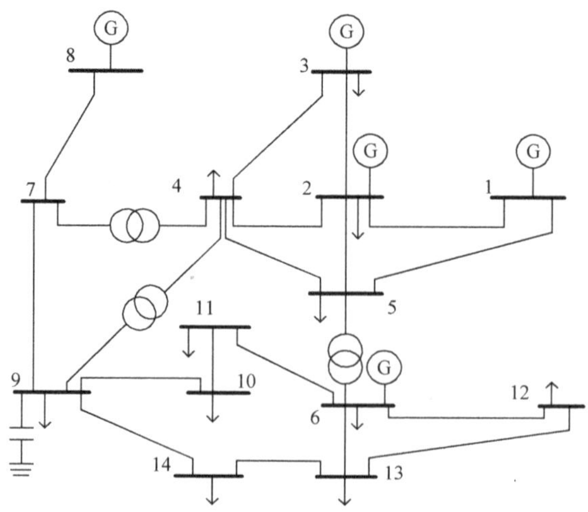

Figure 5. The IEEE-14-bus power system.

Table 3. Active and reactive power injections into system buses.

Bus Number	Generation Active (MW)	Generation Reactive (Mvar)	Load Active (MW)	Load Reactive (Mvar)
1	2.3824	−0.1490	0.0000	0.0000
2	0.4000	0.4904	0.2170	0.1270
3	0.0000	0.2744	0.9420	0.1900
4	0.0000	0.0000	0.4780	−0.0390
5	0.0000	0.0000	0.0760	0.0160
6	0.0000	0.2960	0.1120	0.0750
7	0.0000	0.0000	0.0000	0.0000
8	0.0000	0.3092	0.0000	0.0000
9	0.0000	0.0000	0.2950	0.1660
10	0.0000	0.0000	0.0900	0.0580
11	0.0000	0.0000	0.0350	0.0180
12	0.0000	0.0000	0.0610	0.0160
13	0.0000	0.0000	0.1350	0.0580
14	0.0000	0.0000	0.1490	0.0500

6.1. Comparison of Estimation Effects with WLS, EKF, and AIEKF

Next, the estimation effect of AIEKF proposed in this paper is compared with the standard WLS and EKF before injecting false data. As shown in Figures 6 and 7, bus 11 was randomly selected to compare the estimation performance of the three algorithms in 60 min. It is clear to see from the figures that although the bus voltage and phase angle fluctuate up and down with time, the AIEKF achieves superior performance relative to WLS and EKF in state estimation. To further demonstrate the estimation capability of the proposed algorithm, the estimation results of the voltage amplitude and phase of each bus under the three algorithms after stabilization are shown in Figures 8 and 9.

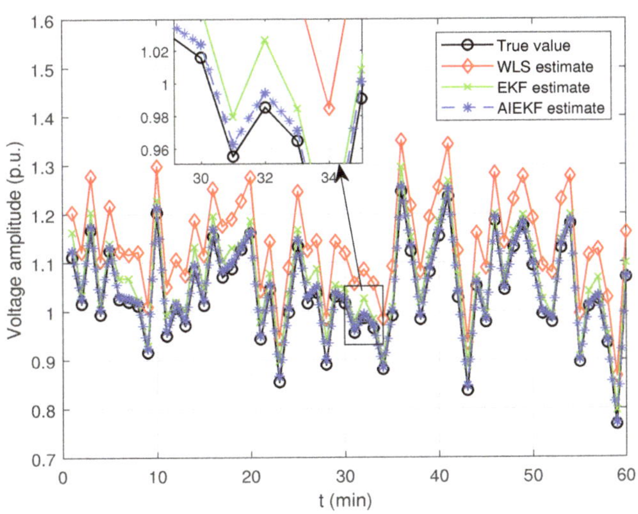

Figure 6. Bus 11 voltage amplitude estimation.

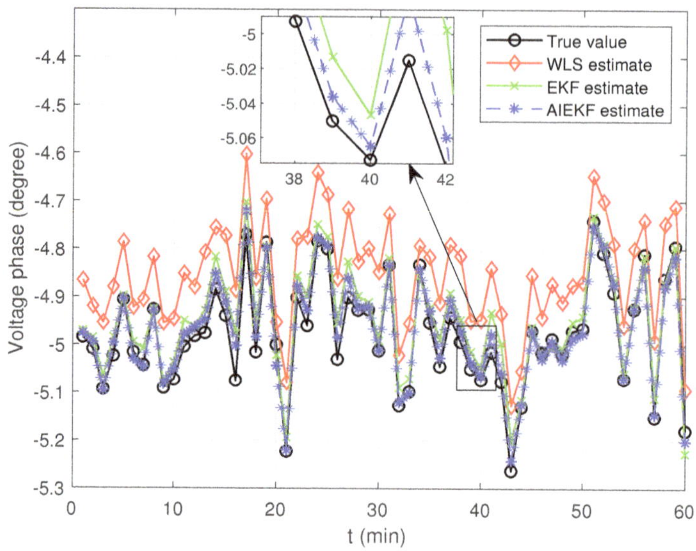

Figure 7. Bus 11 voltage phase estimation.

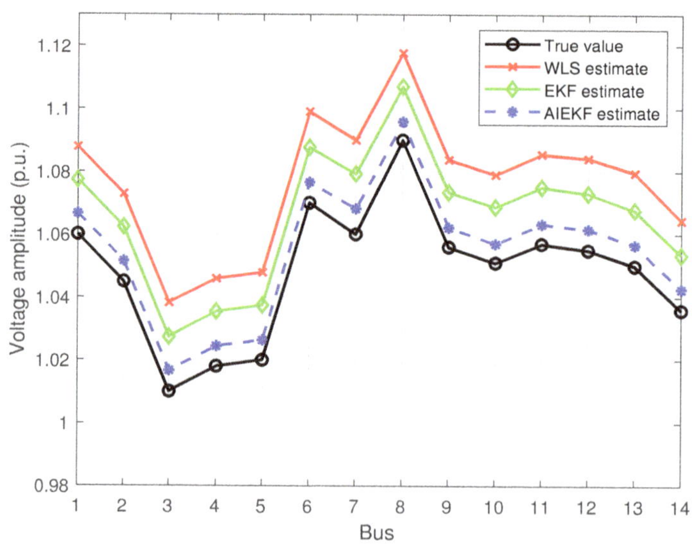

Figure 8. Bus voltage amplitude estimation.

To validate the efficacy of the AIEKF algorithm introduced in this paper for state estimation, we use the root mean square error (RMSE) as a metric to assess the accuracy of the algorithm's estimations. The RMSE calculation formula is provided below.

$$RMSE = \sqrt{\frac{1}{N}\sum_{i=1}^{N}(x_i - \hat{x}_i)^2} \tag{36}$$

where x_i is the ith component of the true value of the state variable, $\hat{x}_i$ is the ith component of the estimation of the state variable, and N is the dimension of the state variable.

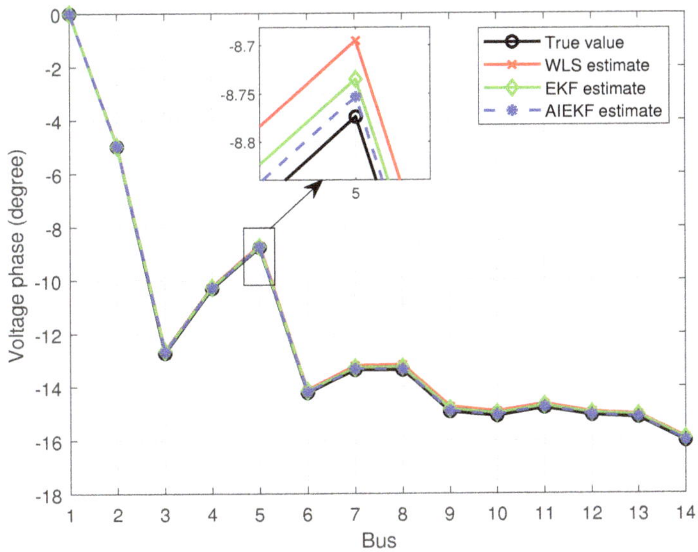

Figure 9. Bus voltage phase estimation.

The RMSE performance metric is calculated in the IEEE-14-bus system, and the results are shown in Table 4. As shown in Table 4, the RMSE of the AIEKF algorithm is the smallest of the three algorithms. Compared with WLS and EKF, the RMSE of the AIEKF algorithm decreases by 79% and 67%, respectively.

Table 4. Comparison of the RMSE of the three algorithms.

Algorithm	RMSE
WLS	0.0969
EKF	0.0616
AIEKF	0.0201

6.2. Estimation of the State Variable Before and After FDIA

To assess the viability of the false data injection attack vector strategy, in this paper, we uses the IEEE-14-bus standard test system for simulation and analysis. An attack on a local subnetwork, e.g., an attack vector ($\boldsymbol{a}$), is injected into each bus measurement value. Meanwhile, it is necessary to ensure that the internal power of the subnetwork is conserved and that the subnetwork boundary voltage and the transmission power between the subnetwork and the external network remain unchanged. The introduction of line blocking constraints leads to the response of the grid security analysis system so that the attacked measurement value ($\boldsymbol{z} + \boldsymbol{a}$) is a valid attack value. Under this condition, the attack vector ($\boldsymbol{a} = [P_3, Q_3, P_{1-2}, P_{2-3}, P_{4-2}, Q_{1-2}, Q_{2-3}, Q_{4-2}]^T$) is selected as $\boldsymbol{a} = [0.0020, -0.2029, 0.0084, -0.0073, -0.0059, -0.4874, 0.1329, -0.0723]^T$, and the increment of the rest of $\boldsymbol{z}$ is zero. Figure 10 shows the change in the measurement distribution of the system before and after the attack.

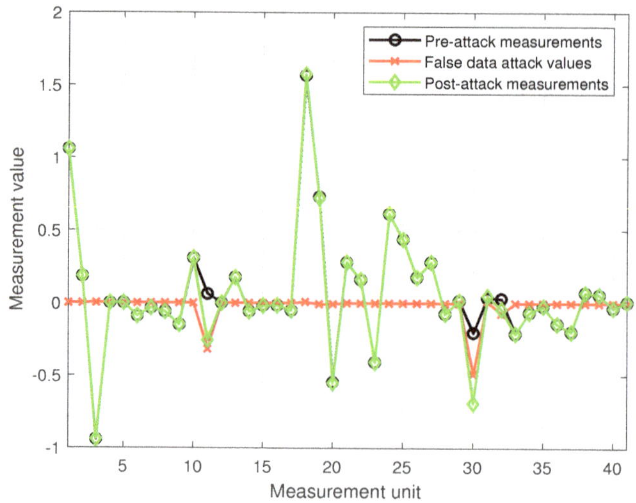

Figure 10. Measurement distribution before and after an attack on the IEEE-14-bus power system.

Once the measurements are tampered with, the state variable (x) changes. Assuming that the system is subjected to a false data injection attack at 75 min, bus 11 is selected to observe the change in bus voltage magnitude and phase angle before and after the false data injection attack occurs. State estimation of system buses using the AIEKF algorithm is performed to improve the stability and accuracy of the detection algorithm. The state estimation results of the two algorithms are shown in Figures 11 and 12. As shown in the figures, in the first 75 min without an attack, AIEKF outperforms WLS in terms of estimation. The system is attacked by false data injection in the 75th minute, and the two algorithms converge to the state expectation at different moments. It is clear that AIEKF converges slowly and with small fluctuations, while WLS is affected by a sudden change in the measurements and converges quickly to the new state value.

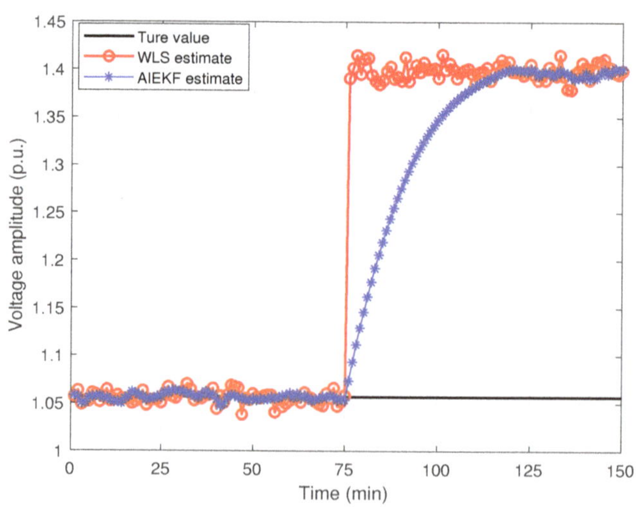

Figure 11. The voltage amplitude change of bus 11 before and after an attack.

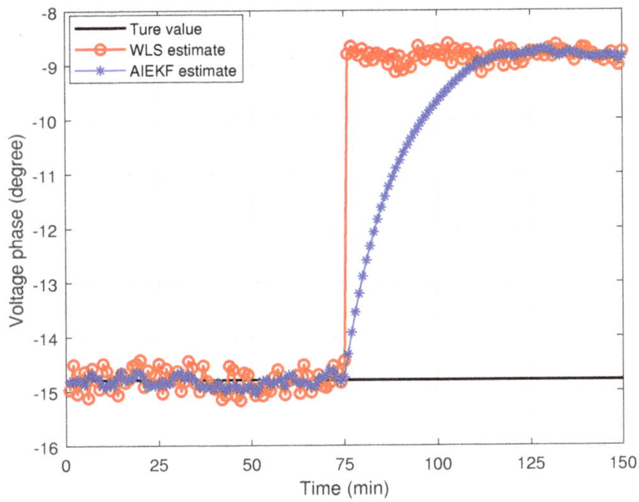

Figure 12. The voltage phase change of bus 11 before and after an attack.

6.3. Detection of FDIAs

Normal operation of the power information physics system produces a certain amount of error, but the residuals caused by measurement noise and system noise are often very small—much smaller than the threshold allowed for the detection of undesirable data errors—so that undesirable data can be prevented from interfering with the system. The values of $J(\hat{x})$ before and after an attack determined using weighted least squares are shown in Figure 13. Before the attack, the value of $J(\hat{x})$ is 4.5728. When the false data attack is injected into the system, the value of $J(\hat{x})$ is 4.7041. It is not difficult to find that the residual does not change much before and after an attack. The IEEE-14-bus system has a total of 41 measurements, the redundancy is $k = m - n = 41 - 27 = 14$, and the significance level (α) is 0.05. According to the statistical chi-square distribution table, the threshold of bad data detection is 23.685. The residual of the injected attack is within the threshold. However, the voltage amplitude and voltage phase are changed. The false data attack vector successfully achieves the attack.

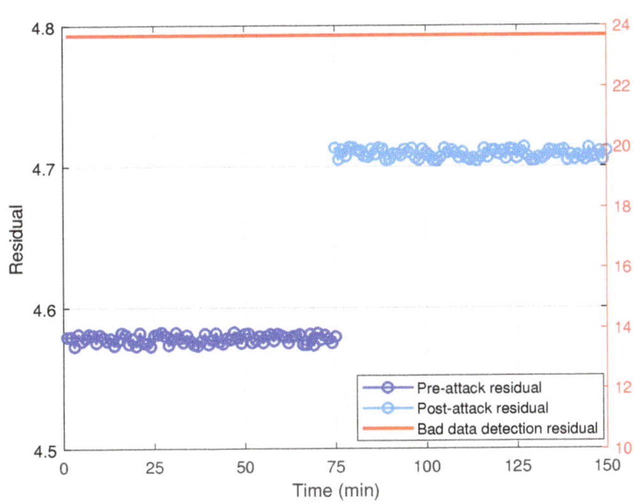

Figure 13. The residual of WLS estimation before and after an attack.

This paper proposes a detection method based on the computation of the Euclidean distance between two points in the state space to detect FDIAs. Using Monte Carlo simulation with 1000 independent experiments, we can obtain the normal-case Euclidean distance distribution. The maximum value is taken as the detection threshold, i.e., $\max\{d(1), \ldots, d(n), \ldots\} = 1.847$. The detection margin ($\mu$) is set to 0.03, and according to Equation (30), the detection threshold can be derived as $\tau_D = 1.85$. After an attack, the Euclidean distance changes to 17.9586. Figure 14 shows the Euclidean distance distribution based on the two algorithms before and after an attack.

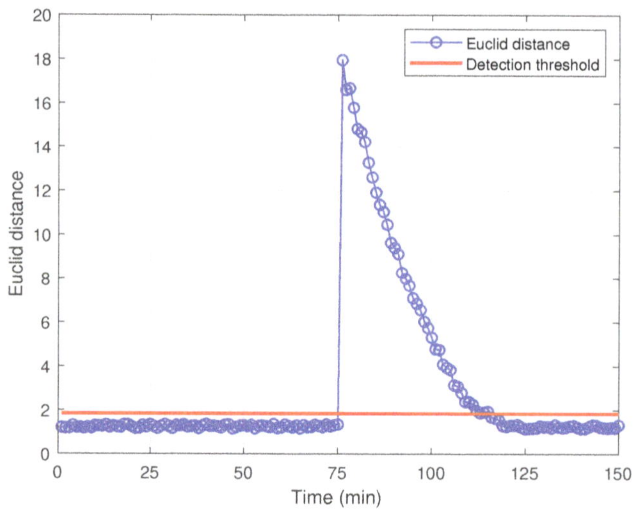

Figure 14. The Euclidean distance before and after an attack.

As can be seen from the figure, during the first 75 min, when the system is not under attack, the Euclidean distance between the state points stays within a certain range below the predefined detection threshold, which indicates that the system does not detect an attack according to the judgment conditions. When the system is attacked after the 75th minute, the two algorithms converge to the new state values at different moments. At this moment, the Euclidean distance of the voltage state estimate fluctuates considerably with the attack and exceeds the predefined detection threshold. Therefore, FDIAs can be detected, which triggers the attack alarm system.

7. Conclusions

In this research, we introduce an approach that combines weighted least squares with an adaptive interpolation extended Kalman filter to detect FDIAs in power systems. AIEKF effectively reduces the nonlinear errors associated with the extended Kalman filters, leading to enhanced accuracy in estimating the state of the power system. When a power system is subject to false data injection attacks, the state estimation weighted least squares statistic is characterized by a real-time nature, where changes in state variables are instantaneous, whereas adaptive interpolation extended Kalman filtering is characterized by hysteresis, and a change in state variables requires a process. Based on the difference between the two algorithms, the Euclidean distance is introduced as a metric for detecting whether the system is injected with false data or not. Additionally, the relevant detection threshold is obtained using Monte Carlo simulation. The experiments show that the method is effective in detecting false data injection attacks.

Subsequent research will consider the study of the localization of FDIAs and the development of a new joint estimation algorithm that can simultaneously achieve the detection and localization of false data injection attacks.

Author Contributions: Conceptualization, G.Z.; methodology, Y.L. and W.G.; software, G.Z. and X.G.; validation, J.Z.; formal analysis, X.G.; resources, W.G.; data curation, P.H.; writing—original draft preparation, G.Z.; writing—review and editing, W.G. and Y.L.; visualization, P.H. and X.G.; supervision, W.G.; project administration, Y.L.; funding acquisition, W.G. All authors have read and agreed to the published version of the manuscript.

Funding: This research was supported in part by the National Natural Science Foundation of China (NSFC) (U21A20146), the Collaborative Innovation Project of Anhui Universities (GXXT-2020-070), the Open Research Fund of Anhui Province Key Laboratory of Detection Technology and Energy Saving Devices (JCKJ2022C02, JCKJ2022A10), and the Open Research Fund of the Key Laboratory of Advanced Perception and Intelligent Control of High-end Equipment of the Ministry of Education (GDSC202208).

Data Availability Statement: Not applicable.

Acknowledgments: We thank the anonymous reviewers for their valuable comments.

Conflicts of Interest: The authors declare no conflict of interest.

References

1. Marashi, K.; Sarvestani, S.S.; Hurson, A.R. Consideration of Cyber-Physical Interdependencies in Reliability Modeling of Smart Grids. *IEEE Trans. Sustain. Comput.* **2018**, *3*, 73–83. [CrossRef]
2. Guo, H.; Pang, Z.H.; Sun, J.; Li, J. An Output-Coding-Based Detection Scheme Against Replay Attacks in Cyber-Physical Systems. *IEEE Trans. Circuits Syst. II* **2021**, *68*, 3306–3310. [CrossRef]
3. Ghosh, S.; Sampalli, S. A Survey of Security in SCADA Networks: Current Issues and Future Challenges. *IEEE Access* **2019**, *7*, 135812–135831. [CrossRef]
4. Zhou, J.; Chen, B.; Yu, L. Intermediate-Variable-Based Estimation for FDI Attacks in Cyber-Physical Systems. *IEEE Trans. Circuits Syst. II* **2020**, *67*, 2762–2766. [CrossRef]
5. Gao, Y.; Ma, J.; Wang, J.; Wu, Y. Event-Triggered Adaptive Fixed-Time Secure Control for Nonlinear Cyber-Physical System with False Data-Injection Attacks. *IEEE Trans. Circuits Syst. II* **2023**, *70*, 316–320. [CrossRef]
6. Wang, Y.; Gu, D.; Peng, D.; Chen, S.; Yang, H. Stuxnet Vulnerabilities Analysis of SCADA Systems. *Commun. Comput. Inf. Sci.* **2012**, *345*, 640–646. [CrossRef]
7. Liang, G.; Weller, S.R.; Zhao, J.; Luo, F.; Dong, Z.Y. The 2015 Ukraine Blackout: Implications for False Data Injection Attacks. *IEEE Trans. Power Syst.* **2016**, *32*, 3317–3318. [CrossRef]
8. Lu, K.D.; Wu, Z.G. Multi-Objective False Data Injection Attacks of Cyber–Physical Power Systems. *IEEE Trans. Circuits Syst.* **2022**, *69*, 3924–3928. [CrossRef]
9. Yu, W.; Bu, X.; Hou, Z. Security Data-Driven Control for Nonlinear Systems Subject to Deception and False Data Injection Attacks. *IEEE Trans. Netw. Sci. Eng.* **2022**, *9*, 2910–2921. [CrossRef]
10. Liu, Y.; Reiter, M.K.; Ning, P. False data injection attacks against state estimation in electric power grids. In Proceedings of the 2009 ACM Conference on Computer and Communications Security (CCS), Chicago, IL, USA, 9–13 November 2009; pp. 1–33. [CrossRef]
11. Yang, Q.; Yang, J.; Yu, W.; An, D.; Zhang, N.; Zhao, W. On False Data-Injection Attacks against Power System State Estimation: Modeling and Countermeasures. *IEEE Trans. Parallel Distrib. Syst.* **2014**, *25*, 717–729. [CrossRef]
12. He, Y.; Mendis, G.J.; Wei, J. Real-Time Detection of False Data Injection Attacks in Smart Grid: A Deep Learning-Based Intelligent Mechanism. *IEEE Trans. Smart Grid* **2017**, *8*, 2505–2516. [CrossRef]
13. Musleh, A.S.; Chen, G.; Dong, Z.Y. A Survey on the Detection Algorithms for False Data Injection Attacks in Smart Grids. *IEEE Trans. Smart Grid* **2020**, *11*, 2218–2234. [CrossRef]
14. Moslemi, R.; Mesbahi, A.; Velni, J.M. A fast, decentralized covariance selection-based approach to detect cyber attacks in smart grids. *IEEE Trans. Smart Grid* **2018**, *9*, 4930–4941. [CrossRef]
15. Chen, Y.; Huang, S.; Liu, F.; Wang, Z.; Sun, X. Evaluation of Reinforcement Learning-Based False Data Injection Attack to Automatic Voltage Control. *IEEE Trans. Smart Grid* **2019**, *10*, 2158–2169. [CrossRef]
16. Zhao, J.; Zhang, G.; Scala, L.M.; Dong, Z.Y.; Chen, C.; Wang, J. Short-Term State Forecasting-Aided Method for Detection of Smart Grid General False Data Injection Attacks. *IEEE Trans. Smart Grid* **2017**, *8*, 1580–1590. [CrossRef]
17. Li, X.; Wang, Z.; Zhang, C.; Du, D.; Fei, M. A Novel Dynamic Watermarking-Based EKF Detection Method for FDIAs in Smart Grid. *IEEE/CAA. J. Autom. Sinica* **2022**, *9*, 1319–1322. [CrossRef]
18. Manandhar, K.; Cao, X.J.; Hu, F.; Liu, Y. Combating False Data Injection Attacks in Smart Grid using Kalman Filter. In Proceedings of the 2014 International Conference on Computing, Networking and Communications (ICNC), Honolulu, HI, USA, 3–6 February 2014; pp. 16–20. [CrossRef]
19. Shi, W.; Wang, Y.; Jin, Q.; Ma, J. PDL: An Efficient Prediction-Based False Data Injection Attack Detection and Location in Smart Grid. In Proceedings of the 2018 IEEE 42nd Annual Computer Software and Applications Conference (COMPSAC), Tokyo, Japan, 23–27 July 2018; pp. 676–681. [CrossRef]

20. Abreu, O.A.; Messina, F.; Vega, L.R. Stealth Attacks on the SADI with Prior Information on the State Covariance Matrix. In Proceedings of the 2022 IEEE Biennial Congress of Argentina (ARGENCON), San Juan, Argentina, 7–9 September 2022; pp. 1–7. [CrossRef]
21. Huang, K.; Xiang, Z.; Deng, W.; Yang, C.; Wang, Z. False Data Injection Attacks Detection in Smart Grid: A Structural Sparse Matrix Separation Method. *IEEE Trans. Netw. Sci. Eng.* **2021**, *8*, 2545–2558. [CrossRef]
22. Khalaf, M.; Youssef, A.; El-Saadany, E. Joint Detection and Mitigation of False Data Injection Attacks in AGC Systems. *IEEE Trans. Smart Grid* **2019**, *10*, 4985–4995. [CrossRef]
23. Kurt, M.N.; Yilmaz, Y.; Wang, X. Distributed Quickest Detection of Cyber-Attacks in Smart Grid. *IEEE Trans. Inf. Forensics Secur.* **2018**, *13*, 2015–2030. [CrossRef]
24. Cheng, G.; Lin, Y.; Zhao, J.; Yan, J. A Highly Discriminative Detector Against False Data Injection Attacks in AC State Estimation. *IEEE Trans. Smart Grid* **2022**, *13*, 2318–2330. [CrossRef]
25. Wang, Y.; Shi, W.; Jin, Q.; Ma, J. An Accurate False Data Detection in Smart Grid Based on Residual Recurrent Neural Network and Adaptive threshold. In Proceedings of the 2019 IEEE International Conference on Energy Internet (ICEI), Nanjing, China, 27–31 May 2019; pp. 499–504. [CrossRef]
26. Mousavian, S.; Valenzuela, J.; Wang, J.H. Real-time data reassurance in electrical power systems based on artificial neural networks. *Electr. Pow. Syst. Res.* **2013**, *96*, 285–295. [CrossRef]
27. Mahi-al-rashid, A.; Hossain, F.; Anwar, A.; Azam, S. False Data Injection Attack Detection in Smart Grid Using Energy Consumption Forecasting. *Energies* **2022**, *15*, 4877. [CrossRef]
28. Yu, J.J.Q.; Hou, Y.; Li, V.O.K. Online False Data Injection Attack Detection With Wavelet Transform and Deep Neural Networks. *IEEE Trans. Industr. Inform.* **2018**, *14*, 3271–3280. [CrossRef]
29. Wang, Y.F.; Zhang, Z.H.; Ma, J.H.; Jin, Q. KFRNN: An Effective False Data Injection Attack Detection in Smart Grid Based on Kalman Filter and Recurrent Neural Network. *IEEE Internet Things J.* **2022**, *9*, 6893–6904. [CrossRef]
30. Jorjani, M.; Seifi, H.; Varjani, A.Y. A Graph Theory-Based Approach to Detect False Data Injection Attacks in Power System AC State Estimation. *IEEE Trans. Ind. Informat.* **2021**, *17*, 2465–2475. [CrossRef]
31. Muscas, C.; Pegoraro, P.A.; Sulis, S.; Pau, M.; Ponci, F.; Monti, A. New Kalman Filter Approach Exploiting Frequency Knowledge for Accurate PMU-Based Power System State Estimation. *IEEE Trans. Instrum. Meas.* **2020**, *69*, 6713–6722. [CrossRef]
32. Deng, R.L.; Zhuang, P.; Liang, H. False Data Injection Attacks Against State Estimation in Power Distribution Systems. *IEEE Trans. Smart Grid* **2019**, *10*, 2871–2881. [CrossRef]
33. Yuan, C.; Zhuo, Y.; Liu, G.; Dai, R.; Lu, Y.; Wang, Z. Graph Computing-Based WLS Fast Decoupled State Estimation. *IEEE Trans. Smart Grid* **2020**, *11*, 2440–2451. [CrossRef]
34. Manousakis, N.M.; Korres, G.N. Application of State Estimation in Distribution Systems with Embedded Microgrids. *Energies* **2021**, *14*, 7933. [CrossRef]
35. Radhoush, S.; Vannoy, T.; Liyanage, K.; Whitaker, B.M.; Nehrir, H. Distribution System State Estimation and False Data Injection Attack Detection with a Multi-Output Deep Neural Network. *Energies* **2023**, *16*, 2288. [CrossRef]
36. Ganjkhani, M.; Fallah, S.N.; Badakhshan, S.; Shamshirband, S.; Chau, K.-W. A Novel Detection Algorithm to Identify False Data Injection Attacks on Power System State Estimation. *Energies* **2019**, *12*, 2209. [CrossRef]
37. Akhlaghi, S.; Zhou, N.; Huang, Z. A Multi-Step Adaptive Interpolation Approach to Mitigating the Impact of Nonlinearity on Dynamic State Estimation. *IEEE Trans. Smart Grid* **2018**, *9*, 3102–3111. [CrossRef]

energies

MDPI

Article

Measuring and Modeling the Skin Effect for Harmonic Power Flow Studies

Eduardo Tavares Silvério * and Jose Rubens Macedo Junior

Faculty of Electrical Engineering, Federal University of Uberlandia, Uberlandia 38408-100, Brazil; jrubens@ufu.br
* Correspondence: eduardot.silverio@ufu.br

Abstract: This research aims to quantify the skin effect and estimate expressions that well represent the phenomenon for harmonic power flow studies. The primary focus is to validate the behavior of the skin effect at harmonic frequencies ranging from 60 Hz to 960 Hz, while considering various amplitudes of electric current. The investigation not only examines the measurement of the skin effect, but also considers the temperature of the tested conductors, aiming to analyze the increase in resistance resulting from temperature rise and resistivity changes. The measurement outcomes demonstrate notable increments in electrical resistance, with resistivity increases of up to 1.9% observed throughout the measurement process. Finally, based on the results obtained through laboratory measurements, mathematical expressions were estimated as a function of frequency. In order to evaluate the simulation time reduction by the proposed expressions, OpenDSS (version: 9.4.1.2; Electric Power Research Institute, Knoxville, TN, USA) software was used, which aims at quantifying the impact of the skin effect on the technical losses. The results from these simulations demonstrate that the proposed expressions to account for the skin effect in conductors reduce the simulation time by around 17% for harmonic power flow.

Keywords: measurement; power distribution system; power loss; skin effect

Citation: Silvério, E.T.; Macedo Junior, J.R. Measuring and Modeling the Skin Effect for Harmonic Power Flow Studies. *Energies* **2023**, *16*, 7913. https://doi.org/10.3390/en16237913

Academic Editor: Ying-Yi Hong

Received: 5 November 2023
Revised: 25 November 2023
Accepted: 26 November 2023
Published: 4 December 2023

1. Introduction

The current harmonic and interharmonic components, which encompass an inseparable reality of electrical systems, have long been studied and quantified by researchers and engineers around the world. The presence of these components, particularly those operating at higher frequencies, on any electric circuit, causes an increase in the resistance of conductors due to the skin effect phenomenon.

Subsequently, scientific publications focused on investigating the skin effect by conducting measurements to establish a relationship between the resistance of the conductor at direct current (RDC) and the resistance of the conductor at alternating current (RAC), considering variations in frequency, such as [1]. Scientific literature that mathematically covers the phenomenon [2,3] was consolidated by utilizing Bessel's equation as a faithful representation of the skin effect, where the particular behavior of the magnetic field for different conductors was later clarified, such as in hard wires and flexible cables [4].

Furthermore, laboratory experiments have addressed specific issues related to the measurement of electrical resistance in conductors at different frequencies [5]. These experiments involved the development of an electrical configuration capable of quantifying resistance under varying electric currents. In parallel, computer simulations have become essential tools for scientific advancements [6]. The use of finite element-based simulated meshes has enabled detailed analyses of the electromagnetic performance of various conductors and electrical equipment geometries.

Building upon previous research, recent scientific studies have focused on practical applications of the phenomenon, such as power losses in arc furnace buses [6]. Additionally, the scientific community has extensively examined the reliability of mathematical solutions provided by computer simulations [7].

However, a shared characteristic observed in the various studies conducted [6–9] is the lack of measurements and real-world applicability of the data to analyze the impact of the skin effect on electric power systems. This knowledge gap highlights the need for a better understanding of quantifying the skin effect. In line with this observation, the study conducted in [10] focused on the practical implications of the skin effect on electrical equipment, such as transformers, resulting in increased power losses.

Similarly, the study in [11] examined the behavior of the skin effect in a 69 kV conventional transmission line with ground return, offering practical insights into the phenomenon, albeit without fully considering the complex nature of the entire electrical system. Despite the innovative mathematical modeling presented, the study primarily addresses the individual impact of the skin effect on the specific electrical device, neglecting the comprehensive analysis of electrical power distribution systems as a whole.

While previous studies have explored the behavior of the skin effect in transformers and transmission lines, there are studies, such as [12,13], that leverage this phenomenon to develop a control logic for protecting a railway system with direct current rails. Specifically, this study presents a precise methodology for analyzing the skin effect in the time domain, accurately reproducing fault currents in the rails. In contrast to earlier research, this study satisfactorily considers the intricate complexities of the local power distribution system. In addition to the mentioned papers, [14] also addresses the phenomenon, pointing out that eddy currents and the skin effect play an important role in electrical machines and power systems.

The impact of harmonic components on eddy current losses in the core of soft magnetic materials is discussed in [13]. The study involves the creation of a 3D model based on the magnetic behavior of the material. Furthermore, the quantification of losses resulting from the skin effect induced by harmonic components was carried out. Ref. [15] presents an approach for analyzing the skin effect in a wire using finite elements. Although this study makes significant contributions to calculating conductor resistance as a function of frequency, it lacks laboratory exploration and measurement, which is an aspect addressed in the current study. Ref. [16] presents a singular contribution on cables losses via a computational method based on a probabilistic approach that predicts some features on litz cables. Despite the singular contribution that this study makes to quantifying losses on conductors, it lacks a focus on electrical power system conductors that the current study aims to fulfill.

The study presented herein has as its objective to present the quantification of resistance increase due to the skin effect in conductors used on power distribution networks. To undertake this study, it is crucial to have a comprehensive understanding of the physical and mathematical aspects of the problem, as well as to conduct laboratory measurements. This allows for the quantification of the skin effect in the electrical resistance of various types of conductors in power distribution lines. Additionally, the expressions estimated for representing the skin effect in conductors were applied to the IEEE 8500-node test feeder utilizing the OpenDSS software, which assesses the impact that the suggested approach has in simulation time and the influence of the skin effect on power distribution losses. It is important to point out that not only are the results from these simulations displayed on item 4.2, but also a detailed approach regarding the electrical characteristics of IEEE 8500-node test feeder. The subsequent section will delve into the physical and mathematical aspects related to the phenomenon.

2. Theoretical Background

In essence, the skin effect consists of the non-uniformity of the electric current density in the cross-section of an electrical conductor. In other words, circular conductors, commonly used in power distribution systems, present a greater flow of electric current on the surfaces of the conductor than on its inner part, and this effect is amplified with increases in electrical current. This phenomenon occurs when a current $i(t)$ varying in time runs through the conductor and, as a result, a magnetic field $\vec{H}(t)$ is generated around it according to Ampere's law. However, in accordance with the law of electromagnetic induction, any time-varying magnetic field generates an internal electromotive force (EMF) in the conductor that is also time-varying, but in the opposite direction to the $\vec{H}(t)$ from which it arises.

As such, a circulating current $i_e(t)$ arises due to the potential difference that is created between the innermost part of the conductor and its surface; as such, $\vec{H_e}(t)$ is generated by virtue of Ampère's law. Figure 1 illustrates the previously mentioned electrical magnitudes. Additionally, Figure 1 displays a solid conductor exclusively for simplicity and clarity of the physical conceptualization of the before mentioned phenomenon. All of the conductors discussed throughout this paper are stranded conductors.

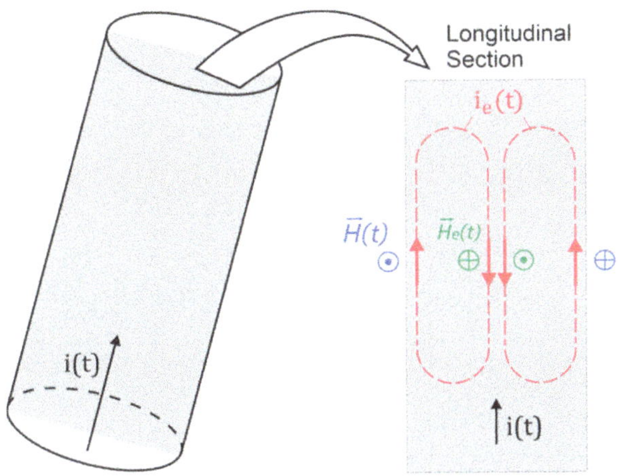

Figure 1. Skin effect on a straight and round conductor under high frequencies.

The law of electromagnetic induction guarantees energy conservation in the flow of electric charge within a conductor. It ensures that, despite the presence of a time-varying magnetic field surrounding the conductor, the resulting electric current remains continuous and does not alter the quantity of charges passing through the conductor's cross-sectional area. On the other hand, it modifies the current density along this cross-section, thus distinguishing the physical process that characterizes the skin effect [17]. The currents that arise inside the conductor $i_e(t)$ are denominated as eddy currents, as shown in Figure 1. In the skin effect, the electric current density of the conductor is not uniform. The inner part the resulting current intensity is $i(t) - i_e(t)$, while closer to the surface, the current is $i(t) + i_e(t)$, and this is the main cause of the non-uniform distribution of the current density in its cross-section.

Mathematical Approach

To mathematically contemplate the skin effect, it is necessary to address the behavior of the magnetic field generated by the electric current that runs through the conductor. In this sense, the magnetic field is time-varying and can be expressed by (1), just as the electric field can be represented by (2).

$$\vec{H}(z,t) = H_y(z)Re\left[e^{j\omega t}\right]\vec{a_y} \tag{1}$$

$$\vec{E}(z,t) = E_x(z)Re\left[e^{j\omega t}\right]\vec{a_x} \tag{2}$$

Therefore, by applying (1) and (2) to Ampere's law and the law of electromagnetic induction, respectively, while considering that the electric field also varies in time and is perpendicular to the direction of $\vec{H}(z,t)$, one arrives at (3) and (4).

$$\nabla \times \vec{E} = \frac{\partial E_x(z)}{\partial z}\vec{a_y} = -j\omega\mu H_y(z)\vec{a_y} \tag{3}$$

$$\nabla \times \vec{H} = -\frac{\partial H_y}{\partial z}\vec{a_x} = j\omega\varepsilon E_x(z)\vec{a_x} \tag{4}$$

Under the intent of obtaining a differential equation that relates only to the magnetic field, expression (4) is rotated.

$$\nabla \times \left(\nabla \times \vec{H}\right) = j\omega\varepsilon\nabla \times \left(E_x(z)\vec{a_x}\right) \tag{5}$$

Substituting the Laplacian vector Equation (6) into (5), one arrives at (7).

$$\nabla^2\vec{H} = \nabla\left(\nabla\cdot\vec{H}\right) - \nabla \times \left(\nabla \times \vec{H}\right) \tag{6}$$

$$\nabla\left(\nabla\cdot\vec{H}\right) - \nabla^2\vec{H} = j\omega\varepsilon\nabla \times \vec{E} \tag{7}$$

As $\vec{H}(z,t)$ is a vector field that has a zero divergence, it is possible to rewrite (7) according to Gauss' law, considering the calculation of the magnetic field gradient, and replacing (3) in (7); one arrives at (8).

$$\frac{\partial^2 H_y(z)}{\partial z^2} = -\omega^2\mu\varepsilon H_y(z) \tag{8}$$

Since expression (8) is the Helmholtz equation in phasor form [18], one can specify the Helmholtz equation as (9) and define the propagation constant of the magnetic field (10) utilizing a conductive material.

$$\frac{\frac{\partial^2 H_y(z)}{\partial z^2}}{H_y(z)} = -\lambda^2 = -\omega^2\mu\varepsilon \tag{9}$$

$$\lambda = \omega\sqrt{\mu\varepsilon} = \omega\sqrt{\mu\varepsilon\prime\left(1 - j\frac{\sigma}{\omega\varepsilon\prime}\right)} \tag{10}$$

Noteworthy is that for conducting materials $\frac{\sigma}{\omega\varepsilon\prime} \gg 1$ as such, (10) is summarized as the following:

$$\lambda = (1+j)\sqrt{\pi f\mu\sigma} = \alpha + j\beta \tag{11}$$

In this sense, the solution to the Helmholtz equation presents the behavior of the magnetic field. It enhances the context to clarify the physical significance of the propagation constant for the skin effect. The solution to the Helmholtz equation is given by the following:

$$\vec{H}(z,t) = H_0 e^{-\alpha} e^{j(\omega t - \beta z)}\vec{a_y} \tag{12}$$

In (12), the term $e^{j(\omega t - \sqrt{\pi f\mu\sigma z})}$ represents the oscillatory feature of the magnetic field, where β proposes the phase shift. However, the term $e^{-\sqrt{\pi f\mu\sigma z}}$ presents exponential decay α of the original magnitude of the vector field (H_0) as the wave travels in the direction +z. Through these equations, it is possible to define the skin effect by means of skin depth. The skin depth δ corresponds to the distance traveled by the magnetic field in the material medium in which its original magnitude is decreased in the order of e^{-1}. In other words, skin depth elucidates the degree of penetration of the magnetic field and of the circulating internal currents in a conductor with losses [18], as shown in Figure 2. Therefore, for the condition presented in (12), z should be equal to $1/\sqrt{\pi f \mu \sigma}$ for attenuation on the order of e^{-1}, where the skin depth is defined as the following:

$$\delta = \frac{1}{\sqrt{\pi f \mu \sigma}} \tag{13}$$

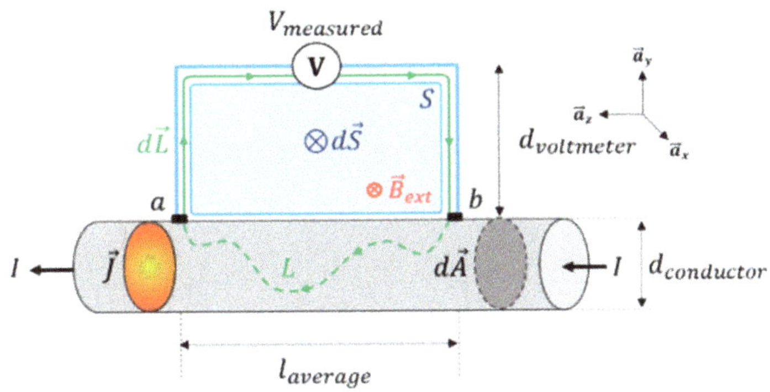

Figure 2. Setup for voltage probe connection.

Hence, the skin effect can be mathematically defined as the skin depth in a conductor, as represented by (13), and it physically signifies a higher concentration of current in the cross-section of the conductor, as depicted in Figure 2. Therefore, it is crucial to investigate how the current density varies based on the conductor's geometry. With this in mind, significant emphasis is placed on studying the current distribution of various types of conductors. As a result, in (14), two different forms are used to calculate the current through a conductor, such as the surface integral of J and Ampère's law.

$$Hr = \int_0^r Jrdr \tag{14}$$

To understand how the geometry of the conductor influences the current density of the conductor, one derives (14) concerning the radius of the conductor. Furthermore, it is also pertinent to derive (14) in time, since the magnetic field is a function of this variable, resulting in the following:

$$\frac{dJ}{dt} = \frac{d^2H}{drdt} + \frac{1}{r}\frac{dH}{dt}. \tag{15}$$

The result of (15) provides the variation of the current density in the conductor for these two variables, space (r) and time (t), according to the variations of the magnetic field. In this way, the following developments will be directed to finding mathematical expressions for the derivatives of the magnetic field in time and space in simpler forms for the result of (15). Noteworthy still is that $\vec{E} = \rho \vec{J}$, thus $\mu dr \frac{dH}{dt} = \rho \vec{J}$, where one arrives at (16) and substitutes (16) into (15), one obtains (17), which is the differential equation that models the phenomenon presented in Figure 2, in a way that specifies the density of the electric current along the cross-section of the conductor.

$$\frac{dH}{dt} = \frac{\rho}{\mu}\frac{d\vec{J}}{dr} \tag{16}$$

$$\frac{d^2J}{dr^2} + \frac{1}{r}\frac{dJ}{dr} - \frac{\mu}{\rho}\frac{dJ}{dt} = 0 \tag{17}$$

Moreover, in the phasor form, where $d/dt \to j\omega$ and $k^2 = j\omega\mu\sigma$, one has the following:

$$r^2\frac{d^2J}{dr^2} + r\frac{dJ}{dr} - r^2k^2J = 0. \tag{18}$$

The differential equation shown in (18) has the term r^2k^2 arranged to facilitate the mathematical development to reach an elegant solution. The expression (19) is the electric current density as a function of the radius of the conductor under analysis, expressed by Bessel's equations with 0 and 1 indexes (I_0 and I_1) [18,19]. The differential equation shown in (18) is a function of r_0, which is the conductor radius, and r represent the variable, which determines the intensity of the electric current in the cross-section of the conductor, as shown in (19). It is important to note that on (18) the term r^2k^2, during the mathematical development to reach a solution, turns into kr within the Bessel's expressions of indexes 0 and 1.

$$J(r) = \frac{Ik}{2\pi r_0}\frac{I_0(kr)}{I_1(kr_0)} \tag{19}$$

Considering that the current density, when a direct current runs through the conductor is $J_{DC} = I/\pi r_0^2$, then $J(r)/J_{DC}$ is given by the following:

$$\frac{J(r)}{J_{DC}} = \frac{\left((1+j)\left(\frac{r_0}{\delta}\right)\right)}{2}\frac{I_0\left((1+j)\left(\frac{r_0}{\delta}\right)\left(\frac{r}{r_0}\right)\right)}{I_1\left((1+j)\left(\frac{r_0}{\delta}\right)\right)} \tag{20}$$

Notably, the current density on the outer part of the conductor is greater in conductors with larger cross-sections, thus underscoring the impact of the skin effect [18,19]; this is evident from (20) as well. Therefore, the electrical resistance of conductors under such conditions increases, which implies an increase in power losses due to the transport of active power. Consequently, to evaluate the impact of such a phenomenon, the proposal is put forward for the study of the behavior of R_{AC}/R_{DC}, in similar fashion to the approach already used for current density (J_{AC}/J_{DC}).

When considering the resistance in alternating current (R_{AC}), one uses Ohm's law, as seen in (21), with the ratio being R_{AC}/R_{DC} as presented in (22) [19,20].

$$Z = \frac{\rho l}{\pi r_0}\frac{k}{2}\frac{I_0(kr_0)}{I_1(kr_0)} \tag{21}$$

$$\frac{R_{AC}}{R_{DC}} = Re\left\{\frac{J(r_0)}{J_{DC}}\right\} \tag{22}$$

It is observed in (22) that as the ratio $\frac{r_0}{\delta}$ increases, the electrical resistance of the conductor increases proportionally for a conductor with radius r_0 and with a skin depth that decreases (increasing the ratio $\frac{r_0}{\delta}$). A trivial condition concerning the increase in ratio $\frac{r_0}{\delta}$ occurs when harmonic components of electric current flow through the conductor. Under such conditions, harmonic components imply a decrease in the value of δ, in accordance with (13). Indirectly, this causes an increase in the resistance of the conductor, as it increases the ratio $\frac{r_0}{\delta}$ [18,21].

It is well established that harmonic components are inherent in electrical power systems, particularly in power distribution systems. In such systems, it is common to encounter harmonic currents (and correspondingly, voltages) with frequencies up to 960 Hz (16th harmonic order). Beyond this frequency range, the amplitudes of these components become insignificant, exerting minimal impact on the skin effect.

In the specific case of electrical power distribution systems, the skin effect assumes a very applicable role in harmonic power flow studies, especially on medium- and low-voltage power networks. In this sense, the physical–mathematical development carried out so far aimed at establishing physical foundations for understanding the phenomenon, concerning justifying any sensible increase in resistance of the conductor under non-sinusoidal conditions, as well as justifying the investigation into the quantification of this increase. From this standpoint, the objective of this study is to assess the electrical resistance of various conductors employed in power distribution systems, in order to quantify the rise in resistance attributed to the aforementioned phenomenon. The subsequent section outlines the methodology employed to measure the electrical resistance of conductors affected by the skin effect.

3. Measurement Methodology

In pursuance of measuring the impact of the skin effect on conductors, it is necessary to understand which electrical magnitudes are verified in the measurement methodology. Given the relatively low magnitude of the measured quantities, it is crucial to consider the uncertainties associated with the measurements. To determine the resistance of the conductor, an indirect measurement approach was employed. This involved passing an electric current through the conductor under examination and using an oscilloscope to measure the corresponding voltage drop. By applying Ohm's Law, the resistance of the conductor can be calculated using Equation (23).

$$R = \frac{1}{l_{average}} Re\left\{ \frac{\dot{V}_{measured}}{\dot{I}} \right\} \times 1000 \left[\frac{\Omega}{\text{km}} \right] \quad (23)$$

However, to obtain a reliable measurement of the voltage drop in the conductor, an assessment is made into the possible interferences admitted by the voltage loop. That said, voltmeters, when measuring the potential difference between two points a and b, are known to consider the voltage arising from the conservative electric field between these two points. This is in addition to the voltage from the non-conservative magnetic field that varies in time and passes through the loop voltage, where (24) is the potential difference measured by the oscilloscope, which includes both previously mentioned vector fields.

$$V_{measured} \approx \int_a^b \vec{E} \cdot d\vec{L} + \frac{\partial}{\partial t} \int_0^d \int_a^b \vec{B}_{ext} \cdot d\vec{S}$$

$$V_{measured} \approx \int_a^b E\, dz + \frac{\partial}{\partial t} \int_0^d \int_a^b B\, dzdy \quad (24)$$

In this context, the measurement methodology used assumes two origins for the potential difference ascertained in the oscilloscope. The first comes from $\vec{E}$ that is independent of the integration path inside the conductor, as well as from the voltage induced by the variation of $\vec{B}_{ext}$ in time bound to the voltage loop, as shown in Figure 2. Additionally, Figure 2 displays a solid conductor exclusively for simplicity and clarity of the physical conceptualization of the before mentioned measurement methodology. All of the conductors measured in this study are stranded conductors.

From Figure 2, it can be noted that $\vec{B}_{ext}$ passes through the voltage loop $\mathbf{S}$, where an induced voltage appears between the conductor points a and b, which is measured with the oscilloscope, and constitutes an effective value of the voltage $V_{measured}$, as shown in (24). It is important to point out that the induced voltage in question greatly impacts the measurement, due to the low order of magnitude of the voltage arising from $\int_a^b \vec{E} \cdot d\vec{L}$ between points a and b. In this case, the smallest possible voltage loop is used to minimize the impact of the voltage induced by $\vec{B}_{ext}$. In accordance with [16], the suggestion is that the height of the voltage loop equals the conductor diameter. More objectively, $d_{voltmeter} = d_{conductor}$ while considering the setup in Figure 2.

In addition to taking electromagnetic factors into account when measuring the electrical resistance of conductors, it is important to consider the presence of nearby conductive materials. These materials can induce eddy currents through the proximity effect, potentially interfering with the measurement procedure. To mitigate this issue, all conductive materials were positioned at a distance of more than one meter from the measurement setup [22,23].

In this manner, Figure 3 illustrates the measurement setup used to measure the resistance of the conductors. Figure 3 shows the programmable source, CMC 256 plus, manufactured by Omicron Electronics Corp. (Houston, TX, USA), used to operate as a current source. In addition, there is a digital oscilloscope, RTH1004, manufactured by Rohde & Schwarz (Teisnach, Germany), with four isolated channels and a 5 GSa/s sampling rate.

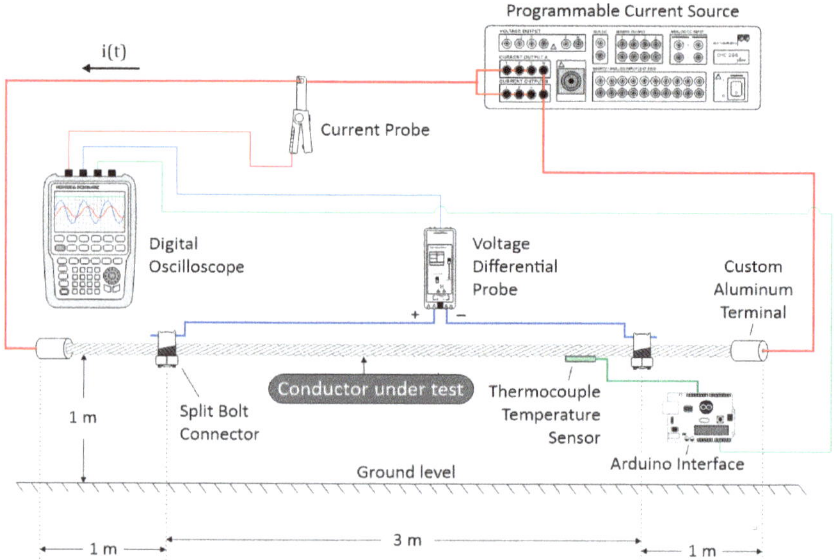

Figure 3. Measurement setup to quantify the skin effect in conductors applied to distribution power systems.

In order to fully describe the measurement setup, the main characteristics are stated as follows:

- The ground is composed of concrete.
- The current return path is 1 m from the conductor under test, close to ground level.
- Wooden racks were used as a non-conductive material to keep the conductor under testing 1 m away from ground level.
- Voltage differential probes cables were placed according to $d_{voltmeter} = d_{conductor}$ for each conductor, as displayed in Figures 2 and 3.

The sensors used for the acquisition of electrical quantities were the A622, manufactured by Tektronix (Portland, OR, USA), employed for the acquisition of electric current, and the TA041, manufactured by PicoTech (Cambridge, UK), for the acquisition of the voltage signal. The configuration displayed in Figure 3 shows the electronic devices used, as well as the three crucial items for reaching the objective of this study.

Additionally, the post-processing voltage signal registered on the oscilloscope corresponds to the flow chart presented in Figure 4. Briefly, the oscilloscope registered a time domain signal that was applied to a Python algorithm developed exclusively to perform

Fourier transforms. Its purpose is to obtain the voltage magnitude and phase angle of the desired frequency, which is the current frequency flowing through the conductor under test. It is important to highlight that the reference phase angle (0° electrical degrees) is related to the current signal, once the programmable power source is a current source. Additionally, the magnitude of the out-of-phase inductive voltage signal is larger than the resistive voltage signal, especially regarding high frequencies. It is an expected response, since the inductive part of the conductor responds to a linear relation to frequency; meanwhile, the resistive part responds according to (22) in relation to frequency. In other words, the measured voltage signals presented a phase angle higher than 80°, indicating that voltage is predominantly inductive. The considered resistive voltage signal was computed (see Figure 4) and applied to (23), obtaining the conductor's electrical resistance.

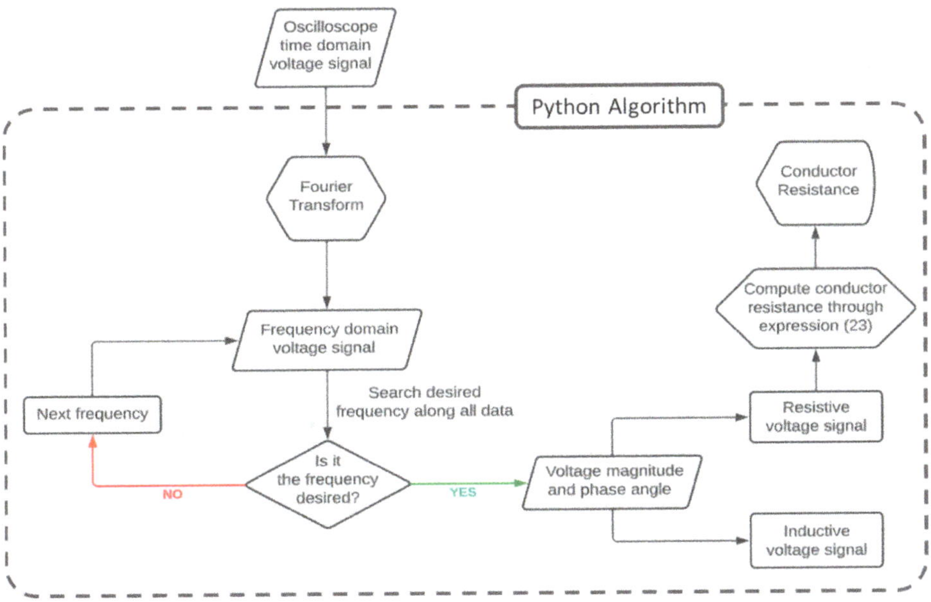

Figure 4. Data measurement post-processing to compute conductor's electrical resistance.

Firstly, the conductor under test in this study includes the main conductors used in electrical power distribution systems, which are 2 AWG—AAC, 1/0 AWG—ACSR, 2/0 AWG—AAC, 4/0 AWG—AAC, 150 mm^2 protected and 336.4 MCM—ACSR. Here, it is noteworthy that for better contact between the power source conductors (red thicker lines, Figure 3) and the conductor under test, it is necessary to use terminals both for current injection in the conductor under test, as well as for measuring the voltage.

As shown in Figure 3, the aluminum terminal ensures that the electrical current is distributed as expected. To this end, its geometric dimensions are shown in Figure 5, where D represents the diameter of the conductor under test.

To ensure that the current distribution is homogeneous in the region where the voltage in the conductor is measured, a distance of 1 m from the input to the output terminals of the electric current was assumed. This implies that $\vec{J}$ is homogeneous in the cross-section of the conductor, guaranteeing an adequate measurement [23].

Finally, there are split bolt terminals for connecting the oscilloscope voltage channels, allowing for a better connection, as shown in Figure 6.

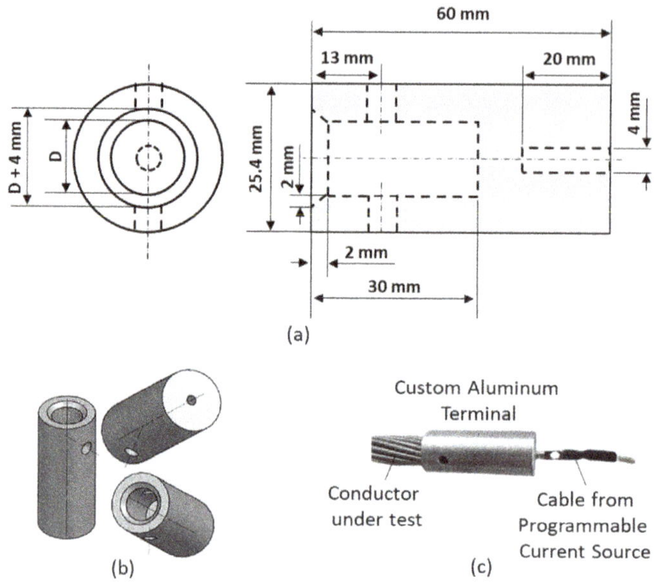

Figure 5. Custom aluminum terminals, considering (**a**) frontal and lateral views, (**b**) 3D view and (**c**) actual terminal connection.

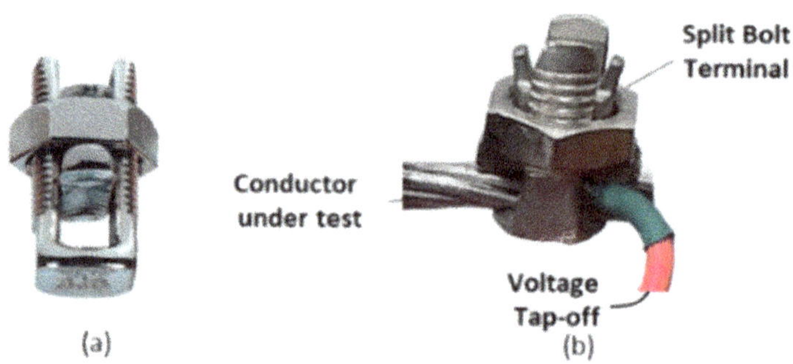

Figure 6. (**a**) Split bolt terminal and (**b**) actual terminal connection.

In addition to the previous recommendation regarding the voltage loop during the measurement of conductor electrical resistance, reference [23] also mentions the potential influence of temperature on the accurate quantification of resistances considering the skin effect.

This aspect is of great importance, since the electric current heats up the conductor and, consequently, increases the resistivity (ρ) of the conductor. Additionally, the resistance of the conductor under test alters owing to its shape, with the coefficients of linear and superficial expansion of the conductor being physical aspects to be investigated due to an increase in temperature along the measurement process. In this sense, the coefficients of linear expansion of aluminum and steel are of the order of 10^{-6} m/°C and 10^{-5} m/°C, respectively [24]. As such, errors associated with linear expansion are negligible, with the contribution arising from linear or superficial expansion being irrelevant to the order of magnitude of the electrical resistance. Therefore, in order to assess the impact of temperature on the resistivity of the conductor, a device was developed for the acquisition of the temperature using an Arduino UNO and temperature sensor DS18B20, in conjunction with the acquisition of 12-bit data and a resolution from the measurement of 0.0625 °C, as shown in Figure 3. To commence the measurement process, all devices involved were certified to be at 20 °C. Thermally conductive aluminum adhesive tape was used in order to guarantee proper heat transfer from the conductor to the sensor. Consequently, the thermal analysis aims to determine the error linked to the skin effect at each frequency, as presented in Table 1 and consolidated in graph form in Figure 7.

Table 1. Temperature variation $\left(\Delta T = T_f - 20\ ^\circ\text{C}\right)$ and resistivity (ρ) increase considering current (40 A) flowing through the conductor at several frequencies.

Harmonic	2 AWG—AAC		1/0 AWG—ASCR		2/0 AWG—AAC		4/0 AWG—AAC		150 mm²—AAC		336.4 MCM—ACSR	
	ΔT (°C)	$\frac{\rho}{\rho_{20^\circ C}}$ (%)	ΔT (°C)	$\frac{\rho}{\rho_{20^\circ C}}$ (%)	ΔT (°C)	$\frac{\rho}{\rho_{20^\circ C}}$ (%)	ΔT (°C)	$\frac{\rho}{\rho_{20^\circ C}}$ (%)	ΔT (°C)	$\frac{\rho}{\rho_{20^\circ C}}$ (%)	ΔT (°C)	$\frac{\rho}{\rho_{20^\circ C}}$ (%)
DC	4.406	1.775	4.527	1.824	3.473	1.399	2.602	1.048	1.724	0.694	1.395	0.562
1	4.410	1.777	4.535	1.827	3.499	1.410	2.625	1.058	1.757	0.708	1.402	0.565
2	4.447	1.792	4.572	1.843	3.510	1.414	2.636	1.062	1.798	0.724	1.432	0.577
3	4.470	1.801	4.595	1.852	3.516	1.417	2.642	1.065	1.835	0.739	1.462	0.589
4	4.485	1.808	4.610	1.858	3.521	1.419	2.647	1.067	1.868	0.753	1.493	0.602
5	4.497	1.812	4.622	1.863	3.524	1.420	2.650	1.068	1.899	0.765	1.523	0.614
6	4.507	1.816	4.632	1.867	3.527	1.422	2.653	1.069	1.927	0.777	1.553	0.626
7	4.516	1.820	4.641	1.870	3.530	1.423	2.656	1.070	1.953	0.787	1.583	0.638
8	4.523	1.823	4.648	1.873	3.532	1.423	2.658	1.071	1.978	0.797	1.613	0.650
9	4.530	1.825	4.655	1.876	3.534	1.424	2.660	1.072	2.002	0.807	1.643	0.662
10	4.535	1.828	4.660	1.878	3.536	1.425	2.662	1.073	2.025	0.816	1.673	0.674
11	4.540	1.830	4.665	1.880	3.537	1.425	2.663	1.073	2.049	0.826	1.704	0.687
12	4.545	1.832	4.670	1.882	3.539	1.426	2.665	1.074	2.073	0.836	1.734	0.699
13	4.550	1.833	4.675	1.884	3.540	1.427	2.666	1.074	2.099	0.846	1.764	0.711
14	4.554	1.835	4.679	1.885	3.541	1.427	2.667	1.075	2.126	0.857	1.794	0.723
15	4.557	1.837	4.682	1.887	3.542	1.427	2.668	1.075	2.156	0.869	1.824	0.735
16	4.561	1.838	4.686	1.888	3.543	1.428	2.669	1.076	2.188	0.882	1.854	0.747

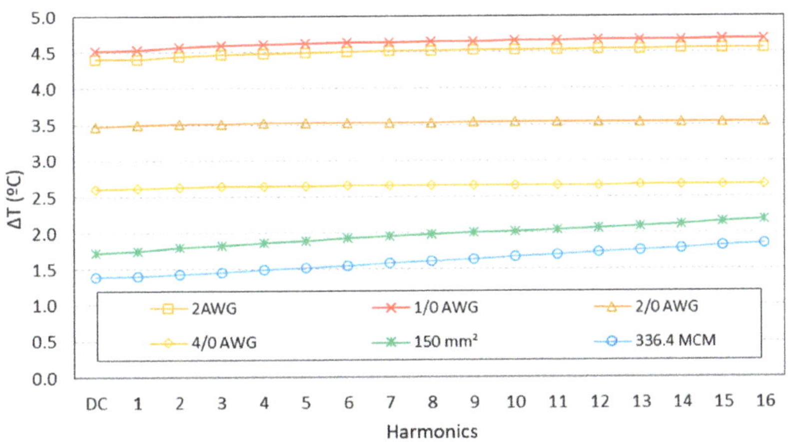

Figure 7. Temperature variation $\left(\Delta T = T_f - 20\ ^\circ\text{C}\right)$ considering current (40 A) flowing through the conductor at several frequencies.

To examine the impact of temperature on the resistivity of the conductor (ρ), the linear relationship between these two quantities was adopted as in reference [24]. Furthermore, reference [24] highlights that at lower temperatures, the correlation between temperature and electrical resistivity in aluminum alloys remains linear. Similarly, reference [25] also addresses this aspect. In the cases examined within this study, the maximum temperature increase was estimated to be approximately 5 °C, starting from an initial temperature of 20 °C. Additionally, it was observed that for thicker conductors, a temperature rise was noted as current frequency increased, as Figure 7 shows for 150 mm^2 and 336.4 MCM conductors. Meanwhile, the remaining conductors displayed a notable lower temperature increase with current frequency, which suggests that the skin effect results in significant heating for thicker conductors, considering current frequency increases.

It should be emphasized that the increase in temperature observed for harmonic components, compared to the temperature rise for DC current flowing through the conductor, is primarily caused by the skin effect. The recorded data regarding temperature rise for harmonic components are presented in Table 2.

Table 2. Temperature rise due exclusively to skin effect.

Conductor	$\Delta T = T_{16harmonic} - T_{DC}$(°C)	$\frac{\rho}{\rho_{DC}}$(%)
2 AWG—AAC	0.155	0.062
1/0 AWG—ASCR	0.159	0.064
2/0 AWG—AAC	0.070	0.028
4/0 AWG—AAC	0.067	0.027
150 mm^2—AAC	0.464	0.187
336.4 MCM—ASCR	0.429	0.172

To specifically quantify the error resulting from temperature rise associated with the skin effect, Table 2 indicates that the largest error is observed for 150 mm^2—the AAC conductor, with an error of 0.187% corresponding to a temperature rise of 0.464 °C. Another noteworthy aspect is that the skin effect is more prominent for thicker conductors, and Table 2 as well as Figure 7 showcase this result, as expected.

Therefore, based on the analysis of aluminum alloys used in the examined conductors, it is concluded that the relationship between temperature and electrical resistivity can be considered linear [23]. Furthermore, it is appropriate to model the resistivity by means of (25) to specify the percentage increment for resistivity $\frac{\rho}{\rho_{20^\circ C}} \times 100\%$.

$$\rho_T = \rho_{Al_{(20^\circ C)}}[1 + \alpha(T - T_0)]\left[\frac{\Omega}{\mathrm{m}}\right] \quad (25)$$

Considering that $\alpha = 0.00403$ ($1/°C$) [13] and $\rho_{Al_{(20^\circ C)}} = 61\%\ IACS$ (Ω·m), as indicated by the manufacturer of the conductors [24], one can quantify how much the resistance of the conductors is influenced by the increase in resistivity due to temperature rise.

The results indicate that the temperature increase observed during the process of measuring conductor resistance, in the worst-case scenario (conductor 1/0 AWG—ACSR), leads to an error of 1.888% in the quantification of the resistance value. As expected, the 1/0 AWG—ACSR conductor has lower ampacity, and therefore experiences greater power dissipation through Joule's effect. Additionally, its geometry includes steel reinforcement, which has a resistivity 20 times greater than typical aluminum alloys [24], resulting in greater heating during the measurement process.

To summarize the measured data, Table 1 shows an increase in electrical resistivity due to temperature rise. For DC, all of the conductors exhibited a maximum resistivity increase of 1.9%. This percentage remained consistent for both the fundamental frequency and all harmonic components.

Furthermore, Table 2 estimates an approximate 0.2% increase in electrical resistivity for the skin effect measurement due to temperature rise. However, this heating effect has a negligible impact on the conductor's resistance measurement. In other words, the increase in electrical resistivity caused by the RMS values of the fundamental and harmonic components contributes to a 1.9% rise in resistivity, while the skin effect causes only a 0.2% increase, which is considered practically negligible.

Having discussed the measurement configuration for the skin effect and the variables that can introduce errors, the subsequent section will delve into the analysis of the measured data and their application to reduce simulation time for harmonic power flow studies.

4. Results and Discussion

4.1. Measurement Results—Skin Effect

Through performing tests, an increase in the resistance of the studied conductors was obtained, which considered the skin effect. Noteworthy here is that verification was sought for the DC resistances values of the conductors made available by the manufacturers [26,27] and as such, was confirmed by the measurement methodology proposed in this study. Therefore, the measured reference value was used to specify the start of the measurement process. From this value, Table 3 presents the measured resistance values in DC resistances.

Table 3. Electrical resistances of conductors in direct current at 20 °C.

Conductor	$R_{measured}$ (Ω/km)	$R_{datasheet}$ (Ω/km)	Error (%)
2 AWG—AAC	0.8652	0.8541	1.31
1/0 AWG—ASCR	0.5285	0.5360	1.40
2/0 AWG—AAC	0.4517	0.4550	0.72
4/0 AWG—AAC	0.2616	0.2676	2.21
150 mm^2—AAC	0.2046	0.2060	0.64
336.4 MCM—ASCR	0.1726	0.1703	1.36

The carried-out measurements showed errors of less than 2.5%, thus ensuring that the measurement process is adequate, these being the reference values used in Figure 8. Based on the measurement results, the theory is corroborated, in which thicker conductors are subject to a greater influence of the skin effect, as illustrated in Figure 8 for the 4/0 AWG, 150 mm^2 and 336.4 MCM conductors. In these conductors, a greater increase in electrical resistance was noted, which results from the skin effect. Furthermore, one notes that the 1/0 AWG conductor, although belonging to the conductors being tested with lower thickness, exhibited a relevant impact against the skin effect. In comparison, the 2 AWG conductor together with the 2/0 AWG, composed only of aluminum wires (AAC), exhibited the lowest influence from the skin effect. From Figure 8, verification was made that the conductors that presented the greatest heating, such as 2 AWG and 1/0 AWG, stand out for showing an increasing tendency as the electrical current increases, indicating, through progressive 3D line graphs, the impact of temperature on the measurement process. Meanwhile, this behavior is not observed in the remainder of the conductors, since the range of test currents (20–40 A) is much lower than the ampacity of the conductors, highlighting, therefore, significantly lower heating of these conductors.

The comparison between the measurement results and the theoretical expectations is significant, as it provides insights into the behavior of the measured data, as displayed in Figure 9. Additionally, the magnitude of the resistance increase is an important aspect to consider. It was observed that for electric currents at 960 Hz, running through the conductor, an increase of around 60% was verified in the electrical resistanc, in comparison to DC resistance for the 1/0 AWG, 4/0 AWG, 150 mm^2 and 336.4 MCM conductors.

Against this backdrop, the following topics aim to quantitatively assess the impact that the skin effect has on power simulation studies, while considering distorted conditions of electric current and the measurement values obtained at this stage of the study.

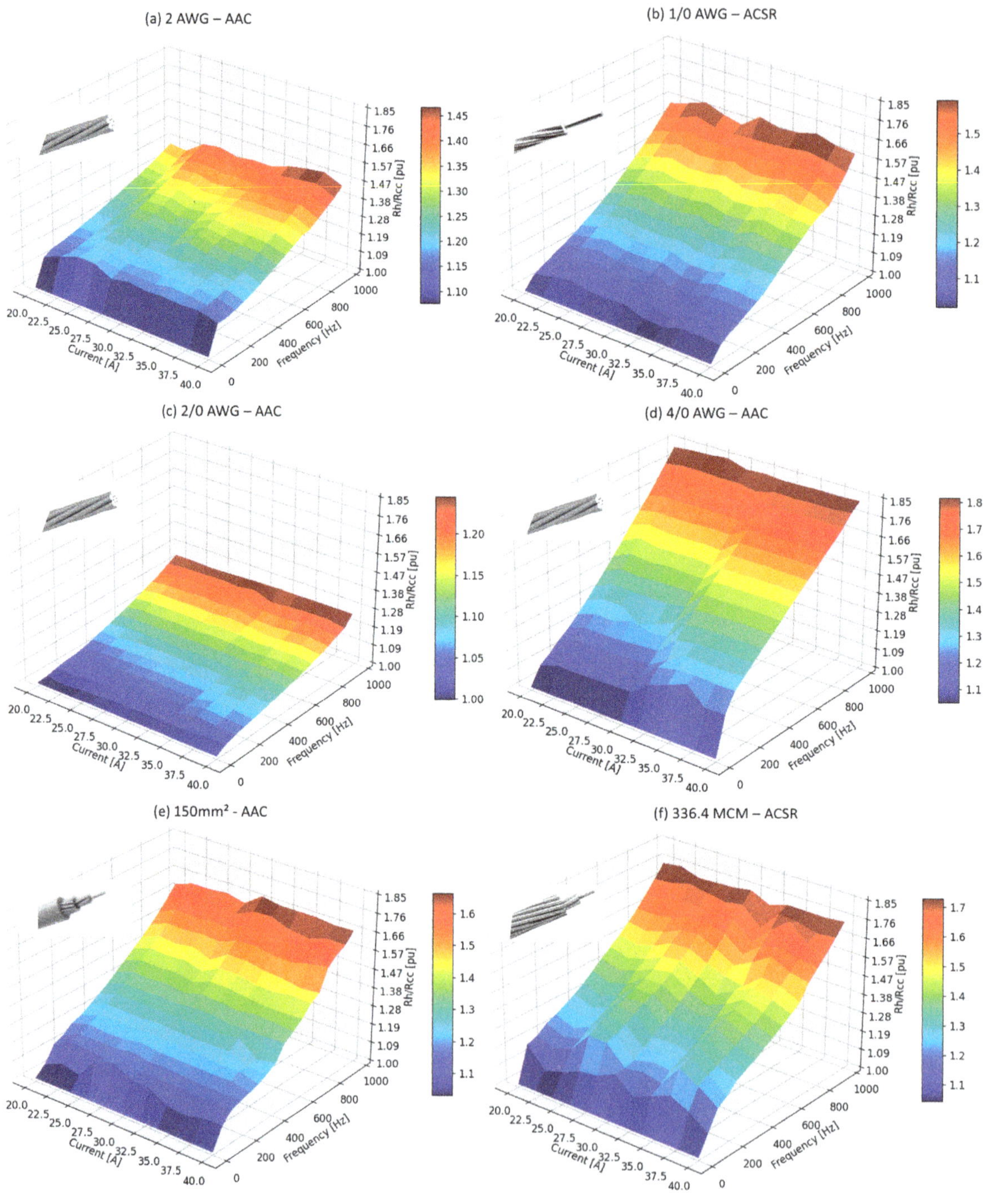

Figure 8. Measured electrical resistances of conductors, considering the skin effect.

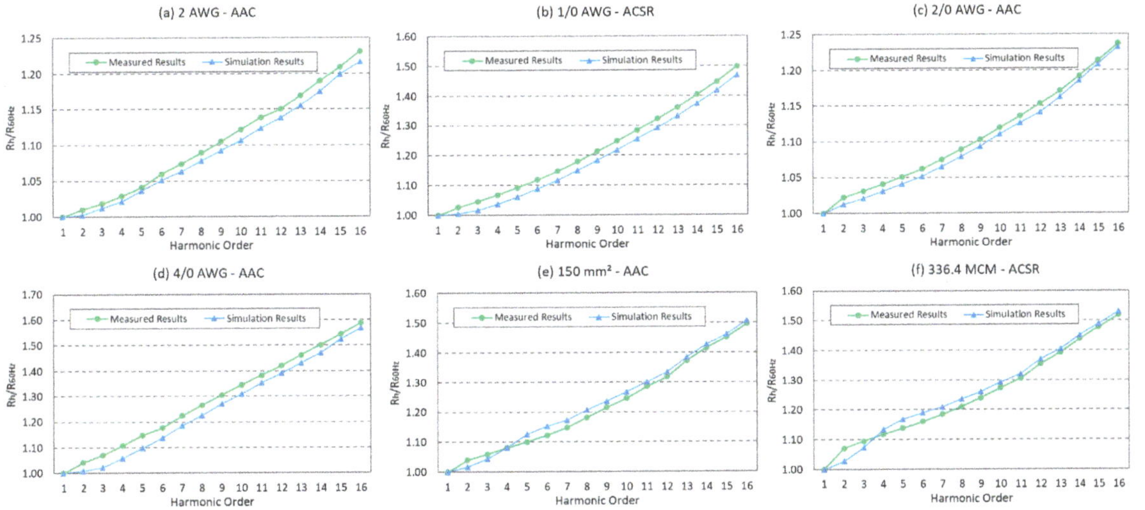

Figure 9. Measured data after (26) is applied versus simulated results of conductor's electrical resistance, considering the skin effect.

4.2. Statistical Simplified Method to Estimate Skin Effect for Low Frequencies

To simplify the process and obtain a single resistance value per frequency, an averaging method was employed considering the variation in current in the conductor. For each frequency, the resistance values were synthesized to be considered, as pointed out in (26).

$$R_h = average(R_{20A}, R_{22A}, \cdots, R_{40A}) \quad (26)$$

Based on this compilation, Figure 10 provides the values that were computed as a result of applying expression (26) to the data. Figure 10 shows that the conductors under test are appropriately estimated through a linear equation considering low frequencies. Not only are the expressions presented in Figure 10 simpler, but they also reduce the processing time when computing the skin effect. It should be emphasized that for harmonic power flow studies, especially those covering the distribution power network of an entire operator region, they may take up to 48 h of simulation time. To compare the time processing between Bessel's proposed expression (22) adapted to 60 Hz resistance as a reference against the Figure 10 expressions, an algorithm in Python (PyCharm) was implemented. The results show that Bessel's expression (22) required 53.85 ms to compute the skin effect for all conductors for the before mentioned frequency range, as opposed to 41.88 ms of simulation time for the Figure 10 expressions. Therefore, the reduction in simulation time is approximately 20% using the proposed method.

For the purposes of an objective assessment of the study, the IEEE 8500 node feeder [28], which is known throughout the scientific community, was used to calculate the power losses under distorted conditions. The studied feeder is shown in Figure 11, in which the line thickness for the segments corresponds to the regions where energy losses are more expressive. It is important to point out that information with greater detail on this feeder is available in [28].

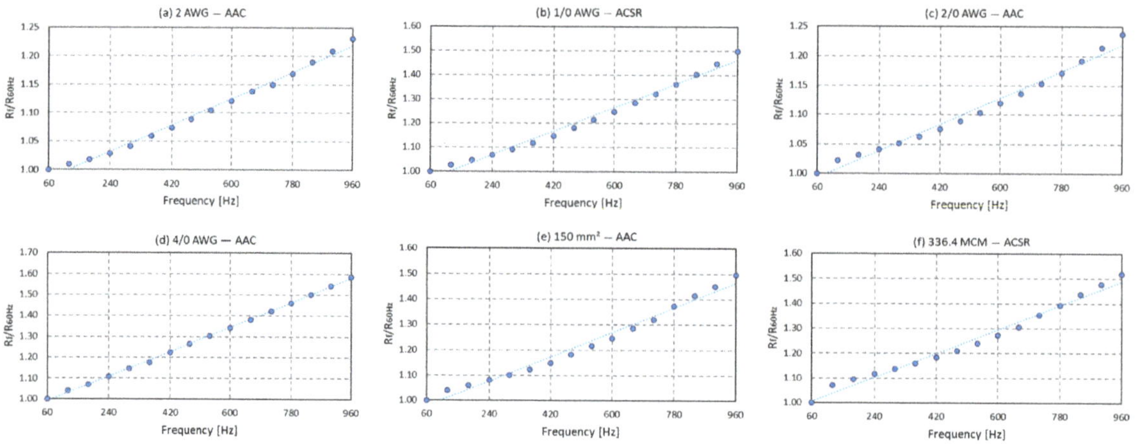

Figure 10. Sampled data after (26) is applied with linear regression expressions and R^2 values.

Figure 11. IEEE 8500 nodes feeder—line thicknesses relate to power loss intensities.

Simulations were performed using OpenDSS and COM Interface with Python (PyCharm). To clearly meet the objective of this study, it was crucial to define the harmonic emissivity spectrum of the non-linear loads of this feeder. To this end, the use of input data from [29] was established for the frequency range of 60 Hz to 960 Hz. Another noteworthy aspect to mention is that additional conductors beyond the ones measured are present at the IEEE 8500 node feeder. To address this, an assumption was made to consider the behavior of the closest laboratory-tested conductor for these untested conductors.

To correctly estimate the time reduction for the harmonic power flow study, the two following separated conditions were performed:

Case 1: Bessel's expression (22), adapted to 60 Hz resistance as a reference, as a correction factor for the skin effect $\left(\frac{R_f}{R_{60\text{Hz}}}\right)$ applied to the resistance of line parameters, according to cables used at the transmission line, as expressed in (27).

Case 2: Figure 10 expressions as a correction factor for the skin effect applied $\left(\frac{R_f}{R_{60\text{Hz}}}\right)$ to the resistance of line parameters, according to the cables used at the transmission line, as expressed in (27).

$$R_{conductor_f} = \frac{R_f}{R_{60\text{Hz}}} R_{conductor}[\Omega] \tag{27}$$

The simulation times of both conditions previously mentioned are shown in Table 4.

Table 4. Simulation times spent for harmonic power flow study case.

Case 1	Case 2	Time Reduction	Percentage Reduction
279.64 s	231.25 s	48.39 s	17.30%

Another noteworthy aspect is the reliability of the results provided from the proposed method compared to Bessel's expressions. Also, from a power systems perspective, power losses are segregated into several components, as illustrated in Figure 12. Skin effect losses are responsible for 4.00% of additional losses on the IEEE 8500 node test feeder.

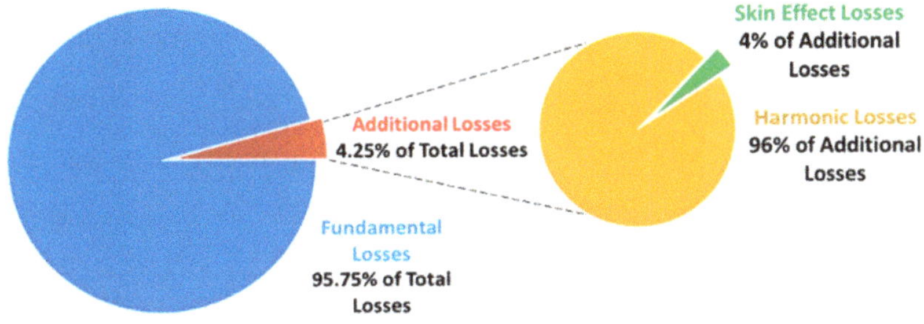

Figure 12. Power losses computed for IEEE 8500 nodes, considering distorted conditions and the skin effect.

5. Conclusions

The research conducted in this study offers a new perspective on the impact of the skin effect on conductors used in energy distribution networks, particularly in relation to Joule losses associated with this phenomenon. The study aimed to provide a mathematical understanding of the non-uniform current distribution in the cross-section of the conductor, thereby explaining its physical origin. A notable aspect of this research is the methodology developed for measuring the skin effect, which was applied to six commonly used conductors in energy distribution networks. The evaluation criteria used for measuring the electrical resistance of the conductors in direct current showed an error of less than 2.5% compared to the manufacturer's specified values.

Furthermore, the study examined the behavior of conductor resistance under harmonic electric currents ranging from the 2nd to the 16th harmonic order, in order to assess the significance of the skin effect. It was observed that the impact of temperature rise on resistivity during the measurement process was 1.9% due to the heating effect caused by the current. Conversely, the temperature rise caused by the skin effect was less than 0.2%, which is considered negligible when it comes to resistance measurements.

After measuring the resistance of the conductors under each magnitude (20–40 A) and frequency (60–960 Hz), the IEEE 8500 node feeder was used to perform a harmonic power flow study. In this regard, it was noted that the proposed expressions provided a reduction in simulation time of approximately 17%, and the skin effect was responsible for a percentage of 4.00% in additional losses.

Hence, the studies conducted in this research aim to provide a practical perspective on a well-known phenomenon within the scientific community, specifically addressing the proposed expressions as a method to reduce simulation times for harmonic power flow studies.

Author Contributions: Review and editing, E.T.S. and J.R.M.J.; conceptualization, E.T.S. and J.R.M.J.; writing—original draft, E.T.S.; simulations, E.T.S.; methodology, E.T.S. and J.R.M.J.; formal analysis, E.T.S. and J.R.M.J. All authors have read and agreed to the published version of the manuscript.

Funding: The research is supported by the Coordination of Superior Level Staff Improvement and National Electrical Energy Agency under funding number 23117.060101/2020-80.

Data Availability Statement: All data are displayed in graphs and figures in the study.

Conflicts of Interest: The authors declare no conflict of interest.

References

1. Cabral, S.H.L.; Bertoli, S.L.; Medeiros, A.; Hillesheim, C.R.; De Souza, C.K.; Stefenon, S.F.; Nied, A.; Leithardt, V.R.Q.; Gonzalez, G.V. Practical Aspects of the Skin Effect in Low Frequencies in Rectangular Conductors. *IEEE Access* **2021**, *9*, 49424–49433. [CrossRef]
2. Gassab, O.; Chen, Y.; Shao, Y.; Li, J.; Wen, D.-E.; He, F.; Su, Z.; Zhong, P.; Wang, J.; Zhao, D.; et al. Accurate Formulation of The Skin and Proximity Effects in High-Speed Cable System. *IEEE Access* **2022**, *10*, 100682–100699. [CrossRef]
3. Davis, E.W.; Dana, A.S.; Dwight, H.B. Skin Effect and Proximity Effect in Tubular Conductors. *Am. Inst. Electr. Eng.* **1922**, *41*, 1131–1140. [CrossRef]
4. Dai, D.; Zhang, X.; Wang, J. Calculation of AC Resistance for Stranded Single-Core Power Cable Conductors. *IEEE Trans. Magn.* **2014**, *50*, 6301104. [CrossRef]
5. Yin, W.; Tang, J.; Lu, M.; Xu, H.; Huang, R.; Zhao, Q.; Zhang, Z.; Peyton, A. An Equivalent-Effect Phenomenon in Eddy Current Non-Destructive Testing of Thin Structures. *IEEE Access* **2019**, *7*, 70296–70307. [CrossRef]
6. Li, S.; Han, Y.; Liu, C. Coupled Multiphysics Field Analysis of High-Current Irregular-Shaped Busbar. *IEEE Trans. Compon. Packag. Manuf. Technol.* **2019**, *9*, 1805–1814. [CrossRef]
7. Kusiak, D. The magnetic field and impedances in three-phase rectangular busbars with a finite length. *Energies* **2019**, *12*, 1419. [CrossRef]
8. Cruciani, S.; Campi, T.; Maradei, F.; Feliziani, M. Finite-Element Modeling of Conductive Multilayer Shields by Artificial Material Single-Layer Method. *IEEE Trans. Magn.* **2020**, *56*, 7502504. [CrossRef]
9. Brito, A.I.; Machado, V.M.; Almeida, M.; das Neves, M.G. Skin and proximity effects in the series-impedance of three-phase underground cables. *Electr. Power Syst. Res.* **2016**, *131*, 30–37. [CrossRef]
10. Lopera, J.M.; Prieto, M.J.; Díaz, J.; Jorge, G. A Mathematical Expression to Determine Copper Losses in Switching-Mode Power Supplies Transformers Including Geometry and Frequency Effects. *IEEE Trans. Power Electron.* **2015**, *30*, 5670–5681. [CrossRef]
11. Monteiro, J.H.A.; Costa, E.C.M.; Pinto, A.J.G.; Kurokawa, S.; Gatous, O.M.O.; Pizzolato, J. Simplified skin-effect formulation for power transmission lines. *IET Sci. Meas. Technol.* **2014**, *8*, 76–83. [CrossRef]
12. He, J.; Yu, L.; Wang, X.; Song, X. Simulation of Transient Skin Effect of DC Railway System Based on MATLAB/Simulink. *IEEE Trans. Power Deliv.* **2012**, *27*, 848–855. [CrossRef]
13. Li, J.; Yang, Q.; Li, Y.; Zhang, C.; Qu, B. Measurement and Modeling of 3-D Rotating Anomalous Loss Considering Harmonics and Skin Effect of Soft Magnetic Materials. *IEEE Trans. Magn.* **2017**, *53*, 6100404. [CrossRef]
14. Guo, W.; Li, D.; Ye, L. A Model of Magnetic Field and Braking Torque in Liquid-Cooled Permanent-Magnet Retarder Accounting for the Skin Effect on Permeability. *IEEE Trans. Veh. Technol.* **2019**, *68*, 10618–10626. [CrossRef]
15. González-Teodoro, J.R.; Kindl, V.; Romero-Cadaval, E.; Asensi, R. Analysis of Skin Effect in Single Wire Resistance by Finite Element Methods. In Proceedings of the 2020 IEEE 14th International Conference on Compatibility, Power Electronics and Power Engineering (CPE-POWERENG), Setubal, Portugal, 8–10 July 2020; pp. 1–6. [CrossRef]
16. Gyimóthy, S.; Kaya, S.; Obara, D.; Shimada, M.; Masuda, M.; Bilicz, S.; Pavo, J.; Varga, G. Loss Computation Method for Litz Cables with Emphasis on Bundle-Level Skin Effect. *IEEE Trans. Magn.* **2019**, *55*, 6300304. [CrossRef]
17. Sim, J.-H.; Kim, D.-Y.; Kim, S.-I.; Hong, J.-P. Analytical Electromagnetic Modeling and Experimental Validation of Vehicle Horn Considering Skin Effect in Its Solid Cores. *IEEE Trans. Magn.* **2017**, *53*, 8001704. [CrossRef]
18. Kazimierczuk, M.K. *High-Frequency Magnetic Components*; John Wiley & Sons: Hoboken, NJ, USA, 2014; ISBN 978-1-118-71779-0.
19. Namadmalan, A.; Jaafari, B.; Iqbal, A.; Al-Hitmi, M. Design Optimization of Inductive Power Transfer Systems Considering Bifurcation and Equivalent AC Resistance for Spiral Coils. *IEEE Access* **2020**, *8*, 141584–141593. [CrossRef]
20. Lakhdari, A.E.; Cheriet, A.; El-Ghoul, I.N. Skin effect based technique in eddy current non-destructive testing for thickness measurement of conductive material. *IET Sci. Meas. Technol.* **2019**, *13*, 255–259. [CrossRef]
21. Fridman, B.E.; Medvedev, M.V. Skin Effect in Coaxial Conductors of Pulse Facilities. *IEEE Trans. Plasma Sci.* **2019**, *48*, 482–490. [CrossRef]
22. Hayt, W.; Buck, J. *Engineering Electromagnetics*; McGraw Hill: New York, NY, USA, 2011; ISBN 978-0073380667.
23. Suchantke, R. Alternating Current Loss Measurement of Power Cable Conductors with Large Cross Sections Using Electrical Methods. Ph.D. Dissertation, Berlin University, Berlin, Germany, 2018.
24. Callister, W.D., Jr. *Fundamentals of Materials Science and Engineering*; John Wiley & Sons: Hoboken, NJ, USA, 2001; ISBN 0-471-39551-X.
25. Brandt, R.; Neuer, G. Electrical Resistivity and Thermal Conductivity of Pure Aluminum and Aluminum Alloys up to and above the Melting Temperature. *Int. J. Thermophys.* **2006**, *27*, 805–815. [CrossRef]
26. Basic Properties for Conducting Materials. Available online: https://www.nexans.com.br/.rest/eservice/dam/v1/file/223721/BasicProperties_br.pdf (accessed on 20 November 2022).
27. Catalog and Electrical Characteristics for Conductors. Available online: https://www.nexans.com.br/en/products/Transmission.and.Distribution/OHL.AWG/AWG/Aluminium.528.html (accessed on 15 December 2022).

28. Arritt, R.F.; Dugan, R.C. The IEEE 8500-node test feeder. In Proceedings of the IEEE PES T&D 2010, New Orleans, LA, USA, 19–22 April 2010. [CrossRef]
29. Neves, V.R., Jr. Determination of the Typology of Load Curves at Harmonic Frequencies for Low Voltage Residential Consumers. Master's Thesis, Universidade Federal de Uberlândia, Uberlândia, Brazil, 2022. [CrossRef]

energies

Article

Research on a Two-Layer Optimal Dispatching Method Considering the Mutual Aid of Peak Regulating Resources among Regional Power Grids

Tianmeng Yang [1], Suhua Lou [2,*], Meng Zhang [3], Yanchun Li [1], Wei Feng [1] and Jicheng Liu [1]

1 Northeast Branch of State Grid Corporation of China, Shenyang 110180, China; 18240369315@163.com (T.Y.); mty18842620317@163.com (Y.L.); 13971415920@163.com (W.F.); m202271914@hust.edu.cn (J.L.)
2 State Key Laboratory of Advanced Electromagnetic Engineering and Technology, Wuhan 430074, China
3 China Energy Engineering Group Liaoning Electric Power Design Institute Co., Ltd., Shenyang 110180, China; mzhang4947@ceec.net.cn
* Correspondence: shlou@mail.hust.edu.cn; Tel.: +86-132-9661-3265

Abstract: Since the power generation structures and load characteristics in each province in China are quite different, the distribution of peak regulating resources and demands are extremely imbalance. Restricted by a low power marketization degree, peak regulating resource shortages, and transmission channel blocks, the efficient utilization of new energy is facing greater pressures. In order to improve the mutual aid in regional power grids and to obtain more precise simulation results, this paper proposes a two-layer optimization dispatching model, considering the mutual aid of peak regulation resources between each province. It determines the optimal startup mode and the units' power output in each province and obtains the power output arrangements for all the units and the technical and economic indicators. The model and the solution method are original and innovative. And it effectively solved the unequal distribution problem between the peak regulating demands and resources of each provincial power grid. Finally, taking an actual regional power grid in China as an example, the simulation results show that the proposed model can significantly improve the utilization rate of new energy, which verifies the effectiveness and feasibility of the proposed model and methods presented in this paper.

Keywords: regional power grids; optimal dispatching; peak regulation resource mutual assistance; peak regulation capability; two-layer optimization

Citation: Yang, T.; Lou, S.; Zhang, M.; Li, Y.; Feng, W.; Liu, J. Research on a Two-Layer Optimal Dispatching Method Considering the Mutual Aid of Peak Regulating Resources among Regional Power Grids. *Energies* **2024**, *17*, 667. https://doi.org/10.3390/en17030667

Academic Editor: Vítor Monteiro

Received: 26 October 2023
Revised: 22 January 2024
Accepted: 22 January 2024
Published: 30 January 2024

1. Introduction

With the proposal of "carbon peak and carbon neutral", power generation structures are turning to a low-carbon transformation, and renewable energy power is developing rapidly in China [1–3]. In order to make full use of renewable energy, the government has put forward high requirements for the utilization rate [4–6]. Therefore, how to plan the power grid's development and optimize the operation modes, and how to make full use of the limited, flexible regulating resources are critical problems. What is more, how to maximize the utilization rate of renewable energy and reduce the peak regulating costs have become critical problems, which need to be solved for the current renewable energy development process in China [7–10].

Currently, there are papers that have studied the optimal dispatching operation methods based on considering the full utilization of peak regulating resources, but the research objects have been provincial power grids, and these are not representative. In [11], a multi-objective optimal scheduling model, considering the economy and flexibility of deep peak regulation, is constructed, but the power grid under study is not representative. In [12,13], by establishing the peak regulation right transaction of wind power and thermal power units, a combined thermoelectric economic dispatching model is established. It can stimulate the predisposition of thermoelectric units to participate in peak regulation, which can improve the wind power utilization rate. In [14,15], based on a peak regulating energy consumption cost model for the different stages of thermal power units, an economic dispatching model, which gives priority to wind power utilization, is established. In [16,17], based on the Manson–Coffin formula, a thermal power unit loss cost model with variable loads is established. And based on the full acceptance of wind power, an economic dispatching model based on a hierarchical deep peak regulation is established, which aims to minimize the total power generation costs.

In the studies of the optimization dispatching method in [18,19], the feasibility of peak regulation between regional power grids is verified. In [20,21], to solve the possible shortage problems of peak regulating resources, a transaction mechanism for peak regulation auxiliary service between the regional power grids is proposed, and a clearing pricing model is first constructed. In [22,23], based on an analysis of the power source and load characteristics of multiple regions, a typical peak–valley mutual aid operation scheme is given, and the typical operation mode and economic evaluation results are optimized. In [24], a two-stage iterative power exchange optimization method considering the marginal power generation costs is put forward. However, this method only aims at obtaining the minimum operation costs and cannot fully use regional peak regulating resources. In [25], an environmental economy dispatching model, considering the inter-provincial power balance of a regional power grid, is established. However, only the actual power limitation is taken into consideration, and the influence of the inter-provincial connection lines' actual electricity trading is not taken into consideration. In [26], by studying the different power sources' peak regulating characteristics, an optimal dispatching method considering peak regulating resources is studied. However, only the deep peak regulating operation costs of thermal power units are considered, and the influence of other peak regulating power sources is not considered.

In summary, peak regulating resources can be exchanged through inter-provincial transmission lines between provincial power grids in China, so as to make full use of peak regulating resources in the regional power grid. Based on this, this paper analyzes the peak regulation characteristics of different power sources, fully considers the differences between the surplus and gap of peak regulation power in the different provinces, and puts forward a two-layer optimization dispatching model with the lowest operation costs and highest utilization rate of renewable energy for a regional power grid. The upper-layer optimization model's objective function is the lowest operation costs for a regional power grid, and the lower-layer optimization model's objective function is the highest utilization rate of renewable energy. Based on this two-layer optimization model, the inter-provincial transmission curve with the lowest costs and the highest renewable power utilization rate can be re-optimized. And the optimized operation results for the whole regional power grid can be obtained. And last, taking an actual regional power grid as an example, the feasibility and high efficiency of the proposed model and optimization method are verified from the simulation results.

2. System Peak Regulating Resource Analysis

2.1. Analysis of Different Power Supplies

The peak regulation scale for active power and the quantity of electricity is large in China. Now, the main resources that can effectively participate in peak regulation are coal-fired units, gas units, cascade hydro-power units (non-runoff), pumped storage, and other types of large-capacity power storage. Among them, coal-fired units are non-stop peaking units, and gas units and cascade hydro-power units can be used as stop peaking units, thanks to their rapid start and stop characteristics [24,25]. The peak regulating characteristics of each power supply are shown in Table 1.

Table 1. Peak regulating characteristics of different power supplies.

Power Supply	Peak Regulating Ability
Coal-fired units	Conventional thermal units: 40~50% After flexibility transformation: 55~70%
Gas units	Have shutdown ability, peak regulation ability: 100%
Cascade hydro-power unit (non-runoff)	Have shutdown ability, peak regulation ability: 100%
Pump storage units	160%
Other energy storage units	160%
Small hydro-power stations (radial flow)	0%
Wind	0%
Photovoltaic	0%

2.2. Analysis of Coal-Fired Thermal Units' Energy Consumption Costs

Considering the rapid development of renewable energy, the function of coal-fired units has gradually changed from the main electricity source to regulating the power supply. When the coal-fired thermal units change to a regulating function, all the units' operating states change frequently. And the characteristics of the technical and economic costs also significantly change.

Considering the operating states and the energy consumption characteristics of coal-fired units, the coal-fired units' peak regulating processes can be divided into three stages: a basic peak regulation stage (RPR), a non-oil supply peak regulation stage (DPR), and an oil supply peak regulation stage (DPRO). The characteristic curves for the energy consumption costs during each peak regulating period are shown in Figure 1. $P_{\mathrm{max}}^{\mathrm{G}}$ represents the rated power output of the coal-fired unit, $P_{\mathrm{a}}^{\mathrm{G}}$ represents the minimum technical power output of the RPR stage, $P_{\mathrm{b}}^{\mathrm{G}}$ represents the minimum technical power output of the DPR stage, and $P_{\mathrm{c}}^{\mathrm{G}}$ represents the minimum technical power output of the DPRO stage.

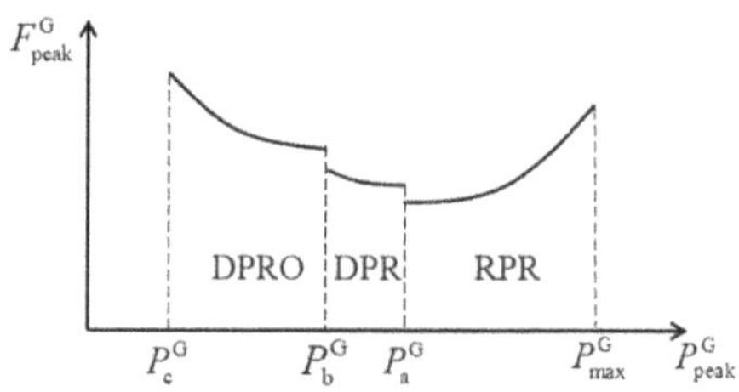

Figure 1. Peak regulating costs of the fuel units.

In recent years, thanks to the technical progress of thermal units, the minimum stable combustion load of the 200 MW and 300 MW units can be reduced to 45% of rated power after flexibility transformation. And the minimum stable combustion load of the 600 MW and 1000 MW units can be reduced to 30~35% after flexibility transformation. The formula of thermal power units' energy consumption costs is shown below.

$$F_{\mathrm{peak}}^{\mathrm{G}} = \begin{cases} F_{\mathrm{coal}}(P_{\mathrm{peak}}^{\mathrm{G}}) & P_{\mathrm{a}}^{\mathrm{G}} < P_{\mathrm{peak}}^{\mathrm{G}} \leq P_{\mathrm{a}}^{\mathrm{max}} \\ F_{\mathrm{coal}}(P_{\mathrm{peak}}^{\mathrm{G}}) + F_{\mathrm{loss}}(P_{\mathrm{peak}}^{\mathrm{G}}) & P_{\mathrm{b}}^{\mathrm{G}} < P_{\mathrm{peak}}^{\mathrm{G}} \leq P_{\mathrm{a}}^{\mathrm{G}} \\ F_{\mathrm{coal}}(P_{\mathrm{peak}}^{\mathrm{G}}) + F_{\mathrm{loss}}(P_{\mathrm{peak}}^{\mathrm{G}}) + F_{\mathrm{oil}}(P_{\mathrm{peak}}^{\mathrm{G}}) & \\ +F_{\mathrm{en}}(P_{\mathrm{peak}}^{\mathrm{G}}) & P_{\mathrm{c}}^{\mathrm{G}} < P_{\mathrm{peak}}^{\mathrm{G}} \leq P_{\mathrm{b}}^{\mathrm{G}} \end{cases} \tag{1}$$

In the formula, $F_{\mathrm{coal}}(P_{\mathrm{peak}}^{\mathrm{G}})$ represents the thermal power units' operating coal consumption costs, which is described by the second-order consumption model in (2).

$$F_{\mathrm{coal}}(P_{\mathrm{peak}}^{\mathrm{G}}) = k_{\mathrm{coal}} \cdot (\alpha {P_{\mathrm{peak}}^{\mathrm{G}}}^2 + \beta P_{\mathrm{peak}}^{\mathrm{G}} + \gamma) \tag{2}$$

In the formula, k_{coal} represents the coal price coefficient, and α, β, γ represent the consumption characteristic coefficients of thermal units, respectively.

$F_{\text{loss}}(P^{\text{G}}_{\text{peak}})$ represents the unit loss costs caused by the excessive thermal stress in the DPR and DPRO stages. The unit loss costs are mainly determined by the cracking cycle of the rotor, as shown in (3).

$$F_{\text{loss}}(P^{\text{G}}_{\text{peak}}) = \frac{1}{2N_{\text{T}}(P^{\text{G}}_{\text{peak}})} \tau k_{\text{unit}} \tag{3}$$

In the formula, $N_{\text{T}}(P^{\text{G}}_{\text{peak}})$ represents the cracking cycle of the rotor, which is interrelated with the power output of units. τ represents the actual operating loss coefficient of the thermal power units, and the value in the DPRO stage is larger than in the DPR stage. K_{unit} represents the investing cost coefficient of thermal power units.

$F_{\text{oil}}(P^{\text{G}}_{\text{peak}})$ represents the fuel consumption costs in the DPRO stage, which is mainly determined by fuel consumption quantity and oil price in the DPRO stage, as shown in (4).

$$F_{\text{oil}}(P^{\text{G}}_{\text{peak}}) = k_{\text{oil}} \cdot E_{\text{oil}} \tag{4}$$

In the formula, k_{oil} represents oil price, and E_{oil} represents fuel consumption quantity in the DPRO stage.

$F_{\text{en}}(P^{\text{G}}_{\text{peak}})$ represents additional environmental punishment fees caused by pollutants' excessive discharge in the DPRO stage, which is shown in (5).

$$F_{\text{en}}(P^{\text{G}}_{\text{peak}}) = k_{\text{punish}} \cdot \lambda_{\text{S}} \tag{5}$$

In the formula, k_{punish} represents the penalty cost coefficient of pollutant discharge, and λ_{S} represents the pollutant emissions.

2.3. Analysis of Pumped Storage Units' Peak Regulating Costs

As a main tool for peak regulating, the pumped storage units have functions including peak load shifting and valley filling, frequency regulating and phase regulating, spinning reserve, and black start. On the one hand, pumped storage units have flexible operating modes with reversible pump turbine units; on the other hand, limited by different operating conditions, pumped storage units have specific operating characteristics and constraints.

The pumped storage unit works the same as the conventional generator in the generating state. In the pumping state, it absorbs power from the power grid. And the operating costs generated by pumped storage units are mainly generated by the conversion between pumping and generating states, which consist of generators' and motors' start-up costs.

$$F^{\text{PH}}_{\text{peak},j,t} = F^{\text{PH-G}}_{j,t} + F^{\text{PH-P}}_{j,t} \tag{6}$$

In the formula, $F^{\text{PH-G}}_{j,t}$ represents the starting costs at the tth period of the jth pumped storage unit working in the generating state, and $F^{\text{PH-P}}_{j,t}$ represents the costs in the pumping state.

3. Two-Layer Optimization Dispatching Model of the Regional Power Grid

3.1. Introduction of the Model

Considering the differences of the power generation structures and load characteristics of each provincial power grid, the distribution of peak regulating demands is extremely imbalanced. Therefore, the regional peak regulating resources should be fully utilized, which can improve the new energy utilization rate of the regional power grid.

Thus, based on the provincial power balance and dispatching management principles in China, a two-layer optimization dispatching model of the regional power grids is proposed in this paper. The model aims to divide the optimal dispatching problem of the regional power grid into two parts: the units' optimal dispatching of each province and the optimization of inter-provincial power transmission plan curves. This two-layer model is extensible and it can be applied into any regional power grid. The flow chart of the two-layer optimization method is shown in Figure 2.

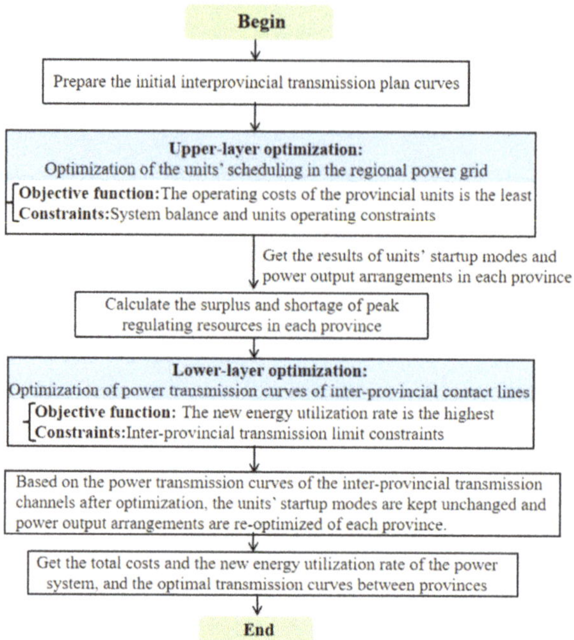

Figure 2. Two-layer optimization dispatching process of regional power grids.

3.2. *Units' Optimization Dispatching Model in Each Province*

3.2.1. Objective Function

The units' optimization dispatching model is for optimizing the startup modes and power output in each province on the basis of the given power transmission plan curves, so that the total operation costs in each province are the lowest. Taking one province as an example, the objective function aims to obtain the minimum operation costs, as shown in Equation (7).

$$\min F = \min(\alpha_1 \cdot F_{\text{oper}} + \alpha_2 \cdot F_{\text{dep}}) \tag{7}$$

In the formula, α_1 and α_2 represent the weight coefficients of operation costs and new energy abandonment penalty costs, respectively. F_{oper} represents the operating costs of all units in the system, including the operating costs of thermal power units and pumped storage units. F_{dep} represents the penalty costs of new energy abandonment.

$$F_{\text{oper}} = F_{\text{peak}}^{\text{G}} + F_{\text{peak}}^{\text{PH}} \\ \begin{cases} F_{\text{peak}}^{\text{G}} = \sum\limits_{t=1}^{T} \sum\limits_{i=1}^{N_{\text{G}}} F_{\text{peak},i,t}^{\text{G}} \\ F_{\text{peak}}^{\text{PH}} = \sum\limits_{t=1}^{T} \sum\limits_{j=1}^{N_{\text{PH}}} F_{\text{peak},j,t}^{\text{PH}} \end{cases} \tag{8}$$

$$F_{\text{dep}} = k_{\text{dep}} \cdot E_{\text{dep}} \tag{9}$$

In the formula, $F_{\text{peak},i,t}^{\text{G}}$ and $F_{\text{peak},i,t}^{\text{PH}}$ represent the peak regulating costs of the ith thermal unit and jth pumped storage unit, respectively, k_{dep} represents the penalty cost coefficient of abandoned power, and E_{dep} represents abandoned electricity of new energy.

As shown above, the peak regulating costs of thermal power units are mainly obtained according to Formula (1), and the peak regulating costs of pumped storage units are mainly obtained according to Formula (6).

3.2.2. Constraints

System Power Balance Constraints

$$P_t^{\mathrm{L}} + P_t^{\mathrm{C}} = P_t^{\mathrm{N}} + \sum_{i=1}^{N_{\mathrm{G}}} P_{\mathrm{peak},i,t}^{\mathrm{G}} + \sum_{j=1}^{N_{\mathrm{PH}}} P_{j,t}^{\mathrm{PH}} - P_{\mathrm{dep},t} \quad (10)$$

In the formula, P_t^{L} represents the load level at the tth period; P_t^{C} represents the reserve capacity at the tth period; P_t^{N} represents the power output of new energy at the tth period; $P_{\mathrm{peak},i,t}^{\mathrm{G}}$ represents the power output of the ith thermal unit at the tth period; $P_{j,t}^{\mathrm{PH}}$ represents the power output of the jth pump storage unit at the tth period, with a positive value representing the generating state and negative value representing the pumping state; $P_{\mathrm{dep},t}$ represents the abandoned power output at the tth period.

Thermal Power Unit Climbing Constraints

$$\mathrm{DP}_i^{\mathrm{G}} \le P_{\mathrm{peak}.i,t}^{\mathrm{G}} - P_{\mathrm{peak},i,t-1}^{\mathrm{G}} \le \mathrm{UP}_i^{\mathrm{G}} \quad (11)$$

In the formula, $\mathrm{DP}_i^{\mathrm{G}}$ and $\mathrm{UP}_i^{\mathrm{G}}$ represent the climbing up and down limitations of the thermal unit, respectively.

Pumped Storage Operation Constraints

According to the actual operating restriction of pumped storage units, the pumped storage unit pumps with rated power in the pumping state. And pumped power generation should meet the storage capacity constraints. Its power output and storage capacity constraints are shown below.

$$P_{j,t}^{\mathrm{PH-P}} = u_{j,t}^{\mathrm{PH-P}} \cdot P_{j,\max}^{\mathrm{PH}} \quad (12)$$

$$0 \le P_{j,t}^{\mathrm{PH-G}} \le u_{j,t}^{\mathrm{PH-G}} P_{j,\max}^{\mathrm{PH}} \quad (13)$$

$$E_{\mathrm{PH,min}} \le E_{\mathrm{PH},t} \le E_{\mathrm{PH,max}} \quad (14)$$

$$E_{\mathrm{PH},t} = E_{\mathrm{PH},t-1} + \Delta T \cdot \left(\sum_{j=1}^{N_{\mathrm{PH}}} \eta_{\mathrm{PH-P}} \cdot P_{j,t}^{\mathrm{PH-P}} - \sum_{j=1}^{N_{\mathrm{PH}}} \frac{P_{j,t}^{\mathrm{PH-G}}}{\eta_{\mathrm{PH-G}}} \right) \quad (15)$$

In the formula, $P_{j,\max}^{\mathrm{PH}}$ represents the rated power output of the jth pumped storage unit; $u_{j,t}^{\mathrm{PH-P}}$ and $u_{j,t}^{\mathrm{PH-G}}$ represent the pumping and generating state variables, respectively, which satisfy $u_{j,t}^{\mathrm{PH-P}} + u_{j,t}^{\mathrm{PH-G}} \le 1$; $P_{j,t}^{\mathrm{PH-P}}$ and $P_{j,t}^{\mathrm{PH-G}}$ represent the pumping power and generating power of the jth unit at the tth period, respectively. $E_{\mathrm{PH,max}}$ represents the rated storage capacity of the pumped storage station; $E_{\mathrm{PH,min}}$ represents the minimum storage capacity of the pumped storage station; $\eta_{\mathrm{PH-P}}$ and $\eta_{\mathrm{PH-G}}$ represent the pumping and generating efficiency of pumped storage units.

Upper and Lower Limits of Thermal Power Units

$$P_{\mathrm{c},i}^{\mathrm{G}} \le P_{\mathrm{peak},i,t}^{\mathrm{G}} \le P_{i,\max}^{\mathrm{G}} \quad (16)$$

In the formula, $P_{\mathrm{c},i}^{\mathrm{G}}$ represents the lowest power output limit of the ith thermal power unit.

3.3. *Analysis of Peak Regulation Amplitude and Gap between Provinces in the Regional Power Grid*

By solving the provincial unit optimization model, the startup plan and power output arrangements can be obtained. They are calculated with the initial inter-provincial power transmission curves. Thus, we calculate the amplitude and gap indexes of peak regulating resources in each province, which can provide the basis for optimizing the transmission curves between provincial power grids.

$$\begin{cases} P_t^{\text{margin}} = \sum_{i=1}^{N_G} \left(P_{\text{peak},i,t}^{\text{G}} - P_{\text{c},i}^{\text{G}} \right) \\ P_t^{\text{insuff}} = P_{\text{dep},t} \end{cases} \quad (17)$$

In the formula, P_t^{margin} and P_t^{insuff} represent the amplitude and gap of peak regulation resource at the tth period, respectively.

3.4. *Optimization Scheduling Model of Inter-Provincial Transmission Curves*

Based on the peak regulation amplitude and gap in each provincial power grid calculated in Section 3.3, the initial inter-provincial transmission curves should be optimized and adjusted. Based on the optimized transmission curves, we can obtain the highest new energy utilization rate of the whole regional power grid. And the peak regulating resources will be called up in a more balanced manner.

3.4.1. Objective Function

The objective function of the inter-provincial transmission curve optimization model mainly considers two aspects; these are the highest new energy utilization rate and the lowest system peak regulation costs. The objective function is shown below.

$$\min E_{\text{dep}} = \min(\beta_1 \cdot k_{\text{dep}} \cdot \sum_{t=1}^{T} P'_{\text{dep},t} + \beta_2 \cdot k_{\text{peak}} \cdot \sum_{l=1}^{N_L} \sum_{t=1}^{T} \Delta P_{l,t}^{\text{D}}) \quad (18)$$

In the formula, β_1 and β_2 represent the weight coefficients of new energy abandonment penalty costs and peak regulating costs; k_{peak} represents the coefficient of system peak regulating costs; $P'_{\text{dep},t}$ represents the abandoned new energy power output at the tth period after optimizing the inter-provincial transmission curves; $\Delta P_{l,t}^{\text{D}}$ represents power differences of the lth inter-provincial transmission lines before and after optimization at the tth period.

Thus, the calculation formulas of the abandoned new energy power output are as follows:

$$\begin{cases} P'_{\text{dep},t} = P_{\text{dep},t} + \sum_{l=1}^{N_L} \Delta P_{l,t}^{\text{D}} \\ \sum_{l=1}^{N_L} \Delta P_{l,t}^{\text{D}} \leq P_t^{\text{insuff}} \end{cases} \quad (19)$$

3.4.2. Constraints

Transmission Limit Constraints

$$P_{l,t}^{\prime\text{D}} = P_{l,t}^{\text{D}} + \Delta P_{l,t}^{\text{D}}$$
$$\begin{cases} P_{l,t}^{\prime\text{D}} \leq P_{l,\text{N}}^{\text{Pos}}, P_{l,t}^{\prime\text{D}} \geq 0 \\ P_{l,t}^{\prime\text{D}} \geq P_{l,\text{N}}^{\text{Neg}}, P_{l,t}^{\prime\text{D}} < 0 \end{cases} \quad (20)$$

In the formula, $P_{l,t}^{\text{D}}$ and $P_{l,t}^{\prime\text{D}}$ represent the transmission power of the lth channel at the tth period before and after optimization, respectively, while $P_{l,\text{N}}^{\text{Pos}}$ and $P_{l,\text{N}}^{\text{Neg}}$ represent the transmission limitation of the lth connection channel in two directions, respectively.

Constraints of Transmission Power Limitation

$$E_l^{\text{D}} = \sum_{t=1}^{T} \sum_{l=1}^{N_L} P_{l,t}^{\prime\text{D}} \quad (21)$$

$$E_l^{\min} \leq E_l^{\text{D}} \leq E_l^{\max} \quad (22)$$

In the formula, E_l^D represents the transmission electricity of the *l*th transmission channel after optimization; E_l^{max} and E_l^{min} represent the upper and lower transmission electricity limitation of the *l*th transmission channel.

4. Case Study

In order to fully verify the effectiveness and feasibility of the proposed model and method, two example power systems are presented for simulation and analysis. They are the IEEE RTS-96 test system and a regional power grid in China, respectively. Since the two-layer model is non-linear, all optimization processes mainly depend on Matlab Software R2016a with CPLEX and the Yalmip optimization engine. Considering the uncertainty of new energy power, the clustering algorithm is used to gather typical scenarios to represent all historical curves.

4.1. Example System Analysis Based on IEEE RTS-96

4.1.1. Overview of the Example System

The example system is modified from the IEEE RTS-96 system, and all parameters of the system are based on the original system. It is divided into three sub-regional power grids. Each region is connected through DC transmission channels. The load characteristics of the three sub-regions are shown in Figure 3. The load level of region 1 and region 2 is lower than that of region 3, and it has the smallest peak–valley difference. The peak regulating abilities of thermal units are shown in Table 2. The wind power capacities are 1300 MW, and the wind power output characteristics in each region are shown in Figure 4. The peak regulating characteristics of wind power output in region 1 are positive, which can reduce the peak regulating pressure, while it is negative in region 2, which aggravates the peak regulating pressure. The characteristics of wind power output in region 3 are stable, which has little influence on peak regulation.

Table 2. Peak regulating capacities of thermal units in each sub-region.

Region	Units Capacities/MW	Wind Power Capacities/MW	Peak Regulating Capacities	Peak Regulating Capacities without Oil	Peak Regulating Capacities with Oil
Region 1	29,410	1300	50%	40%	35%
Region 2	29,410	1300	50%	45%	40%
Region 3	21,100	1300	50%	43%	38%

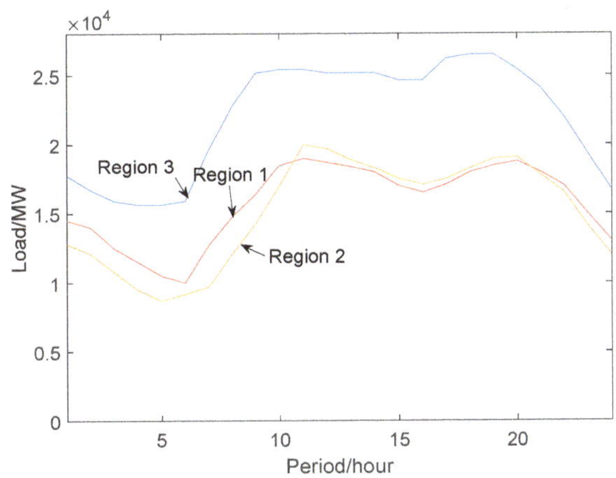

Figure 3. Load characteristics of regional power grid.

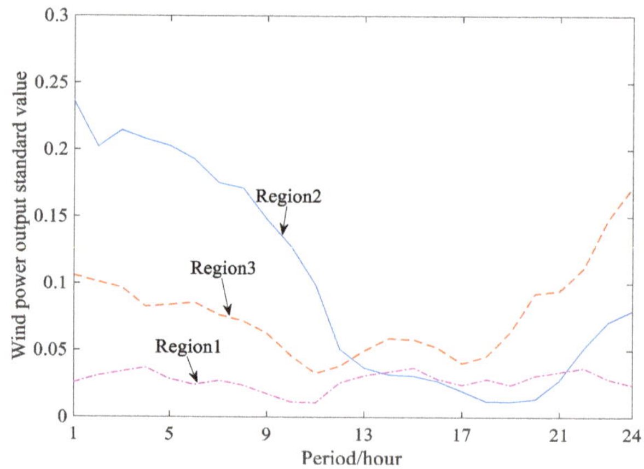

Figure 4. Wind power characteristics of regional power grid.

The weight coefficients α_1 and α_2 are 0.7 and 0.3, and β_1 and β_2 are 0.8 and 0.2.

4.1.2. Analytical Calculation

Regions 1 and 2 are the main power delivery areas with surplus unit capacities and small load level, while region 3 is the main power-receiving area with insufficient unit capacities and large load level. The initial transmission curves between regions take the constant power transmission mode. The results of operating costs in each region before and after optimization are shown in Table 3, and the regional daily transmission curves before and after optimization are shown in Figure 5.

Table 3. Operation costs of each region before and after inter-provincial transmission curve optimization.

	Before Optimization			After Optimization		
Region	**Total Costs/USD 100 Million**	**Operation Costs/USD 100 Million**	**Abandon Punishment Costs/USD 100 Million**	**Total Costs/USD 100 Million**	**Operation Costs/USD 100 Million**	**Abandon Punishment Costs/USD 100 Million**
1	0.28	0.23	0.05	0.24	0.22	0.02
2	0.27	0.23	0.04	0.25	0.23	0.02
3	0.25	0.2	0.05	0.22	0.19	0.03
Total	0.8	0.66	0.14	0.71	0.64	0.07

As can be seen in Figure 5, the power transmission curves of regions 1 and 2 after optimization show great differences. The main reason is that there exist extreme differences between peak regulating capacities and demands in different regions. Since the load peak and valley differences in region 1 are small, the peak regulation demands are the minimum. While the load peak and valley differences in region 2 are large, the peak regulation demands are the maximum, and the peak regulating support ability to other regions is the minimum. Therefore, the optimized transmission curves can fully call up peak regulating resources in the three regions. During low-load periods, region 2 transfers power to region 3, and region 1 transfers power mainly during high-load periods. Thus, it can reduce system operating costs and power abandonment penalty costs, and promote the operating efficiency of each regional power grid.

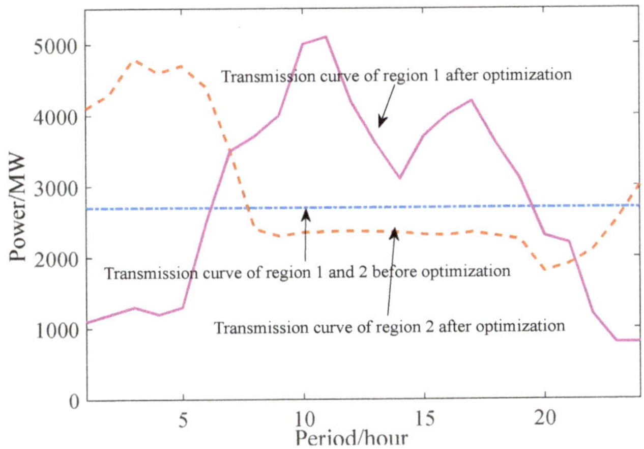

Figure 5. Daily power transmission curves between each regional power grid before and after inter-provincial transmission curve optimization.

4.2. *Example System Analysis Based on an Actual Regional Power Grid*

4.2.1. Overview of the Regional System

In this part, an actual regional power grid in China is taken as an example for analysis and calculation, and this region consists of four provincial power grids. All the parameters of units and load curves are based on the actual power system. In this region, provinces A, B, and C have surplus electricity and province D mainly receives electricity from B and C. The actual power flows are shown in Figure 6. And in actual operation, the power flow direction is mainly opposite to what is shown in Figure 6.

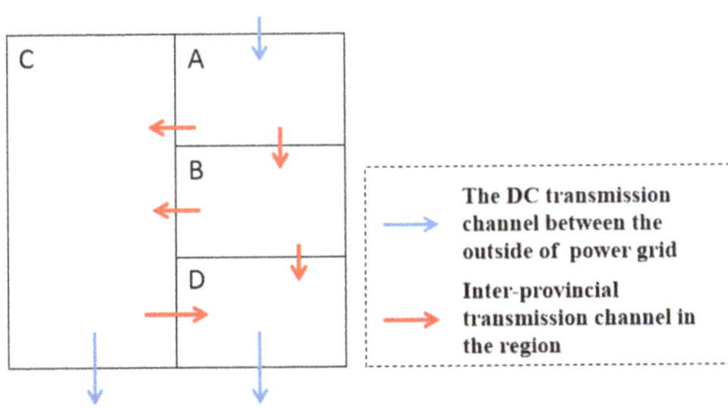

Figure 6. Schematic diagram of power transmission and reception between each province.

Since this region is located at high latitude in China, the load demands and the peak regulating demands are large in winter. Therefore, taking winter load characteristics as an example, the load characteristics of the four provinces are shown in Figure 7. The load demands of provinces A, B, and C are small and the load peak valley differences of province B are the smallest. The power output characteristics of each province are shown in Figure 8.

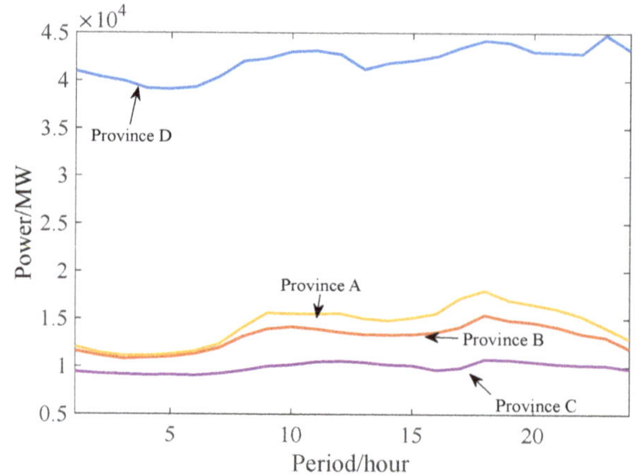

Figure 7. Load characteristics of each province.

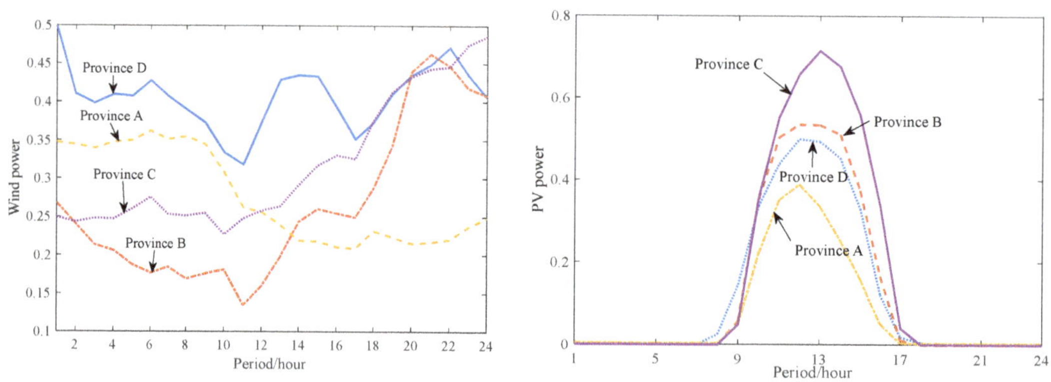

(**a**) (**b**)

Figure 8. (**a**) Wind power output characteristics of each province in regional power grid. (**b**) PV power output characteristics of each province in regional power grid.

4.2.2. Analytical Calculation

Before optimization, the DC power transmission curves outside the region and the initial transmission curves between provinces in the region are based on the historical annual transmission curves. The operating cost results before and after optimization are shown in Table 4. The regional daily transmission curves before and after optimization are shown in Figure 9.

Table 4. Operation costs of each regional power grid before and after inter-provincial transmission plan curve optimization.

	Before Optimization				After Optimization			
Province	Total Costs/USD 100 Million	Operation Costs/USD 100 Million	Abandon Punishment Costs/USD 100 Million	New Energy Utilization Rate%	Total Costs/USD 100 Million	Operation Costs/USD 100 Million	Abandon Punishment Costs/USD 100 Million	New Energy Utilization Rate%
A	0.46	0.396	0.064	87.9	0.419	0.377	0.042	89.9
B	0.545	0.452	0.092	84.8	0.495	0.433	0.061	88.3
C	0.455	0.335	0.121	89.6	0.423	0.315	0.108	90.7
D	0.653	0.599	0.053	93.3	0.617	0.565	0.052	93.5
Total	2.113	1.782	0.33	87.7	1.954	1.69	0.263	90.9

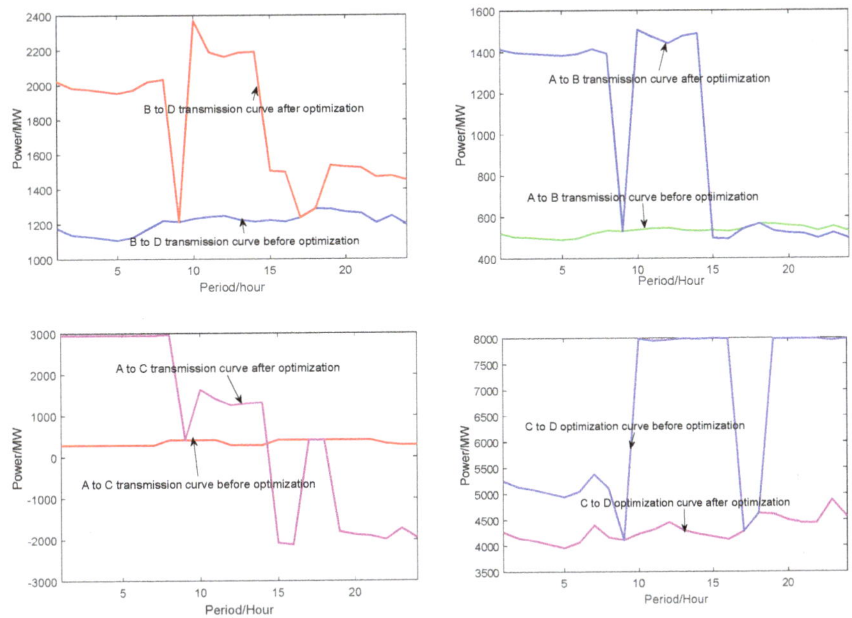

Figure 9. Daily power transmission curves of each regional power grid before and after transmission curve optimization.

As can be seen from Figure 9, the differences in the transmission curves between provinces C and D and between provinces B and D before and after optimization are the largest. The main reason is that the peak regulating capacities and amplitude of province D are the largest. After optimizing the transmission curves between C and D and between B and D, the surplus peak regulation resources can be supported to B and C, which can maximize the new energy utilization rate of the whole regional power grid.

According to the calculation results in Table 4, after the unit startup optimization of the whole system and transmission curve optimization between provinces, the total costs of the regional power grid decrease by 7.6%, the operation costs decrease by 5.2%, and the new energy power abandonment penalty costs decrease by 20.4%. After optimization, the utilization rate of new energy in the whole region increases by 3.2%. And the utilization rate of new energy in province B has the maximum improvement (3.5%).

4.3. Comparison and Analysis

In order to further verify the effectiveness and feasibility of the proposed optimization model and algorithm, the traditional optimization method based on equal peak regulating rate was used for comparison. In the traditional equal peak regulating rate method, it is required for all thermal units in each province in the region to share peak regulating demands equally in accordance. In comparison, the model proposed in this paper focuses on the whole region, which can maximize the mutual aids of peak regulating resources in the whole regional power grid.

The calculation results of the two models are shown in Table 5. As can be seen from the comparison results, the total costs by using the equal peak regulating rate optimization model (model 2) are higher than the proposed model (model 1) in this paper. The new energy utilization rate is lower, and the solution time of the two models is relatively similar. As can be seen from the above comparison results, the proposed optimization model and method in this paper have higher calculation accuracy and solution efficiency. It can fully consider the mutual aid of peak regulating resources and can achieve the highest new energy power utilization rate. What is more, it is more suitable for solving the actual optimization dispatching problems of regional power grids in China.

Table 5. Comparison of calculation results between two optimization models.

Model	Optimization Results	
	Total Costs/USD 100 Million	**New Energy Utilization Rate%**
1	1.953	90.9
2	2.116	87.1

5. Conclusions

Based on deeply studying the peak regulation characteristics of different power supplies, an original two-layer optimal dispatching model is proposed in this paper. The upper layer is mainly the unit startup coordination and optimization scheduling model of each province in the region. The lower layer is mainly the optimal scheduling model of transmission curves in the region. By solving the two-layer optimization model, the lowest operation costs and the highest new energy utilization rate of the regional power grid can be obtained, which can realize full aid and sharing of peak regulating resources in the whole region. Finally, based on the simulation and analysis of the IEEE RTS-96 power system and an actual regional power grid, the feasibility and effectiveness of the proposed model and optimization method are verified.

Author Contributions: T.Y. established the two-layer optimization dispatching model of the regional power grids and established the simulation model for the verification. S.L. established the two-layer optimization dispatching model of the regional power grids. M.Z. corrected the manuscript and put forward many suggestions for improvement. Y.L. put forward some suggestions when building the model and analyzed the simulation results. W.F. checked and corrected the mathematical model of the paper. J.L. corrected the manuscript and put forward many suggestions for improvement. All authors have read and agreed to the published version of the manuscript.

Funding: The work was funded by the science and technology project guided by northeast branch of state grid corporation of China. Project Number: (52992623000G, SGDB0000DJJS2310060).

Data Availability Statement: Data is contained within the article.

Conflicts of Interest: Authors Tianmeng Yang, Yanchun Li, Wei Feng and Jicheng Liu were employed by the company Northeast Branch of State Grid Corporation of China; Meng Zhang was employed by the company China Energy Engineering Group Liaoning Electric Power Design Institute Co., Ltd. The remaining authors declare that the research was conducted in the absence of any commercial or financial relationships that could be construed as a potential conflict of interest.

References

1. Zhang, N.; Zhou, T.; Duan, C. Impact of large-scale wind farm connecting with power grid on peak load regulation demand. *Power Syst. Technol.* **2010**, *34*, 152–158.
2. Zhang, Z.; Zhou, L.; Sun, Y.; Gao, X.; Liu, J.; Dong, Y. Joint Optimal Dispatching Strategy of Pumped Storage and Thermal Power Units with Large-Scale Wind Power Integration. In Proceedings of the 2021 IEEE 5th Conference on Energy Internet and Energy System Integration, Taiyuan, China, 22–24 October 2021.
3. Zhang, Y.; Le, J.; Liao, X.; Zheng, F.; Liu, K.; An, X. Multi-objective hydro-thermal-wind coordination scheduling integrated with large-scale electric vehicles using imopso. *Renew. Energy* **2018**, *128*, 91–107. [CrossRef]
4. Cheng, D.; Xu, B.; Li, W.; Wu, Q.; Cheng, L.; Zhang, Y.; Zhao, W.; Ma, W.; Zhao, J. Optimal Economic Dispatch Strategy of a Hybrid Energy Storage Microgrid Based on Genetic Algorithm. In Proceedings of the 2023 IEEE 18th Conference on Industrial Electronics and Applications, Ningbo, China, 18–22 August 2023.
5. Yang, D.; Yang, Z.; He, G.; Geng, J.; Ligao, J. Wind Power Accommodation Method for Regional Integrated Heat-Power System Considering Generation and Load Uncertainties. *Access* **2021**, *9*, 2169–2180. [CrossRef]
6. Zhu, L.; Chen, N.; Han, H. Key problems and solutions of wind power accommodation. *Autom. Electr. Power Syst.* **2011**, *35*, 29–34.
7. Jiang, P.; Yang, S.; Huo, Y. Static security analysis of power systems considering randomness of wind farm output. *Autom. Electr. Power Syst.* **2013**, *37*, 35–40.

8. Zhang, H.; Yin, Y.; Shen, H. Peak-load regulating adequacy evaluation associated with large-scale wind power integration. *Proc. CSEE* **2011**, *31*, 26–31.
9. Kabouris, J.; Kanellos, F.D. Impacts of large-scale wind penetration on designing and operation of electric power systems. *IEEE Trans. Sustain. Energy* **2010**, *1*, 107–114. [CrossRef]
10. Lin, Z.; Chen, H.; Chen, J.; Huang, J.; Li, M.; Wu, Q. Risk-averse robust interval economic dispatch for power systems with large-scale wind power integration. *CSEE J. Power Energy Syst.* **2021**, *12*, 2096–2106.
11. Ying, Y.; Wang, Z.; Wu, X. Multi-objective strategy for deep peak shaving of power grid considering uncertainty of new energy. *Power Syst. Prot. Control* **2020**, *48*, 34–42.
12. Yang, L.; Liang, X.; Wang, X. Combined heat and power economic dispatching considering peak regulation right trading to improve secondary accommodation capability of wind power. *Power Syst. Technol.* **2020**, *44*, 1872–1880.
13. Tan, Q.; Ding, Y.; Zhang, Y. Optimization model of an efficient collaborative power dispatching system for carbon emissions trading in China. *Energies* **2017**, *10*, 1405. [CrossRef]
14. Lin, L.; Zou, L.; Zhou, P. Multi-angle economic analysis on deep peak regulation of thermal power units with large-scale wind power integration. *Autom. Electr. Power Syst.* **2017**, *41*, 21–27.
15. Yang, B.; Cao, X.; Cai, Z.; Yang, T.; Chen, D.; Gao, X.; Zhang, J. Unit Commitment Comprehensive Optimal Model Considering the Cost of Wind Power Curtailment and Deep Peak Regulation of Thermal Unit. *Access* **2020**, *8*, 505–512. [CrossRef]
16. Lin, L.; Tian, X. Analysis of deep peak regulation and its benefit of thermal units in power system with large scale wind power integrated. *Power Syst. Technol.* **2017**, *41*, 2255–2263.
17. Tian, L.; Xie, Y.; Hu, B.; Liu, X.; Li, F. A deep peak regulation auxiliary service bidding strategy for CHP units based on a risk-averse model and district heating network energy storage. *Energies* **2019**, *12*, 3314. [CrossRef]
18. Huang, H.; Xiong, H.; Jiang, B. Design of day-ahead trading mechanism for trans-province peak-shaving auxiliary service of regional power grid. *Smart Power* **2020**, *48*, 119–124.
19. Lu, S.; Zhang, X.; Xu, Y. Two-stage iterative method to optimize tie-line exchange power based on marginal power generation cost. *Power Syst. Prot. Control* **2021**, *49*, 77–88.
20. Zhang, Y.; Hou, Y.; Hu, T. Regional power network environment/economic power dispatching considering balance of power between provinces. *Eng. J. Wuhan Univ.* **2017**, *50*, 548–555.
21. Ohtake, H.; Uno, F.; Oozeki, T.; Yamada, Y.; Takenaka, H.; Nakajima, T. Outlier events of solar forecasts for regional power grid in Japan using jma mesoscale model. *Energies* **2018**, *11*, 2714. [CrossRef]
22. Fang, B.; Xiao, L.; Sun, C.; Chen, Y.; Xue, Y.; Zhang, D. An optimization dispatch method for regional power grid considering peaking resource mutual support. *Electr. Drive* **2022**, *52*, 54–60.
23. Li, J. An improved quantum-behaved particle swarm optimization method for economic dispatch problems with multiple fuel options and valve-points effects. *Energies* **2012**, *5*, 240–250.
24. Liu, D.; Huang, Y.; Wang, W. Analysis on provincial system available capability of accommodating wind power considering peak load dispatch and transmission constraints. *Autom. Electr. Power Syst.* **2011**, *35*, 77–81.
25. Zhuo, Z.; Zhang, N.; Xie, X. Key technologies and developing challenges of power system with high proportion of renewable energy. *Autom. Electr. Power Syst.* **2021**, *45*, 171–191.
26. Chen, J.; Wu, W.; Zhang, B.; Wang, B.; Guo, Q. A spinning reserve allocation method for power generation dispatch accommodating large-scale wind power integration. *Energies* **2013**, *6*, 5357–5381. [CrossRef]

MDPI AG
Grosspeteranlage 5
4052 Basel
Switzerland
Tel.: +41 61 683 77 34
www.mdpi.com

Energies Editorial Office
E-mail: energies@mdpi.com
www.mdpi.com/journal/energies

www.ingramcontent.com/pod-product-compliance
Lightning Source LLC
Chambersburg PA
CBHW040748250926
51850CB00007BB/1437
* 9 7 8 3 7 2 5 8 1 6 1 5 6 *